Cold Atoms

The Quantum World of Ultra-Cold Atoms and Light

Book I: Foundations of Quantum Optics

Cold Atoms

ISSN: 2045-9734

Series Editor: Christophe Salomon (*Laboratoire Kastler Brossel, École Normale Supérieure, France*)

Vol. 1 Quantum Gases: Finite Temperature and Non-Equilibrium Dynamics
edited by N. Proukakis, S. Gardiner, M. Davis and M. Szymańska

Vol. 2 The Quantum World of Ultra-Cold Atoms and Light
Book 1: Foundations of Quantum Optics
by Crispin Gardiner and Peter Zoller

The Quantum World of Ultra-Cold Atoms and Light

Book I: Foundations of Quantum Optics

Crispin Gardiner
University of Otago, New Zealand

Peter Zoller
University of Innsbruck, Austria

Imperial College Press

ICP

Published by

Imperial College Press
57 Shelton Street
Covent Garden
London WC2H 9HE

Distributed by

World Scientific Publishing Co. Pte. Ltd.
5 Toh Tuck Link, Singapore 596224
USA office: 27 Warren Street, Suite 401-402, Hackensack, NJ 07601
UK office: 57 Shelton Street, Covent Garden, London WC2H 9HE

British Library Cataloguing-in-Publication Data
A catalogue record for this book is available from the British Library.

Cold Atoms — Vol. 2
THE QUANTUM WORLD OF ULTRA-COLD ATOMS AND LIGHT
Book 1: Foundations of Quantum Optics

ISBN 978-1-78326-460-5
ISBN 978-1-78326-461-2 (pbk)
ISBN 978-1-78326-462-9 (ebook: library)
ISBN 978-1-78326-463-6 (ebook: individual)

for

Hanni and Helen

Preface

The quantum world, once seen as inaccessible to direct observation and irrelevant to direct practical application, is now ready to join the mainstream of applied physics and become the home of advanced technologies. Up until recently, the technical underpinning was based on the carefully controlled use of laser light and atoms—the field of quantum optics. More recently, a variety of other techniques have been developed, some using the methods solid state physics used in the semiconductor industry, some using superconductor technology, and others using assemblies of ultra-cold atoms trapped in an optical lattice.

This is the culmination of half a century of exploration of the foundations of quantum mechanics using lasers and atoms, with the almost magical result: *quantum mechanics is correct!* This theoretical structure of quantum mechanics, formulated quite hastily in the mid-1920s by a few scientists to explain atomic spectra, has now been applied to a wide variety of systems—nuclei, elementary particles, solids, superconductors, metals, etc. So far, it has never shown itself to be in disagreement with experiment, even in situations where its predictions appear quite paradoxical.

This book is the first in a trilogy designed to make available to those working in the field, and to advanced students, a full suite of theoretical techniques needed for quantum technologies. These have been developed within the paradigm of quantum optics, a field based on the use of atoms and light. As participants in the theoretical aspects of quantum optics, ultra-cold atoms, and quantum information, we have worked both together and independently on most of the significant themes which have arisen in the last thirty years, and during this period have both taught classes and supervised research projects in the field.

We decided to write this as a three volume set so that we could set out the common theoretical framework in *Book I*, describe the optical and atomic physics aspects in *Book II*, and then develop the study of ultra-cold atoms in *Book III*. We aim to make each volume quite compact, and to write at the level of the advanced undergraduate or early postgraduate. The theoretical viewpoint is based on quantum stochastic processes, that is, the quantum-mechanical master equation and related techniques. These have proved to be very successful in quantum optics, and have been adapted to the study of ultra-cold atoms, as well as to most of the other methods of implementing quantum technologies which are currently under investigation.

The scope of the books is determined by physical and mathematical considerations. This volume, *Book I*, is compact and largely self-contained. It gives a simple grounding in the relevant parts of the theory of classical stochastic processes, and then goes on to develop the basic aspects of quantum Markov processes. As preparation, the reader should need no more than a solid background in quantum mechanics and an acquaintance with Maxwell's equations. Based on that background, this relatively short book covers almost all of the basic techniques, and contains many examples which illustrate both the physics and the mathematics being discussed.

For those who wish to go further, an extensive exposition of the theory of classical stochastic processes can be found in *Stochastic Methods*, in a form and notation consistent with this book. Similarly, in the field of quantum optics, our previous book, *Quantum Noise*, will be valuable to the reader, especially for the clarification of some of the more complex aspects of the field. However, the physics has developed considerably even since the completion of the last edition in 2004, and *Book II* and *Book III* cover significant aspects of physics that have developed since that time.

The next two volumes will not be strongly dependent on each other, but will build on the material in this book in a way appropriate for their respective fields. In *Book II* we develop those basic ideas of quantum optics and ultra-cold atoms which are in the forefront of current research into quantum technologies. This includes a careful grounding in the extended quantum stochastic techniques, in which the details of the mutual influence the quantum stochastic nature of light and that of atoms are linked together. The field of quantum information, based on the possibility of implementing quantum computers and similar devices using ultra-cold atoms and light, provides the main current field of study to which these methods can be applied.

Book III covers the field of ultra-cold atoms using the quantum stochastic paradigm. It includes an overview of some of the fascinating systems which can now be created using optical lattices and ultra-cold atoms; a field in which quantum simulation of quite complex systems is now starting to become realistic. It is a surprising fact that the methods of quantum optics, with their orientation based on a modular paradigm, have nevertheless proved themselves capable of describing the apparently non-modular systems which appear in the field of ultra-cold atoms. *Book III* is designed to show how this is done in a coherent and consistent manner.

Otago and Innsbruck
December 2013

Crispin Gardiner
Peter Zoller

Acknowledgments

This book, and its companion volumes, represent a summation of knowledge developed by ourselves and many others in the field over a period of about forty years, and during this period we have interacted with colleagues all around the world, as well as in our own countries. Our primary debt is to all of these colleagues, who have collectively created the environment within which we have worked. Particular thanks, however, are owed to those with whom we have worked most closely, namely the late Dan Walls, Ignacio Cirac, Rainer Blatt, Rob Ballagh, Matthew Collett, Jeff Kimble, Bill Phillips, Rudi Grimm, Hans Briegel, Immanuel Bloch, Scott Parkins, Ashton Bradley, Blair Blakie, Simon Gardiner and Dieter Jaksch. We are very grateful to Catarina Sahlberg for assistance in proofreading and indexing.

During our careers, we have been supported by many institutions: the University of Otago, the Leopold Franzens University of Innsbruck, the University of Colorado, Victoria University of Wellington, the University of Southern California, and the University of Waikato.

During the time in which we have been writing we have been generously supported by New Zealand through its research funds, the *Marsden Fund* and the *New Economy Research Fund*, and by Austria through *die Österreichische Akademie der Wissenschaften* and *die Standortagentur Tirol.*

Contents

I The Physical Background

1. Controlling the Quantum World

Quantum mechanics, definitively formulated in the 1920s, proceeded within the space of perhaps only a decade, to revolutionize chemistry and physics. The success of quantum mechanics was dramatic—the basic understanding of atoms, molecules, solids, conductivity and semiconductors was essentially complete by the end of the 1930s. The impact of quantum mechanics as a reliable and accurate description of the physical world was succinctly put by *Bardeen* [1] in his Nobel Prize lecture:

> In this lecture we shall attempt to describe the ideas and experiments which led to the discovery of the transistor effect as embodied in the point-contact transistor. ... It was dependent both on the sound theoretical foundation largely built up during the thirties and on improvement and purification of materials, particularly of germanium and silicon, in the forties. ... The general aim of the program was to obtain as complete an understanding as possible of semiconductor phenomena, not in empirical terms, but on the basis of atomic theory. A sound theoretical foundation was available from work done during the thirties:
>
> i) Wilson's quantum-mechanical theory, based on the energy band model, and describing conduction in terms of excess electrons and holes. ...
>
> ii) Frenkel's theories ... in which general equations were introduced which describe current flow when non-equilibrium concentrations of both holes and conduction electrons are present. ...
>
> iii) Independent and parallel developments of theories of contact rectification by *Mott, Schottky, Davydov.*

Research in physics has continued the paradigm noted by Bardeen. The quantum-mechanical framework has been exploited repeatedly to explain and predict phenomena of increasingly subtlety, complexity or fragility. All of these phenomena have become experimentally accessible as the result of the availability of tools of increasing power and precision, most of them a direct result of the precise and apparently universal picture of the world provided by quantum mechanics.

The laser, invented in the 1960s, has its origin in the same kind of paradigm as the transistor. In his Nobel Prize lecture, *Townes* [2] showed how the logic of quantum mechanics led to the construction of the maser, the microwave precursor of the laser:

> My own particular interest came about from the realization that probably only through the use of molecular or atomic resonances could coherent oscillators for very short waves be made, and the sudden discovery in 1951 of a particular scheme which seemed to really offer the possibility of substantial generation of short waves by molecular amplification. ...
>
> The crucial requirement for generation, which was also recognized by *Basov* and *Prokhorov* was to produce positive feedback by some resonant circuit and to ensure that the gain in energy afforded the wave by stimulated molecular transitions was greater than the circuit losses. ...
>
> We called this general type of system the maser, an acronym for *M*icrowave *A*mplification by *S*timulated *E*mission of *R*adiation. ... Maser amplification is the key process in the new field known as quantum electronics—that is, electronics in which phenomena of a specifically quantum-mechanical nature play a prominent role.
>
> It is well known that an amplifier can usually be made into an oscillator, or vice versa, with relatively minor modifications. But it was only after experimental work on the maser was started that we realized this type of amplifier is exceedingly noise-free.

Townes, *Gordon* and *Zeiger* built the first maser in 1953, and the first laser—the extension of the maser principle to wavelengths in the optical domain—was demonstrated by *Maiman* in 1960. By the end of the decade most of the laser types we now know had been demonstrated and lasers were commercially available at a reasonable price.

These two inventions, both founded in quantum mechanics—the transistor and allied solid state devices, and the laser and related optical systems—now dominate our world. It is hard to imagine anyone in the developed world who does not make use of lasers and solid state electronics on a daily basis. For the scientist the influence is even stronger, for nearly every instrument used in a modern laboratory makes use of the solid state and optical devices which have arisen from the transistor and the laser. Some of these are simply convenient devices, like the optical mouse, or the LED computer screen, but in the physics of atoms and light, there are now highly specialized laser and optical devices of extraordinary precision which enable quantum control at an unprecedented level of accuracy and reproducibility.

As well as permeating the everyday world, the laser brought about a revolution in the physics of atoms, molecules and light. The precise intensity and coherence of laser light led to the major paradigms of quantum engineering which are the subject of this trilogy:

i) Exotic quantum states of light, such as *antibunched light, squeezed light* and *quantum correlated light beams* enable optical measurements to be done at very high precision, and as well, provide a framework inside which quantum effects can be used in information technology.

ii) The interactions between atoms and laser light can be used to manipulate the internal states of atoms, enabling the experimenter to put an atom into almost any quantum state.

iii) The mechanical interaction between laser light and atoms can be used to trap and cool, ultimately leading to the production of quantum degenerate matter, such as the Bose–Einstein condensate, and later, assemblies of Fermions at almost zero temperature.

iv) More recently, *engineered* devices have been investigated, in which the concepts of quantum optics are implemented using devices designed and manufactured specifically for that purpose. The most promising idea is to emulate quantum optics on solid state devices, which would be both robust and cheap, and therefore provide a realistic platform for quantum technology.

1.1 Quantum Optics

For similar reasons to those noted by Townes for the maser, the light from a laser is very different from that emitted by thermal sources, such as the the mercury arc lamp which, up to time of the laser, provided the most intense source of light with high spectral purity. The light is not only spectrally pure, but in an ideal laser, it is also quantum-mechanically pure, that is, the light from an ideal laser can be regarded as being almost in a quantum-mechanical pure state. From this feature began the field of quantum optics—that is, optics in which the quantum nature of light would play an essential role.

1.1.1 The Development of the Laser

a) Laser Theory: The difference between laser light and light from an ordinary lamp was not immediately appreciated. However, the development of accurate theoretical descriptions of the lasing process, particularly by *Lax, Haken* and *Lamb* [3–7], demonstrated that laser light was highly coherent and that associated with this, the statistics of photon counting from an ideal laser would be Poissonian, as opposed to the power-law distribution expected from thermal light. The laser initiated a new paradigm in quantum statistical mechanics, as a source of light in a quantum-mechanically pure state, but one in which this pure state arose in a driven open quantum system, dominated by dissipation.

The paradigm until then was that of thermodynamic equilibrium, in which the purest states were ground states, achieved only at the lowest temperatures. The most analogous example is that of the superconductor, whose theoretical explanation by *Bardeen, Cooper* and *Schrieffer* [8] was one of the great triumphs of twentieth-century theoretical physics. The ground state of a superconductor is formed from a linear combination of normal state configurations in which electrons are virtually excited in pairs of opposite spin and momentum. This state, now known as the BCS state, is lower in energy than the normal state by small amount $\hbar\Delta$. When $k_B T < \hbar\Delta$, the number of electron pairs in the ground state becomes a macroscopic number, and the resulting behaviour is that of a superconductor. Thus, the BCS state only appears *below* the critical temperature, and its appearance happens in thermodynamic equilibrium.

In contrast, laser light only appears *above* a critical pumping rate, and is thus a genuine non-equilibrium phenomenon. The theoretical treatment of the laser which arose was that of a driven open quantum system, dominated by dissipation, and involving many atoms. This gave rise to a description in terms of quantum Markov processes, in which both classical and quantum noise played an essential role.

b) Practical Implementations of Lasers: Although lasers were initially only laboratory instruments, it took very little time to develop the full range of lasers now used. Initially, lasers produced only pulses of light, and many of the most useful lasers are still pulsed lasers, available with pulse lengths of picoseconds, femtoseconds and even attoseconds. But steady state gas lasers (generally called CW, or continuous-wave lasers) soon became available, and then dye lasers, pumped by gas lasers, provided continuously tuneable laser beams. Methods for stabilizing these were introduced and continuously improved.

Nowadays a wide range of solid state laser devices exist, producing laser light for almost any need, and lasers are available in an enormous range of power levels. They can be used, at one extreme, for cutting, welding, or even initiating nuclear reactions, and at the other extreme, for the precision manipulation of individual ions and atoms.

1.1.2 The Development of Quantum Optics

Lasers enabled light of an unprecedented intensity to be directed at optical materials and atomic vapours. The high intensities achieved did not necessarily mean that the laser beam had very high power, but rather that the power was concentrated in relatively few optical modes. Light of this kind behaves essentially like a classical electromagnetic field when it interacts with matter, and it is from this property that the practical utility of laser light arises as a research tool.

a) Materials and Atoms: Three main research methods arose from the availability of high-quality laser light, as follows:

i) *Non-linear Optics*: One of the first consequences was the development of the field of non-linear optics, in which the high intensities of laser light now available enabled the effects of an intensity-dependent refractive index to be extensively explored. The refractive index arises fundamentally from the interaction of the light with individual atoms in the material. Nevertheless, because there are very large numbers of atoms within even a highly focused laser beam, the traditional description in terms of a refractive index is normally still possible. In fact, within this traditional optical point of view, phenomena which depend critically on quantum mechanics can still be observed and described.

ii) *Manipulation of Atomic States*: The influence on atomic spectroscopy was revolutionary. Using the coherence and spectral purity of laser light, it became possible to manipulate the quantum states of atoms with precision and reliability. The traditional way of looking at atomic transitions as incoherent processes driven by a stream of photons arriving independently was replaced by the picture of a two-level system driven by a coherent optical field. This point of view was similar to that which had been introduced by *Rabi* using radio frequency fields on Zeeman sublevels of atoms, and which had been further developed by *Bloch, Purcell* and others, leading to the field of nuclear magnetic resonance.

iii) *Optomechanical Effects*: Electromagnetic fields exert forces on atoms, but it is only with the intensities available from lasers that these forces are measurable and significant in a laboratory situation. With the development of stable and reliable lasers it has become possible to trap atoms in a strong light field, and to cool them to temperatures of the order of microKelvins or less, leading ultimately to the creation of the world's first Bose–Einstein condensate in 1995 [9], and the inauguration of a whole new field of physics, that of ultra-cold atoms, the subject of *Book III* of this trilogy.

iv) *Ultra-Cold Atoms in Optical Lattices*: By setting up standing waves of very intense light, lattices of optical traps can be constructed, in one, two and three dimensions. The very low kinetic energies available from ultra-cold atoms at Bose–Einstein condensation temperatures make it feasible to load an optical lattice with one atom (or more) per site. These atoms have weak interactions with each other, and provide a lattice implementation of quantum field theory. Formulated by *Jaksch, Bruder, Cirac, Gardiner* and *Zoller* [10], and implemented experimentally by *Greiner, Mandel, Esslinger, Hänsch* and *Bloch* [11], the physics provided by atoms thus trapped in an optical lattice provides a complete toolbox for quantum engineering, such as was never available using traditional condensed matter systems.

b) Non-classical Light: The concept of *non-classical light* arose in the early period of quantum optics, stimulated by the realization that the theoretical methods developed for lasers permitted many more kinds of light than arose from either the laser itself or from thermal sources. Here two principal kinds of non-classical light were theoretically predicted and then observed, namely *antibunched light* and *squeezed light*—the situation at the time was elegantly summarized by two review articles by *Walls* [12, 13].

In squeezed light the statistics are modified in a way which is equivalent to the reduction of the quantum noise, and can be viewed as light in which pairs of photons are quantum-mechanically correlated. It can also be used to increase the precision of interferometric measurement techniques. Since it is typically produced with a non-linear crystal, squeezed light is an example of the non-linear optics point of view. Squeezed light can be created in two optical modes, and in that case is equivalent to the production of entangled states (see Sect. 1.2.2) of correlated photons. Such entangled states have applications in quantum cryptography and quantum communication.

In antibunched light, the photons arrive at a detector more regularly than in the case of coherent light from a laser, for which the arrival times obey Poissonian statistics. Typically, antibunched light is produced by the interaction of light on a specific atomic transition in individual atoms, and thus is an example of the atomic spectroscopy point of view. Single photon sources, now being used for quantum cryptography, are equivalent to a kind of antibunched light.

1.1.3 The Elementary Objects of Ideal Quantum Optics

In addition to *ideal laser fields*, the two elementary objects out of which quantum-optical systems can be modelled are *harmonic oscillators* and *two-level atoms*. Both of these are characterized by a single transition frequency ω. The harmonic oscillator has an infinite number of energy levels with level spacing $\hbar\omega$, while the two-level atom has only two levels, spaced $\hbar\omega$ apart.

The real quantized electromagnetic field is described by an infinite number of harmonic oscillators, one for each mode of the field, and by the use of appropriate resonators, such as a Fabry–Pérot cavity, specific modes can be isolated, and a very good approximation to an ideal harmonic oscillator can be constructed. The real atoms used have very many levels, with a variety of spacings between them. The transitions between these energy levels have a linewidth, and this linewidth is normally very much less than the level spacing. This means that transitions between pairs of levels can be easily isolated, and the dynamics of a single pair of levels can then be modelled quite well by a two-level atom.

Quantum optics is the science founded on coupling these kinds of systems together, using real atoms and light. In practice it is often necessary to extend this simple picture. This can involve introducing other energy levels to the two-level system, or using non-linear optical materials to give very useful modifications of the harmonic oscillator Hamiltonian. The full range of phenomena available for study or application then becomes very rich.

By *ideal* quantum optics, we mean the theoretical physics of these simplified systems, in contrast to *real* quantum optics, in which this physics is implemented using *real* atoms and light. There are disadvantages in implementing ideal quantum optics using real atoms and light. In particular, the essentially three-dimensional nature of the light field to which an atom couples is a problem. One would like to couple to tightly defined light beams, but what is the best technology for doing this has not yet been established. Similarly, atoms in quantum optics have to be trapped by rather complicated devices, involving intense light fields or strong magnetic fields. Furthermore, the coupling between atoms and light can only be made very strong using devices which are in practice rather difficult to construct.

There are good reasons to investigate whether the kind of technology used in the semiconductor industry, based on carefully engineered devices constructed microscopically, could be used instead of atoms and light to implement the model two-level atoms and harmonic oscillators of ideal quantum optics. Currently there is a range of alternative technologies under investigation, and some of these are now able to emulate almost all of the capabilities of the quantum optics of real atoms and light. In the next decade, we can expect some of these alternative quantum-optical technologies to mature and yield useable devices, initiating an era in which quantum technology enters the mainstream of modern high-tech engineering.

1.2 Quantum Information

The concepts and formalism of quantum optics will certainly remain with us as the fundamental set of tools to describe quantum technologies. The capability of the ideal quantum optics to describe a useful quantum technology is without question, and the theoretical tools—the object of this trilogy—are very well developed and very powerful. Using these tools, the field of *quantum information* has been developed over the past fifteen or so years as the paradigm for the technological application of the quantum world.

During the development of quantum mechanics and its applications, there were issues of interpretation and doubt that seemed to some to be profound, and to others to be irrelevant. Even though within ten years of its formulation in 1925, quantum mechanics had successfully explained the physics of atoms, the chemical bond, the basic structure of the solid state, and the general behaviour of nuclei, doubts about the completeness or consistency of quantum mechanics were being raised. The most profound of these were expressed in the *Einstein–Podolsky–Rosen* paradox (now known universally as the EPR paradox) [14], and the case of the *Schrödinger* cat [15].

1.2.1 The Einstein–Podolsky–Rosen Paradox

The EPR paradox considers the quantum correlations of very distant objects, such as the spins of two particles initially bound in a state of zero spin. The bound state can decay in such a way as to conserve spin angular momentum, after which two particles can then depart from each other to a state of remote separation. Then on measuring the spin projection of one particle, we immediately know that the spin of the other particle has the opposite sign. Viewed this way, one might see this as little different from a classical situation, in which one spin always had value $+1$ while the other had value -1. However many years later, in 1964 *Bell* [16], on reviewing the paradox, showed that this was not so. The spin zero state is never of this kind, and can be formulated independently of the axis of quantization. Further he showed that any classical system must satisfy certain inequalities, now known as the *Bell inequalities*, which would be violated in

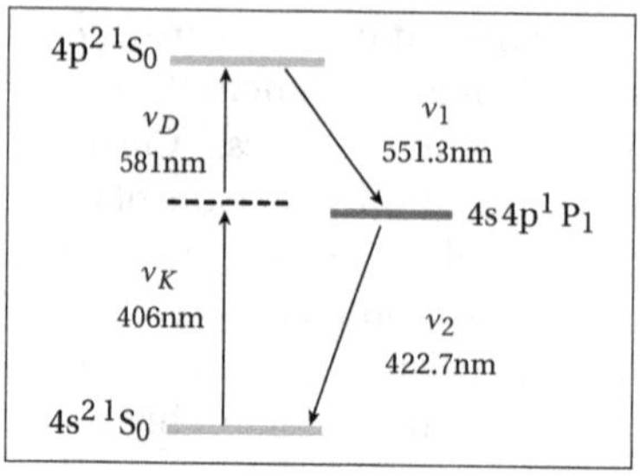

Fig. 1.1. Relevant levels of calcium, as used in the EPR experiment of Aspect, Grangier and Roger [17]. The atom, selectively pumped to the upper level by the non-linear absorption of ν_K and ν_D emits photons ν_1 and ν_2 correlated in polarization. From measurements on these photons it is found that the Bell inequalities are violated, thus verifying that the EPR paradox exists in reality.

certain quantum-mechanical situations. The first confirmation that the Bell inequalities were indeed violated was given by the experiment of *Aspect, Grangier* and *Roger* [17], which involved correlated photons from calcium atoms pumped with appropriate lasers as illustrated in Fig. 1.1.

1.2.2 Schrödinger's Cat and Entangled States

Schrödinger's cat paradox, in which he considers how a cat might be put into a quantum superposition state of death and life, is fundamentally based on the same aspect of quantum mechanics as the EPR paradox, namely, that of *entanglement.* The simplest example of entanglement comes from a state of two non-interacting particles A and B, for each of which there is a basis set, which we can enumerate as $|A,i\rangle$ and $|B,j\rangle$, where i and j are indices which label the individual basis states. *All linear superpositions of the product basis are permitted by quantum mechanics,* and among these are states like $|E\rangle = |A,1\rangle|B,2\rangle - |A,2\rangle|B,1\rangle$. The EPR argument shows that these states cannot be eliminated from quantum mechanics, and indeed the state created from the decay of a bound state of zero spin is exactly of this kind. Furthermore there is now indisputable evidence that such entangled states are not only possible, but are both pervasive and useful.

The essence of the Schrödinger cat paradox lies in the idea that a macrosopic object (such as a living cat) might be able to be put into a superposition state of "alive" and "dead". Experience tells us that, for macrosopic systems, this kind of phenomenon does not happen with sufficient probability to ever be experienced. Thus, any technology which wishes to exploit quantum-mechanical entanglement will necessarily require the use of atomic-sized systems, or some carefully constructed very small quantum devices.

1.2.3 The Quantum Computer

In 1982 *Feynman* [18] expressed the view:

> ...the full description of quantum mechanics for a large system of R particles is given by a function $\psi(\boldsymbol{x}_1, \boldsymbol{x}_2, \ldots, \boldsymbol{x}_R, t)$ ...and therefore, because it has too many variables, it *cannot be simulated* with a normal computer.

While Feynman's reasoning is not absolutely watertight, there is no doubt that a quantum system is in practice very much more difficult to simulate than a similar classical system. The logical next step was to investigate the possibility of using fully quantum-mechanical logic gates to implement a *quantum computer*, and in 1986 Feynman [19] showed how to implement such gates abstractly using the rules of quantum mechanics, in a way which requires the use of the entangled states discussed above. These concepts were given a more rigorous and systematic form by *Deutsch* [20, 21], who also suggested that a quantum computer might be more powerful than a conventional computer, even for problems not of a specifically quantum-mechanical nature. However it still remained to be shown how to implement a quantum computer in practice, and how to use it for any practical tasks.

1.2.4 Shor's Algorithm and the Cirac–Zoller Ion Trap Quantum Computer

Two breakthroughs happened in 1994. Firstly, *Shor* [22] devised a quantum computation algorithm for factorizing integers into primes, a problem of central importance to encryption technologies, and he showed this algorithm was more efficient than any known for a conventional computer. This was the definitive verification of Deutsch's conjecture—a quantum computer *can* be more powerful than a classical computer for a classical problem.

Secondly, *Cirac* and *Zoller* [23] showed that a system of trapped ions, interacting by their electric fields and manipulated and addressed using lasers, provided a complete implementation of all of the necessary attributes of a quantum computer in a form feasible in the laboratory. Such a quantum computer requires in practice extensive use of solid state electronics, highly stabilized lasers, and the implementation of quantum entanglement. At the date of writing, laboratory systems of trapped ions have been used to demonstrate the basic building blocks of a scalable quantum computer, including high-fidelity gates between qubits [24–26] and the transport of ions in a charge-coupled device between memory and processor regions [27].

1.2.5 Quantum Computing with Trapped Ions

Let us look more closely at the details of an ion trap quantum computer, since this provides an example of how quantum-optical techniques can be used in practice to implement quantum information processing, as well as some of the practical disadvantages of an approach based on traditional techniques of this kind.

Because they are charged, small numbers of ions can be trapped by combinations of electric and magnetic fields, yielding a highly controllable system, a technique pioneered by *Dehmelt* [28] and *Paul* [29]. Typically nowadays tens of ions can be trapped in such a way as to interact with each other quite strongly through their Coulomb fields. In Fig. 1.2 we illustrate the Innsbruck quantum processor [30], an ion trap quantum computer, configured to simulated damped quantum sytems.

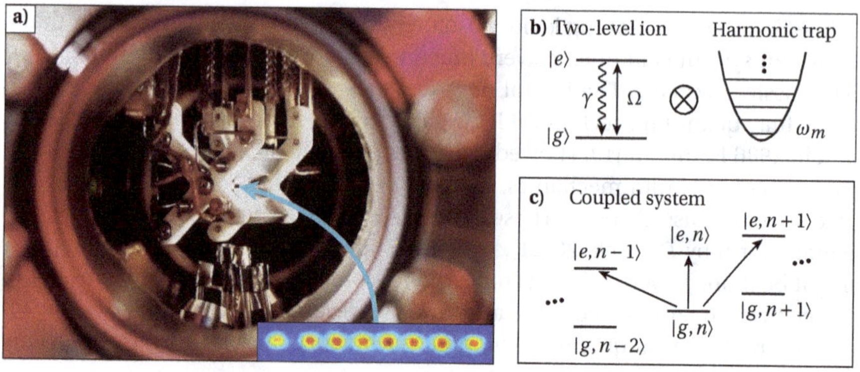

Fig. 1.2. The Innsbruck quantum processor. **a)** A Paul trap containing individually addressed $^{40}Ca^{+}$ ions; after laser cooling, the trapped ions align into a string—a typical image of an ion string (of length about 70μm) is shown in the inset. **b)** The energy levels of a two-level ion (left) and one mode of the ion's motion (right). On the left are the ion's ground state $|g\rangle$ and excited state $|e\rangle$, interacting with radiation characterized by the Rabi frequency Ω and decaying with the rate γ. On the right are the harmonic oscillator potential and equally spaced energy levels for one mode of motion (the electrode arrangement provides an almost harmonic three-dimensional well). **c)** The level structure of the coupled system of the ion and the harmonic oscillator. States are described by the spin ($|g\rangle$ and $|e\rangle$) and motional ($|1\rangle$, $|2\rangle$, ... , $|n\rangle$) degrees of freedom. Arrows indicate the transitions that are possible when appropriately tuned radiation is applied; dotted lines indicate levels not shown. (*Image kindly supplied by Prof. R. Blatt.*)

The practical implementation of the ion trap quantum computer has been pioneered by *Blatt* in Innsbruck and *Wineland* in NIST (Boulder) who, in their review on ion traps and entanglement [31], point out:

> To study entanglement, it is desirable to have a collection of quantum systems that can be individually manipulated, their states entangled, and their coherences maintained for long durations, while suppressing the detrimental effects of unwanted couplings to the environment. This can be realized by confining and laser cooling a group of atomic ions in a particular arrangement of electric and/or magnetic fields.
>
> With such "traps", atomic ions can be stored nearly indefinitely and can be localized in space to within a few nanometres. Coherence times of as long as ten minutes have been observed for superpositions of two hyperfine atomic states of laser-cooled, trapped atomic ions.
>
> In the context of quantum information processing, a typical experiment involves trapping a few ions by using a combination of static and sinusoidally oscillating electric potentials that are applied between the electrodes of a linear quadrupole, an arrangement known as a Paul trap [29]. When the trapped ions are laser cooled, they form the linear "string", in which the spacings are determined by a balance between the horizontal (axial) confining fields and mutual Coulomb repulsion. Scattered fluorescence, induced by a laser beam, can be imaged with a camera. The use of tightly focused laser beams allows the manipulation of individual ions.

Using such traps, the groups of both Wineland and Blatt have been able to entangle ions, and implement the basic operations of the quantum gates needed for quantum information processing in the ion trap quantum computer. In particular, they have demonstrated:

i) High-fidelity gates between qubits;

ii) The transport of ions in a charge-coupled device between memory and processor regions.

1.2.6 Engineered Systems for Quantum Optics

A realistic quantum computer must be *scalable*, that is, it must be possible to make a device as large as necessary by connecting a sufficient number of basic devices to each other. While the ion trap quantum computer is at present the most developed implementation of a quantum computer, and is in principle scalable, it is nevertheless a very complex device. It can only be considered a prototype of the practical quantum computer of the future, whose implementation may use an entirely different technology, as noted in Sect. 1.1.3. There are many proposals for different technological implementations of quantum information and quantum computing, and all of them use the same basic quantum-optical paradigm. However, instead of the familiar setup of atoms trapped in free space, connected by light beams travelling in three dimensions, they create (in a variety of ways) engineered analogues, with the intention of creating implementations of ideal quantum optics in a way which is more effective than the traditional benchtop optical paradigm. For example:

i) Two-level systems can be emulated using superconductor technology using a "Cooper pair box" [32]. The transition frequency is in the microwave regime, and the relevant electromagnetic field can be transmitted along a superconducting transmission line.

ii) "Quantum dots", namely very small crystals of semiconductor, embedded in a matrix of a different kind of semiconductor can also emulate two-level systems [33]. In this case the transitions are in the infrared optical regime, and strong coupling can be implemented using engineered optical resonators in the semiconductor matrix.

iii) Rare-earth atoms can be doped in a crystalline substrate, such as YAG[1], yielding effective two-level systems which are very well isolated from the environment. These function in the optical regime, and the quantum information technology is implemented using photon-echo techniques [34].

This is not an exhaustive list, but serves to give some idea of the ingenious proposals under investigation.

---

[1]YAG is the acronym for yttrium aluminium garnet, $Y_3Al_5O_{12}$. It is widely used as a host material in solid state lasers.

2. Describing the Quantum World

In any attempt to harness the opportunities offered by the quantum world, the fundamental fragility of the kinds of devices which might be utilized has to be faced. Those properties of quantum mechanics, such as coherence and entanglement, which can be exploited for the development of quantum devices, are very easily degraded by extraneous influences. For example, the ion trap quantum computer, as described in Sect. 1.2.4 and Sect. 1.2.5, must deal with technical noise, such as magnetic field fluctuations and thermal noise, as well as genuinely quantum-mechanical noise. In fact it can be reasonably said that the issue of noise is of overwhelming importance in the practical implementation of quantum devices. For the laboratory, this means extraordinary care must be taken to isolate the devices from the outside world, but this is not enough. As well as superb experiments, reliable and efficient theoretical descriptions of quantum and classical noise must be available in order to design the devices.

The early development of theoretical quantum optics soon brought out the need to have a theory which could provide a description of the optical physics in such a way as to include fluctuations and noise having both a thermal and a quantum-mechanical origin. Furthermore, it was necessary to consider systems, such as a gas laser, composed of subsystems, such as the gas providing the laser active medium and the electromagnetic field modes into which light is emitted. Each of these is not very dissimilar to a system in thermal equilibrium. However, in the gas the internal atomic energy levels are pumped so as to be far from the Boltzmann distribution, while the translational modes are not. The cavity modes of the electromagnetic field are highly occupied, but there are few of these in comparison to the many other modes of the electromagnetic field into which the atoms may radiate. Such non-equilibrium behaviour occurs in classical systems, such as heat engines, which are also composed of subsystems out of equilibrium with each other and linked to a moving mechanical system, which is the analogue of the laser light field. The challenge is to get a description which respects the quantum-mechanical framework, and makes it available for study.

The initial phase of theoretical quantum optics was dominated by a number of *ad hoc* methods, all relying on the same basic physical approximations, but adapted to the particular problem under study. Because the necessary structures were all similar, this gave birth to the concept of a *quantum Markov process*, in which the appropriate quantum analogue of the classical Markov process was formulated. By the late 1990s, the field had matured to the extent that all of the concepts of classical stochastic processes had their quantum counterparts. However, in addition to that, the quantum-mechanical wave-particle duality leads to a duality in quantum stochastic systems, where certain stochastic wave properties, such as optical coherence, are seen to be intimately connected to stochastic particle properties, such as the Poissonian photon counting statistics of laser light.

2.1 Classical Stochastic Processes

One of the most significant achievements of twentieth-century science was the formulation and application of the idea of a system which evolved in time in a manner governed by probability laws—such a time evolution is nowadays called a *stochastic process.* There is an enormous variety of possible stochastic processes, in fact the possibilities are many more than can find any reasonable application. *Markov* investigated the simplest possible form of non-trivial stochastic process, that in which one could determine the future time evolution of the system from the knowledge of the state of the system in the present—this is of course known today as a *Markov process.* This kind of stochastic process dominates the modelling of physical systems to this day.

Markov processes provide the most convenient way with which to describe fluctuating classical systems, such as electrical circuits, Brownian motion, and light fields. But more importantly, they are also very relevant to the study of fluctuating *quantum systems*, even if they cannot provide an adequate description of quantum probability. To understand the nature of those quantum Markov processes which are necessary to describe the quantum world, a grounding in classical stochastic processes is indispensable, and *Stochastic Methods* provides a thorough exposition of the topic. However, the bare minimum needed is not nearly so extensive, and Part II of this book provides an outline of the most relevant aspects of stochastic processes for this field of study.

2.1.1 Probabilities, Paths and Correlations

A classical stochastic process can be viewed as the motion of a point in some abstract or possibly real physical space. We can look at this from two points of view.

a) Sample Paths: The motion of a point traces out a path in the space, and a single realization of this is called a *sample path.* In some sense, this corresponds most directly to what one might measure; an experiment would ideally consist of a number of observations of such sample paths.

Correspondingly, one can develop equations of motion with randomly fluctuating terms which give a theoretical description of the motion of a point, and in the case that the motion is described by a Markov process, this is called a *Langevin equation* or a *stochastic differential equation.*

b) Probability Distributions: An alternative point of view is to look at the probability distributions of the position of the point at a number of times, and develop equations of motion for these probability distributions. In the case that the motion is both Markovian and continuous, these lead to partial differential equations for the probability distributions, known as *Fokker–Planck equations.*

Discontinuous motion is also possible. Such motion is normally piecewise continuous, with jumps occurring at random times, and the time-development equation for the probability distribution, called a *master equation*, is no longer a partial differential equation, and instead involves a consideration of all possible jumps.

2.1.2 Markov Processes and the Conditional Probability

For *Markov processes* of all kinds, with continuous or discontinuous motion, the fundamental object is the *conditional probability* $P(x_1, t_1 \mid x_0, t_0)$. This gives the probability, conditioned on starting at the point x_0 at time t_0, of arriving at the point x_1 at a *later* time t_1. This is all that is needed to determine the complete structure of a Markov process, and this is true for all Markov processes.

2.1.3 Measurement and What to Measure

Unlike quantum mechanics, classical mechanics allows the concept of measurement without affecting the system to be measured. Because classical mechanics involves only variables which have simple numerical values, there is not much structure imposed on the variables to be measured. This is in contrast to the quantum theory of measurement, in which the physical effect of the measurement process on the system has to be taken into account, and in a way that is consistent with quantum mechanics.

2.1.4 Correlation Functions and Spectra

Measurements that can be made on a fluctuating system need to be analysed, and the most common tool is to determine the influence of present values on the distribution of future values. The conditional probability is the appropriate mathematical construct, but it is not easily determined in practice. The *autocorrelation function* measures the average value of the product $X(t_1)X(t_2)$ of the variable at two different times. The Fourier transform of the autocorrelation function gives the corresponding *spectrum*, which gives insight into the possible oscillation frequencies in the system, and the degree to which they are excited.

Thus, the correlation functions and spectra arising from a fluctuating system are experimentally straightforward to measure, and reflect the underlying dynamics taking place. They are also straightforwardly formulated theoretically. So, while they do not characterize the system completely, they do encapsulate the most significant physical properties.

2.2 Theoretical Quantum Optics

The quantum-optical systems, such as those used in the devices described in Chap. 1, are *modular* systems, that is, systems composed of distinct parts coupled to each other. For example, the ion trap quantum computer illustrated in Fig. 1.2 is composed of a number of well-separated atoms, which interact with each other weakly through their electric charges, giving rise to coupled vibrational modes. The atoms are individually addressed and manipulated using laser beams, and the quantum information can be transferred to other systems by optical means.

To describe such systems requires a fully quantized theory, with a theoretical description of the individual elements and of the phononic and photonic connections between them. In practice, this description must be a version of quantum field theory.

2.2.1 Quantum Fields and "Stripped-Down" Quantum Electrodynamics

The successful quantization of the electromagnetic field came at the end of a long struggle with the problem of divergences, and yielded the rather daunting formalism of quantum electrodynamics. This formalism is fully relativistic, gauge invariant, and provided a description of photons and electrons of such accuracy that its basic truth could not be doubted. However, for the description of optical and electronic systems, the subtleties of renormalization are not particularly important, while in contrast, the compact description of quantized electromagnetic fields with very large numbers of photons, such as those that are produced by lasers, is of immense advantage in practical calculations.

Therefore, for quantum optics and quantum electronics, a stripped-down version of this formalism was developed, in which the complexities of renormalization and the very intricate Feynman diagram formalism were put aside in favour of a formulation more closely related to electrical circuit theory. The most important simplification arises from the restricted range of frequencies which are in practice required. For example, in the description of the decay of an excited state of an atom to a lower state plus a single photon, only field components with frequencies very close to the transition frequency play any significant role in the decay constant. Nevertheless, because decay takes place in free space, there are an infinite number of frequencies in this narrow range, and this infinity of frequencies is necessary to yield the irreversible nature of atomic decay. There is also always a frequency shift associated with a decay process, and in this case this amounts to the *Lamb shift*, whose explanation does require the full apparatus of quantum electrodynamics. The stripped-down version of quantum electrodynamics cannot give the right value for the Lamb shift, but for practical purposes this is quite unimportant. The Lamb shift is very small, and does not change the basic nature of the system, unlike atomic decay. If, in spite of its smallness, we wish to account for it, we can simply adjust the frequency of the transition between the two levels by an appropriate very small amount.

In summary, the full apparatus of quantum electrodynamics is absolutely necessary to account for phenomena like the Lamb shift and the correction to the magnetic moment of the electron. However, since these corrections are all very small they have little relevance to quantum engineering, so "stripped-down" quantum electrodynamics is usually adequate, and indeed it is also very powerful. The kinds of approximations needed, and thus the formulation of the concept of "stripped-down" quantum electrodynamics, are given in Chap. 12.

2.2.2 The Building Blocks of Quantum-Optical Systems

Quantum optics has a restricted set of simple building blocks, abstracted from the quantum electrodynamics of atoms, light and electrons, as we have already noted in Sect. 1.1.3. Mirrors are formed into optical cavities, which can be abstracted into harmonic oscillators, and atoms are stripped down into a few relevant energy levels between which transitions are optically induced. The most extreme simplification is the *two-level system*, and even when more levels are treated, the basic understanding is of the levels treated two at a time.

2.2.3 The Quantum-Mechanical Master Equation

These constructs, which were necessary for a theoretical description of such systems, were developed as they were needed. The description of the light used to couple the systems was not a major issue.

In the early years, when the only kind of light being used was either thermal or coherent, attention was focused on a description of the quantum state of one or many atoms interacting with light tuned to particular transitions. Aspects of damping and noise were unavoidable, and by the 1980s the standard description found to be of most use was one in which the quantum-mechanical *density matrix* of a system with a few energy levels was governed by a simple equation of motion—the quantum-mechanical *master equation*. This equation is able to describe the irreversible evolution of the density matrix in a way which is consistent with all quantum-mechanical requirements. However, the properties of the incident and emitted light were calculated from the solutions of the master equation using particular procedures developed for each particular problem.

2.2.4 The Input-Output Formalism

It was only with the advent of squeezed light, in which some aspects of the noise properties of light are reduced below those of the vacuum, that a detailed accounting of the quantum nature of the light interacting with an atom or another quantum system (such as a non-linear medium inside an optical cavity) became necessary, and the appropriate formalism was provided by the input-output formalism of *Collett* and *Gardiner* [35, 36]. In this formalism, the nature of the light incident on a system is specified, and from this a corresponding master equation for the system can be deduced and solved. The light emitted from the system can then be described in terms of the system operators and the input light. The methodology is fully modular. In an extension of the formalism developed independently by *Carmichael* [37] and *Gardiner* [38], the method of *cascaded quantum systems* showed how to treat the output of one system as the input of another system within a master equation formalism. Thus, arbitrary numbers of modular subsystems can be coupled *unidirectionally* to each other.

2.3 Quantum Stochastic Methods

The mixture of quantum mechanics and classical probability inherent in the master equation leads naturally to a description of the formalism as that of *quantum Markov processes.* In classical Markov processes, the central object is the *conditional probability,* that is, the probability of a certain configuration, given the knowledge of an initial configuration at a previous time. The density matrix is the natural quantum generalization of the conditional probability. From these elementary beginnings, for both classical and quantum Markov processes, a very rich structure can be developed, which can be used to describe almost any situation encountered in practice.

2.3.1 Quantum Markov Description

The formalism, while very powerful, is based on certain restrictions related directly to the modular nature of quantum-optical systems. The modular components, such as atoms, cavities and detectors, are weakly coupled via optical fields, and the optical fields are themselves essentially monochromatic, with a relatively narrow spectral linewidth.

The description in terms of *quantum Markov processes,* which arises directly out of these restrictions, is fully quantum-mechanical, and hence it can be described in a number of apparently quite different formalisms, related to the well-known choice of the Schrödinger equation or the Heisenberg equations of motion in quantum mechanics.

i) *The Master Equation*: This is simply an equation for the density matrix of a quantum system driven by light of a known kind, as described above.

ii) *Operator Quantum Stochastic Differential Equations*: Here the Heisenberg equations of motion for the operators of the electromagnetic field and the quantum system are transformed in such a way as to make the input optical field become a *quantized noise source* which augments the Heisenberg equations of motion for the system operators. The narrow bandwidth of the driving fields and the weak coupling of the system lead to a description in which the optical fields can be regarded as *quantum white noise*, which is a concept equivalent to that of a quantum Markov process. In fact, the master equation can be derived from this description. The mathematical structure which results is fundamentally based on the "input-output" method as described above.

iii) *The Quantum Stochastic Schrödinger Equation*: A description using the quantum white noise limit but in the Schrödinger picture can also be developed. In this case the noise sources so derived drive the Schrödinger equation for the wavefunction of the system and its light field. The formalism which arises is equivalent to that of quantum stochastic differential equations, and in some ways more unwieldy. However, it is well adapted to the description of entanglement and its evolution, since it is directly related to the wavefunction needed to describe such entanglement.

iv) *The Stochastic Schrödinger Equation*: This is a methodology based on the description of the wavefunction of the system as the photons in the light beams are counted. The description is in terms of a wavefunction for the system (excluding the light fields) driven by a classical stochastic process, corresponding to the various outcomes possible for each measurement. It is derivable from the quantum stochastic Schrödinger equation.

    This procedure arose in the early 1990s as a practical algorithm for the numerical computation of laser cooling, where the number of atomic energy levels necessary for an accurate description can become quite large. The stochastic wavefunction, with N elements, is then a much smaller object than the density matrix, with N^2 elements. Stochastic simulations based on many sample runs of the equations of motion then become quite practical.

2.4 Ultra-Cold Atoms

With the experimental realization of Bose–Einstein condensation in 1995, using methods based on lasers and laser cooling, a field developed in which the methods of quantum optics and those of condensed matter physics were both necessary. In the laboratory, the methods of magnetic and optical trapping of thousands to millions of atoms is done using all the tools of lasers and optics which led to the field of quantum optics. The temperatures reached are in the nanoKelvin range, and the gases were extremely dilute vapours of typically ^{87}Rb and ^{23}Na.

Traditional condensed matter experiments, for which the theory of Bose–Einstein condensation provided a very imperfect model, were based on truly macroscopic numbers of atoms, contained in cryostats and cooled to milliKelvin temperatures. The typical materials studied were ^{4}He and ^{3}He, and these were normally in a liquid form, whose theoretical description is complex, difficult and even today not fully understood.

The field of *ultra-cold atoms* which arose was therefore essentially new, both in the laboratory and in theoretical structure, notwithstanding the fact that some basic tools had already been developed between 1940 and 1960 by condensed matter physicists, in their quest to get tractable models which could describe the basic properties of quantum degenerate matter. To get a full understanding of ultracold atoms, a blend of the methods of quantum optics with those more traditional in condensed matter was needed.

2.4.1 Field Operators and Hamiltonians

The description of ultra-cold atoms requires a quantum field theory which works in the same kind of regime as the stripped-down version of quantum electrodynamics used in quantum optics. In both cases, only a relatively small proportion of the available modes are necessary, but some of these modes may be highly occupied.

a) The Hamiltonian for Ultra-Cold Bosons: Bosonic matter can be described by a matter wave field operator $\psi(\boldsymbol{x})$ with commutation relations,

$$\left[\psi(\boldsymbol{x}),\psi^\dagger(\boldsymbol{x}')\right]=\delta(\boldsymbol{x}-\boldsymbol{x}'),\qquad \left[\psi(\boldsymbol{x}),\psi(\boldsymbol{x}')\right]=\left[\psi^\dagger(\boldsymbol{x}),\psi^\dagger(\boldsymbol{x}')\right]=0. \tag{2.4.1}$$

The basic Hamiltonian used to describe a system of Bose atoms of mass m, interacting with each other by a potential $v(\boldsymbol{x})$ and trapped in a potential $V(\boldsymbol{x})$ takes the form

$$\begin{aligned} H_{\text{BEC}} &= \int d^3\boldsymbol{x}\,\psi^\dagger(\boldsymbol{x})\left(-\frac{\hbar^2\nabla^2}{2m}+V(\boldsymbol{x})\right)\psi(\boldsymbol{x}) \\ &\quad +\tfrac{1}{2}\int d^3\boldsymbol{x}\,d^3\boldsymbol{x}'\,v(\boldsymbol{x}-\boldsymbol{x}')\psi^\dagger(\boldsymbol{x}')\psi^\dagger(\boldsymbol{x})\psi(\boldsymbol{x})\psi(\boldsymbol{x}'). \end{aligned} \tag{2.4.2}$$

b) Pseudopotentials and Local Interaction Models: This Hamiltonian, with the explicit inclusion of the interaction potential $v(\boldsymbol{x}-\boldsymbol{x}')$ is much more detailed than is actually required for sytems of ultra-cold atoms, whose de Broglie wavelength is normally very much longer than the range of the interaction potential. All that is needed is a description valid on a distance scale appropriate for the temperature, and this normally involves only the low energy scattering properties of the interatomic interaction, in particular, the S-wave scattering length a_s. There are many subtleties in how to do this, which are explained in *Book III*, but crudely speaking, this means we can make the substitution

$$v(\boldsymbol{x}-\boldsymbol{x}')\to\frac{4\pi a_s\hbar^2}{m}\delta(\boldsymbol{x}-\boldsymbol{x}'), \tag{2.4.3}$$

into the Bose–Einstein condensate Hamiltonian, provided the description is used only for long wavelengths of the matter field.

c) The Gross–Pitaevskii Equation: When the number of quanta becomes very large, a description in terms of a c-number matter wave field $\phi(\boldsymbol{x}, t)$ becomes possible, and this matter wave field obeys the *Gross–Pitaevskii equation*

$$i\hbar\frac{\partial\phi}{\partial t} = \left(-\frac{\hbar^2\nabla^2}{2m} + V(\boldsymbol{x}) + \frac{4\pi a_s\hbar^2}{m}|\phi|^2\right)\phi. \tag{2.4.4}$$

This equation is a form of the non-linear Schrödinger equation, an equation which also can be used to characterize laser action under certain circumstances.

The matter wave field $\phi(\boldsymbol{x}, t)$ is closely analogous to a classical electromagnetic field, which also arises when the number of quanta becomes very large. The major difference is that Maxwell's equations, which describe both the classical and the quantized electromagnetic field, do not involve Planck's constant, whereas the Gross–Pitaevskii equation contains $\hbar$ in a way which cannot be avoided. Matter waves are therefore essentially quantum-mechanical, whereas electromagnetic waves are not.

d) Hamiltonian for Ultra-Cold Fermions: The classic system of low temperature Fermions is that of a superconductor, in which pairing of electrons of opposite spin and momentum generates a new quantum state. The most interesting aspects of such a system, and consequently of systems of ultra-cold trapped Fermions, can only appear when the spin states are taken into account, so that the appropriate Hamiltonian takes the form

$$\begin{aligned} H_{\text{Fermion}} &= \sum_i \int d^3\boldsymbol{x}\,\psi_i^\dagger(\boldsymbol{x})\left(-\frac{\hbar^2\nabla^2}{2m} + V(\boldsymbol{x})\right)\psi_i(\boldsymbol{x}) \\ &\quad + \tfrac{1}{2}\sum_{i,j}\int d^3\boldsymbol{x}\,d^3\boldsymbol{x}'\,v(\boldsymbol{x}-\boldsymbol{x}')\psi_j^\dagger(\boldsymbol{x}')\psi_i^\dagger(\boldsymbol{x})\psi_i(\boldsymbol{x})\psi_j(\boldsymbol{x}'), \end{aligned} \tag{2.4.5}$$

with the field operators obeying the *anticommutation relations*

$$[\psi_i(\boldsymbol{x}), \psi_j^\dagger(\boldsymbol{x}')]_+ = \delta_{ij}\delta(\boldsymbol{x}-\boldsymbol{x}'), \tag{2.4.6}$$

$$[\psi_i(\boldsymbol{x}), \psi_j(\boldsymbol{x}')]_+ = [\psi_i^\dagger(\boldsymbol{x}), \psi_j^\dagger(\boldsymbol{x}')]_+ = 0. \tag{2.4.7}$$

As in the case of the Bose–Einstein condensate, this Hamiltonian is far more detailed than necessary, and can be simplified by using similar pseudopotential methods.

2.4.2 Methodologies for Ultra-Cold Atoms

The Hamiltonians of the previous section, even simplified by pseudopotential approximations, describe systems of some considerable complexity. They can be described quite advantageously by a blend of the traditional methods of condensed matter theory and the methods of quantum optics. Some phenomena whose description exemplifies the blend of methodologies needed are the following:

i) The description of the kinetics of the growth of a Bose–Einstein condensate provides an example of the need for both approaches. The gross features of condensate growth are well described by a simple equation very like that used to describe a laser, but the finer details require aspects of physical kinetics better known in the field of condensed matter.

ii) The description of a Bose–Einstein condensate in terms of quantized matter wave field is very like that of quantized laser light in a non-linear refractive medium. But a laser is explicitly a non-equilibrium device, whereas Bose–Einstein condensation leads to a condensed state which is thermodynamic equilibrium.

iii) Systems of ultra-cold atoms are so dilute that interference experiments, analogous for example to the traditional optical double slit experiment, are feasible. The theoretical descriptions of these are similar, but not identical. Photons essentially do not interact with each other, whereas the interactions between atoms make Bose–Einstein condensate interference somewhat more subtle.

iv) The most significant practical difference between a Bose–Einstein condensate and laser light arises because the atoms in a Bose–Einstein condensate are conserved, whereas the photons in laser light are *not* conserved. Thus, it is possible to make an atom laser, in which atoms are all emitted into a single quantum state, but the atoms cannot be created on demand, as photons can. This means that the atom laser is a curiosity, rather than a useful device, because its beam is weak and short-lived.

II CLASSICAL STOCHASTIC METHODS

3. Physics in a Noisy World

Noise is a part of physics, deserving a physical treatment as much as any other physical phenomenon. In a laboratory, one always seeks to minimize noise, but it can never be completely eliminated. The laws of thermodynamics, arising as they do from a foundation of statistical mechanics, are intrinsically noisy in any finite system, and the laws of quantum mechanics are necessarily noisy because of the Heisenberg uncertainty principle.

There is a large body of mathematical and physical techniques for treating both classical and quantum noise. The aim of this part of the book is to give the simplest possible formulation of the techniques used for the theoretical treatment of noise in a physical context, to provide a basis for the other parts of the book, which deal with quantum systems, in which which both classical noise and intrinsic quantum noise are simultaneously present.

3.1 Brownian Motion and the Thermal Origin of Noise

The first observation of noise as a physical phenomenon is rightly attributed to the botanist *Brown*. In 1827 [39], using a microscope incorporating the recently invented achromatic objective, he was examining pollen grains suspended in water in order to elucidate the mechanism by which the grains moved towards the ova when fertilizing flowers. He observed that these pollen grains were in a very animated and irregular state of motion, which he thought might be a manifestation of life, and possibly the mechanism of transport for which he was looking.

Brown did not elucidate the mechanism of the phenomenon, which came to be called *Brownian motion*, and it was only in 1905 that Einstein showed that *thermal motion* of the molecules of water would account for the observed motion. He was able to give quantitative predictions, which were soon experimentally verified. Einstein's explanation depended conceptually on the kinetic theories of *Boltzmann*, *Clausius* and *Maxwell*, which saw what Clausius called *the kind of motion we call heat* [40] as a randomized motion, with an average energy of $\frac{1}{2}k_B T$ for each of the huge number of degrees of freedom in a macroscopic system. Brownian motion was the first example of a *stochastic process* to be introduced in physics, and as reformulated by *Langevin* gave rise to the idea of equations of motion incorporating *noise*, *damping* and *temperature*.

3.1.1 Equation of Motion for Brownian Motion

For a Brownian particle of mass m, in a modern notation, the equation of motion for the position x and velocity u is

$$dx = u\,dt, \tag{3.1.1}$$

$$du = -\gamma u\,dt + \sqrt{f}\,dW(t), \tag{3.1.2}$$

in which

$$f = \frac{2\gamma k_B T}{m}. \tag{3.1.3}$$

Here γ is a friction constant, proportional the viscosity of the fluid and the radius of the particle. The *noise term* $dW(t)$ is nowadays called a *Wiener increment*; it is a Gaussian random variable such that

i) $dW(t)$ is statistically independent of $dW(t')$ for $t \neq t'$.

ii) $\langle dW(t)\rangle = 0, \qquad \langle dW(t)^2\rangle = dt.$

The only dependence on the fluid itself is through the factor $\sqrt{2\gamma k_B T/m}$—this value guarantees that the average kinetic energy of the Brownian particle reaches the value $\frac{1}{2}k_B T$ (in a time of order of magnitude $1/\gamma$), thus coming into thermal equilibrium with the fluid.

3.1.2 Markov Processes

By writing the equation of motion in the form above, we can see that if we know both $x(t)$ and $v(t)$, then at time $t + dt$, we have

$$x(t+dt) = x(t) + u\,dt, \tag{3.1.4}$$

$$u(t+dt) = u(t) - \gamma u\,dt + \sqrt{f}\,dW(t). \tag{3.1.5}$$

This means that

i) If we know $x(t)$ and $u(t)$, we know the *probability distribution* of $x(t+dt)$ and $u(t+dt)$, and hence, by iteration, the probability distribution of $x(t+\tau)$ and $u(t+\tau)$, for all positive τ.

ii) In order to determine the future probability distribution, we do not need to know anything about the values of x and v for times previous to t.

iii) Thus, in a probabilistic sense, the *present* determines the *future*, the *past* is unimportant.

This is the mathematical definition of a *Markov process*, which can be seen as the probabilistic generalization of a set of first-order differential equations. Thus, Newton's laws, formulated as equations of motion for position and velocity, are of first order; the knowledge of all of the positions and velocities of a system of many particles at a given time determines these variables for all future times. Similarly, Hamilton's canonical equations, Maxwell's equations, and even the Schrödinger equation all give further examples.

3.1.3 Mathematical Description of a Markov Process

In a more abstract way, we can consider a Markov process to be a description of a vector of quantities $\boldsymbol{X}(t)$. The Markov property implies that if we know that $\boldsymbol{X}(t_0)$ has the value $\boldsymbol{x}_0$, then there is a probability density $p(\boldsymbol{x},t\,|\,\boldsymbol{x}_0,t_0)$, for $\boldsymbol{X}(t)$ to have the value $\boldsymbol{x}$ at any time $t \ge t_0$.

a) Conditional Probability: The Markov property specifies that this information is complete, that is, no information is needed about the values of $\boldsymbol{X}(\bar{t})$ for $\bar{t} < t_0$, and that even if such information is available, the knowledge that $\boldsymbol{X}(t_0)$ has the value $\boldsymbol{x}_0$ makes this previous information irrelevant. The quantity $p(\boldsymbol{x},t\,|\,\boldsymbol{x}_0,t_0)$ is known as the *conditional probability density*, and knowledge of all of its values is a full specification of the Markov process.

b) The Chapman–Kolmogorov Equation: Consider three times $t_0 < t_1 < t_2$. The probability that $\boldsymbol{X}(t)$ has values within a volume element $d\boldsymbol{x}_1$ of $\boldsymbol{x}_1$ at time t_1 *and* that $\boldsymbol{X}(t)$ has values within a volume element $d\boldsymbol{x}_2$ of $\boldsymbol{x}_2$ at time t_2 is

$$p(\boldsymbol{x}_2,t_2\,|\,\boldsymbol{x}_1,t_1)\,p(\boldsymbol{x}_1,t_1\,|\,\boldsymbol{x}_0,t_0)\,d\boldsymbol{x}_2\,d\boldsymbol{x}_1. \tag{3.1.6}$$

Integrating over all values of $\boldsymbol{x}_1$ clearly gives the probability density to have the value $\boldsymbol{x}_2$ at t_2, given that the initial value at t_0 was $\boldsymbol{x}_0$, the resulting equation is known as the *Chapman–Kolmogorov equation*

$$p(\boldsymbol{x}_2,t_2\,|\,\boldsymbol{x}_0,t_0) = \int d\boldsymbol{x}_1\,p(\boldsymbol{x}_2,t_2\,|\,\boldsymbol{x}_1,t_1)\,p(\boldsymbol{x}_1,t_1\,|\,\boldsymbol{x}_0,t_0), \quad t_0 \le t_1 \le t_2. \tag{3.1.7}$$

The Chapman–Kolmogorov equation is a very strong restriction on the conditional probability, and enables a full classification of all possible Markov processes. For further details, see *Stochastic Methods.*

3.2 Brownian Motion, Friction, Noise and Temperature

The methodology we will now introduce has become a standard method in the study of many physical systems, and has led to the concept of the *stochastic differential equation,* which is now a standard mathematical technique. The physical name for such an equation is a *Langevin equation,* after *Langevin* [41], who first formulated the concept in his treatment of Brownian motion.

3.2.1 The Langevin Equation

We consider a particle moving in a liquid, which is assumed to be subject to viscous drag and to a fluctuating force arising from the impacts with molecules. If u is one component of the velocity, we can write Newton's law as

$$\frac{du}{dt} = -\gamma u + \sqrt{f}\,\xi(t), \tag{3.2.1}$$

where γ = friction coefficient = $6\pi a\eta/m$, with a = radius of the (spherical) particle, m is its mass and η is the viscosity.

The term $\sqrt{f}\,\xi(t)$ represents the force arising from the impacts of the molecules on the particle. It is a rapidly fluctuating quantity, and observations of the motion of the particle, as well as physical intuition, lead us to believe that $\xi(t)$ is independent of $\xi(t')$ if t and t' are significantly different from each other.

We can idealize this by assuming

$$\langle \xi(t) \rangle = 0, \qquad \text{(the average effect of the collisions is null)}, \tag{3.2.2}$$

$$\langle \xi(t)\xi(t') \rangle = \delta(t-t'), \qquad \text{(for different } t \text{ the } \xi\text{'s are independent)}. \tag{3.2.3}$$

a) Interpretation as a Wiener Increment: Let us introduce the interpretation

$$dW(t) = \int_t^{t+dt} \xi(t')\,dt'. \tag{3.2.4}$$

From this definition and the requirements (3.2.2, 3.2.3) it is straightforward to show that $dW(t)$ is a Wiener increment as defined in Sect. 3.1. This means that in some sense we can think of $\xi(t)$ as the derivative of $W(t)$.

b) Interpretation as a Markov Process: This also means that we can rewrite the Langevin equation in the form

$$du = -\gamma u\,dt + \sqrt{f}\,dW(t), \tag{3.2.5}$$

that is, in the same form as (3.1.2). From that, the argument of Sect. 3.1.2 shows that this equation, and hence the Langevin equation also, represents a Markov process.

3.2.2 Solution of the Langevin Equation

The solution of the Langevin equation is obtained by completely conventional means, and is

$$u(t) = u(0)\mathrm{e}^{-\gamma t} + \sqrt{f}\int_0^t dt'\,\mathrm{e}^{-\gamma(t-t')}\xi(t'). \tag{3.2.6}$$

Since $\xi(t)$ is a random function, this is a random quantity, for which it is appropriate to calculate average quantities. Using a stochastic differential equation formalism, as will be explained in Chap. 4, this equation can be used to simulate Brownian motion, as is done in Fig. 3.1.

The mean velocity and the mean square velocity are easily calculated, and have obvious physical meaning. Using the correlation functions of the noise term given in (3.2.2), one finds:

a) Mean Velocity: Since the mean of $\xi(t)$ is zero, averaging the solution (3.2.6) gives

$$\langle u(t) \rangle = u(0)\mathrm{e}^{-\gamma t}. \tag{3.2.7}$$

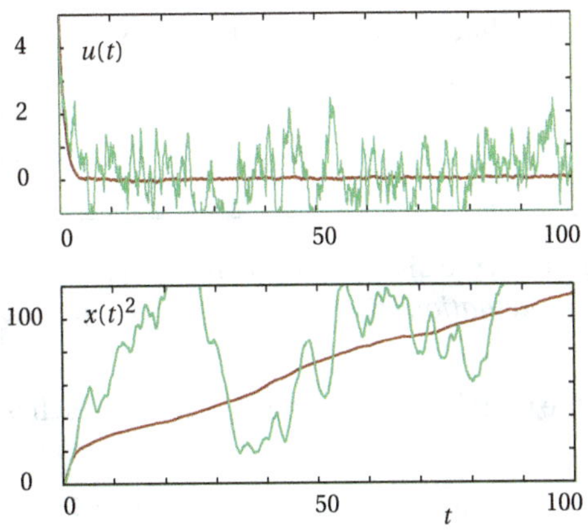

Fig. 3.1. Simulations of Brownian motion as described by (3.2.6), with $\gamma = 1$, $u(0) = 5$, $f = 1$. The green lines are the result of a single simulation, while the brown lines represent the average of 1000 independent trials. Note how the initial behaviour in both graphs is given by the initial non-zero velocity.

b) Mean Square Velocity:

$$\begin{aligned}\left\langle \left[u(t) - u(0)\mathrm{e}^{-\gamma t}\right]^2 \right\rangle &= f\int_0^t dt' \int_0^t dt''\, \mathrm{e}^{-\gamma(t-t')}\mathrm{e}^{-\gamma(t-t'')}\langle \xi(t')\xi(t'')\rangle, \\ &= f\int_0^t dt' \int_0^t dt''\, \mathrm{e}^{-\gamma(2t-t'-t'')}\delta(t'-t''), \\ &= f\int_0^t dt'\, \mathrm{e}^{-2\gamma(t-t')} = \frac{f}{2\gamma}\{1-\mathrm{e}^{-2\gamma t}\}. \end{aligned} \tag{3.2.8}$$

Thus we find

$$\langle u(t)^2\rangle = u(0)^2\mathrm{e}^{-2\gamma t} + \frac{f}{2\gamma}\{1-\mathrm{e}^{-2\gamma t}\}, \tag{3.2.9}$$

$$\mathrm{var}[u(t)] = \frac{f}{2\gamma}\{1-\mathrm{e}^{-2\gamma t}\}. \tag{3.2.10}$$

c) Long and Short Time Limits:

i) For small t,

$$\langle u(t)^2\rangle \approx u(0)^2. \tag{3.2.11}$$

ii) As $t \to \infty$, we have the *stationary limit*

$$\lim_{t\to\infty}\langle u(t)^2\rangle = \frac{f}{2\gamma}. \tag{3.2.12}$$

3.2.3 Fluctuation-Dissipation Theorem

As $t \to \infty$, the Brownian particle must come to thermodynamic equilibrium, and the theorem of equipartition of energy says that mean energy per degree of freedom must be $\frac{1}{2}k_B T$. Putting this together with the limiting result (3.2.12), we deduce that

$$\lim_{t\to\infty}\left\langle \tfrac{1}{2}mu(t)^2\right\rangle = \frac{fm}{4\gamma} = \tfrac{1}{2}k_B T, \tag{3.2.13}$$

so that we have derived the result stated without proof as (3.1.2), namely

$$f = \frac{2\gamma k_B T}{m}. \tag{3.2.14}$$

This connection between the temperature T, the dissipation coefficient γ, and the size f of the noise, is an example of a *fluctuation-dissipation theorem.* Specific points to note are:

i) The left-hand side, f, arises from the correlation function $\langle \xi(t)\xi(t') \rangle$, which is a measure of the *fluctuations.*

ii) The right-hand side is related to the absolute temperature, T, and the friction coefficient γ, which is a measure of how the energy is dissipated.

iii) The physical content is that energy is supplied to the particle by the fluctuating force, and dissipated by the viscous drag. In the steady state, the rate at which energy is dissipated is equal to the rate at which it is supplied.

iv) For a sphere of radius a moving in a liquid with viscosity η, the friction coefficient is given by

$$\gamma = 6\pi a\eta. \tag{3.2.15}$$

v) The specific connection between noise and temperature applies only to systems which come to thermodynamic equilibrium in the steady state. This is a large class of systems, but in the case of a laser, for example, the steady state is not in thermodynamic equilibrium, but is the result of a balance between gain and loss. In that case there is still a connection between dissipation and fluctuation, but it is not simply given in terms of temperature.

3.2.4 Diffusion as a Result of Brownian Motion

There is also a result directly related to the fact that these equations describe the motion of a particle. The velocity is always non-zero, but constantly fluctuating. Under the influence of this fluctuating velocity, the distance moved in the time interval $(0, t)$, is given by

$$x(t) = \int_0^t ds\, u(s). \tag{3.2.16}$$

The mean distance travelled is then the finite value

$$\langle x(t) \rangle = u(0)\frac{1 - e^{-\gamma t}}{\gamma} \xrightarrow[t\to\infty]{} \frac{u(0)}{\gamma}. \tag{3.2.17}$$

This is, of course, the distance it would be expected to travel as a result of its initial velocity and the viscous damping.

Nevertheless, the particle can drift an arbitrary distance from its starting point. Let us calculate the mean squared distance travelled, $\langle x(t)^2\rangle$. Using (3.2.7), and our calculation (3.2.17) of the mean distance travelled, we find that

$$\langle [x(t) - \langle x(t)\rangle]^2\rangle = \int_0^t ds \int_0^t g(s,s')\,ds', \tag{3.2.18}$$

where we have introduced the *autocorrelation function*

$$g(s,s') = \langle \alpha(s)\alpha(s')\rangle, \tag{3.2.19}$$

of the random function

$$\alpha(s) = \sqrt{f}\int_0^s dt\, \mathrm{e}^{-\gamma(s-t)}\xi(t). \tag{3.2.20}$$

a) Time Correlation Functions: The *autocorrelation function* $g(s,s')$ of the random function $\alpha(s)$ is a special case of a *time correlation function*, a quantity of the form $\langle A(s)B(s')\rangle$, where $A(s)$ and $B(s)$ are different random functions. When $s < s'$ this kind of correlation function is a measure of the influence of the variable $A(s)$ on the behaviour of $B(s')$ at the future time s'.

Exercise 3.1 Evaluation of the Autocorrelation Function: Show that

$$g(s,s') = \frac{f}{2\gamma}\left(\mathrm{e}^{-\gamma|s-s'|} - \mathrm{e}^{-\gamma(s+s')}\right), \tag{3.2.21}$$

and hence that

$$\langle x(t)^2\rangle = u(0)^2\left(\frac{1-\mathrm{e}^{-\gamma t}}{\gamma}\right)^2 + \frac{ft}{\gamma^2} - \frac{2f}{\gamma^3}\left(1-\mathrm{e}^{-\gamma t}\right) + \frac{f}{2\gamma^3}\left(1-\mathrm{e}^{-2\gamma t}\right). \tag{3.2.22}$$

b) Long and Short Time Behaviour: In (3.2.22), $u(0)$ is the initial velocity, assumed to be well-defined, i.e., not random. Notice that:

i) The *definition* of the diffusion coefficient of a particle in a fluid is half the mean square distance travelled per unit time for sufficiently long time. We can calculate this directly from (3.2.22) thus:

$$D = \lim_{t\to\infty}\frac{\langle x(t)^2\rangle}{2t} = \frac{f}{2\gamma^2}. \tag{3.2.23}$$

ii) Thus, reinserting the value of f from (3.2.14), we find that

$$D = \frac{k_\mathrm{B}T}{\gamma m} = \frac{k_\mathrm{B}T}{6\pi\eta a}. \tag{3.2.24}$$

iii) In contrast to the long time behaviour, the formula (3.2.22) gives a quadratic short time behaviour in t; namely, for small t

$$\langle x(t)^2\rangle \approx u(0)^2 t^2, \tag{3.2.25}$$

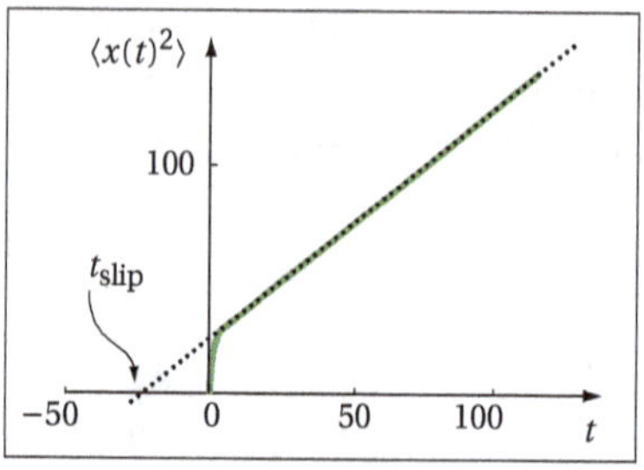

Fig. 3.2. The mean square distance travelled as a function of time, for Brownian motion with non-white noise, showing the asymptotic behaviour of equation (3.2.26). The parameters are $\gamma = 1$, $f = 1$, $u(0) = 5$, as in Fig. 3.1

which is simply the effect of the initial speed of the particle. The full formula (3.2.22) shows that this is damped out. Note that the part not proportional to $u(0)^2$ in fact only increases initially *cubically* with time, as $ft^3/3$, reflecting the need to accelerate the particle initially.

iv) The asymptotic behaviour of $\langle x(t)^2\rangle$ is more precisely given by

$$\langle x(t)^2\rangle \approx 2Dt + \left(\frac{u(0)}{\gamma}\right)^2 - \frac{3f}{2\gamma^3} + \text{exponentially small terms.} \tag{3.2.26}$$

Thus we have an initial non-diffusive behaviour, being taken over by the long-term diffusive behaviour, as illustrated in Fig. 3.2. This leads to an initial "slip time" t_{slip} where the asymptotic formula would correspond to $\langle x(t)^2\rangle = 0$.

3.3 Measurement in a Fluctuating System

The theoretical description provided by a Langevin equation allows one to compute a wide variety of average quantities, in fact in principle, the average of any function of the fluctuating variable as a function of time for any given initial condition. In practice, that is, in any experimental system for which one might seek a stochastic model, the possible kinds of average which can be implemented are limited. The major issue is one of sampling, that in practice some averages depend very strongly on very improbable values of the variable being measured. For example, in Brownian motion the mean square velocity in equilibrium, given by (3.2.13), is $k_B T/m$. A measurement of a quantity like $\exp \lambda u(t)$ for $\lambda^2 \gg m/k_B T$ would sample values of $u(t)$ which occur very infrequently, and would require a very long time to collect enough data to yield a reliable measurement.

3.3.1 Quantities Most Commonly Considered

Since only a limited number of quantities can be measured reliably, it makes sense only to calculate such quantities. If $x(t)$ is the variable of interest, the most common quantities one considers are:

a) Measurements at a Single Time:

i) The mean value $\langle x(t)\rangle$.

ii) The second moment $\langle x(t)^2 \rangle$ and the variance var $[x(t)]$.

iii) The conditional probability $p(x, t \,|\, x_0, t_0)$. Here, accurate measurements will only be available for values of the arguments which yield a probability significantly different from zero. This is not a trivial statement, since sometimes it is possible for very infrequent configurations to be important. This happens, for example, in the study of stochastic bistable systems.

b) Stationary Measurements: The same quantities as at a single time are measured once the system has come to equilibrium, and the corresponding stationary quantities are:

i) The stationary mean value $\langle x \rangle_s$.

ii) The stationary second moment $\langle x^2 \rangle_s$ and the stationary variance var $[x(t)]_s$.

iii) The stationary probability distribution $p_s(x)$.

c) Time Correlation Functions: We can measure $x(t)$ and $x(t')$, and from these construct the time correlation function. Mostly, it is the stationary correlation function that is of interest, and in mathematical terms this is given by

$$\langle x(t)x(t') \rangle_s = \int dx \int dx' \, xx' \, p(x, t \,|\, x', t') \, p_s(x'), \qquad t \geq t'. \tag{3.3.1}$$

When, as in this equation, one correlates the same variable x with itself at two different times, the special terminology *autocorrelation function* is also used. Of course in a mathematical sense, the correlation function can be constructed from the single time and stationary measurements as defined above, but it has a special place in physics because of its connection to the *spectrum*.

d) The Conditional Mean: The mean of $x(t)$ given an initial value at a previous time can be written

$$\langle x(t) | [x', t'] \rangle = \int dx \, x \, p(x, t \,|\, x', t'), \qquad t \geq t'. \tag{3.3.2}$$

This has a strong connection to the correlation function; from (3.3.1) we have

$$\langle x(t)x(t') \rangle_s = \int dx' \, \langle x(t) | [x', t'] \rangle x' \, p_s(x'), \qquad t \geq t'. \tag{3.3.3}$$

3.3.2 The Regression Theorem

The significance of the last result lies in the concept of a *regression theorem*, by which we mean the following: Quite often we find that the conditional mean obeys a linear differential equation such as

$$\frac{d}{dt} \langle x(t) | [x', t'] \rangle = -\Lambda \langle x(t) | [x', t'] \rangle. \tag{3.3.4}$$

If this is so, the relationship (3.3.3) shows that a time correlation function such as $\langle x(t)x(t') \rangle_s$ obeys the same equation, but with initial condition $\langle x^2 \rangle_s$. In one variable this is not a very profound result, but the corresponding result for many variables is of significant use. The extension to a *quantum regression theorem* is of great importance, and this is done in Sect. 13.4.3.

3.3.3 Stationary Processes

When the system is allowed to settle for a sufficiently long time with no external time-dependent influences we speak of it as a *stationary process.* In this case one will normally find:

i) The mean $\langle x(t)\rangle$ approaches a stationary value

$$\langle x(t)\rangle \to \langle x\rangle_s\,. \tag{3.3.5}$$

ii) The autocorrelation function depends only on the time difference $t-t'$; that is, we can define

$$G_s(t-t') \equiv \langle x(t)x(t')\rangle_s\,. \tag{3.3.6}$$

iii) For large time differences $|t-t'|\to\infty$, the autocorrelation function factorizes,

$$\lim_{|t-t'|\to\infty} \langle x(t)x(t')\rangle_s = \langle x\rangle_s{}^2, \tag{3.3.7}$$

and consequent on this, we can define

$$g_s(t-t') \equiv G_s(t-t') - \langle x\rangle_s{}^2 \longrightarrow 0 \quad \text{as } |t-t'|\to\infty\,. \tag{3.3.8}$$

iv) A *stationary process* is defined as one in which the correlation functions of *all orders* depend only on time differences. Clearly, this property is, at the very least, difficult to check in any actual system.

3.3.4 Spectrum and Autocorrelation Function

One of the clearest ways of characterizing a physical system is through the spectrum of a physical observable, which displays the activity of the system at a range of frequencies. Peaks in a spectrum are usually directly related to physically interesting phenomena, and can be measured in a variety of ways. One of the most useful techniques is to measure the autocorrelation function, and then perform a Fourier transform.

a) Conventions for the Definition of the Autocorrelation Function: There are various conventions for the definition of the autocorrelation function. Conventionally, in a physical context both $G_s(t-t')$ and $g_s(t-t')$ are called the "autocorrelation function". In this book we will not attempt to impose a rigid convention, but rather will seek to make it clear in any particular case exactly which definition is being used.

b) The Spectrum: The stationary autocorrelation functions and the concept of a spectrum are directly connected to each other through a Fourier transformation.

Let us define a Fourier transform variable

$$\tilde{x}(\omega + \mathrm{i}\gamma, \tau) = \int_{-\infty}^{0} dt\, \mathrm{e}^{(\mathrm{i}\omega+\gamma)t} x(t+\tau). \tag{3.3.9}$$

This gives a way of assessing the "amount" of fluctuation in a frequency band of width $1/\gamma$ of the variable at the time τ—it is equivalent to applying a filter of bandwidth γ to the fluctuating variable $x(t)$.

The variable of interest is the *spectrum*, which we define as follows. Firstly let us define $S_\tau(\omega)$, (which is not itself the spectrum) by

$$S_\tau(\omega) = \lim_{\gamma\to 0} \frac{\gamma}{\pi} \langle |\tilde{x}(\omega + \mathrm{i}\gamma, \tau)|^2 \rangle_{\mathrm{s}} \tag{3.3.10}$$

$$= \lim_{\gamma\to 0} \frac{\gamma}{\pi} \int_{-\infty}^{0} dt \int_{-\infty}^{0} dt'\, \mathrm{e}^{\mathrm{i}\omega(t-t')+\gamma(t+t')} \langle x(t+\tau)x(t'+\tau)\rangle_{\mathrm{s}}. \tag{3.3.11}$$

We now take account of the the stationarity condition (3.3.8), and also change variables to $t_1 = t - t'$, $t_2 = t + t'$, so that

$$S_\tau(\omega) \to S(\omega) \tag{3.3.12}$$

$$= \lim_{\gamma\to 0} \frac{\gamma}{2\pi} \int_{-\infty}^{\infty} dt_1 \int_{-\infty}^{-|t_1|} dt_2\, \mathrm{e}^{\mathrm{i}\omega t_1 + \gamma t_2} \left\{ \langle x\rangle_{\mathrm{s}}^{2} + g_{\mathrm{s}}(t_1)\rangle \right\}, \tag{3.3.13}$$

$$= \lim_{\gamma\to 0} \frac{\gamma}{2\pi} \int_{-\infty}^{\infty} dt_1\, \mathrm{e}^{\mathrm{i}\omega t_1} \frac{\mathrm{e}^{-\gamma|t_1|}}{\gamma} \left\{ \langle x\rangle_{\mathrm{s}}^{2} + g_{\mathrm{s}}(t_1)\rangle \right\}, \tag{3.3.14}$$

$$= \langle x\rangle_{\mathrm{s}}^{2} \frac{1}{\pi} \lim_{\gamma\to 0} \frac{\gamma}{\gamma^2+\omega^2} + \frac{1}{2\pi} \int_{-\infty}^{\infty} dt\, \mathrm{e}^{\mathrm{i}\omega t} g_{\mathrm{s}}(t). \tag{3.3.15}$$

Thus, the spectrum of fluctuations is given by

$$S(\omega) = \langle x\rangle_{\mathrm{s}}^{2} \delta(\omega) + \frac{1}{2\pi} \int_{-\infty}^{\infty} dt\, \mathrm{e}^{\mathrm{i}\omega t} g_{\mathrm{s}}(t). \tag{3.3.16}$$

i) The *spectrum* of fluctuations is only well defined when the system is in a steady state, or, more formally, is *stationary*. The relationship between the stationary autocorrelation function $g_{\mathrm{s}}(t)$ and the spectrum is known as *Khinchin's* theorem [42].

ii) Both the spectrum and the autocorrelation function can be measured. The spectrum is measured by filtering the signal and measuring the size of the filtered signal; the autocorrelation function is measured by sampling the signal at fine time intervals, and then simply carrying out the requisite arithmetic. Using a fast Fourier transform, it is easy to transform from one to the other.

iii) The spectrum can also be defined by the formulae

$$y(\omega) = \int_{-T}^{0} dt\, \mathrm{e}^{-\mathrm{i}\omega t} x(t), \tag{3.3.17}$$

$$S(\omega) = \lim_{T\to\infty} \frac{1}{2\pi T} |y(\omega)|^2. \tag{3.3.18}$$

In *Stochastic Methods* we show this is the same as our definition (3.3.16).

3.3.5 Fourier Analysis of Fluctuating Functions

We can define a Fourier transform variable by

$$c(\omega) = \frac{1}{2\pi}\int dt\, x(t) e^{-i\omega t}. \tag{3.3.19}$$

If $x(t)$ is a real quantity,

$$c(\omega) = c^*(-\omega). \tag{3.3.20}$$

a) Correlation Functions of Fourier Transform Variables: For convenience, in what follows we assume $\langle x\rangle = 0$. Hence,

$$\langle c(\omega)\rangle = \frac{1}{2\pi}\int dt\,\langle x\rangle e^{-i\omega t} = 0, \tag{3.3.21}$$

$$\langle c(\omega)c^*(\omega')\rangle = \frac{1}{(2\pi)^2}\int\int dt\,dt'\, e^{-i\omega t + i\omega' t'}\langle x(t)x(t')\rangle, \tag{3.3.22}$$

$$= \frac{1}{2\pi}\delta(\omega-\omega')\int d\tau\, e^{i\omega\tau} g_s(\tau), \tag{3.3.23}$$

$$= \delta(\omega-\omega')S(\omega). \tag{3.3.24}$$

Here we find not only a relationship between the mean square $\langle |c(\omega)|^2\rangle$ and the spectrum, but also the result that stationarity alone implies that $c(\omega)$ and $c^*(\omega')$ are uncorrelated, since the term $\delta(\omega-\omega')$ arises because $\langle x(t)x(t')\rangle$ is a function only of $t-t'$.

b) White Noise: When spectrum is independent of ω, that is, when $S(\omega) = \bar{S}$, the Fourier amplitudes satisfy

$$\langle c(\omega)c^*(\omega')\rangle = \bar{S}\delta(\omega-\omega'). \tag{3.3.25}$$

This can be seen as a defining property of *white noise*. The terminology arises because all frequencies are represented equally, as is the case in white light. In this case, the autocorrelation function is given by

$$g_s(t) = 2\pi\bar{S}\delta(t), \tag{3.3.26}$$

so that in white noise, the fluctuations at different times are uncorrelated. This means we can write Gaussian white noise in terms of the Langevin source $\xi(t)$ with the properties (3.2.2, 3.2.3) in the form

$$x_W(t) = \sqrt{2\pi\bar{S}}\,\xi(t). \tag{3.3.27}$$

4. Stochastic Differential Equations

The concept of a Langevin equation, as introduced in Sect. 3.2 is best formalized mathematically under the terminology of a *stochastic differential equation.* The purpose of this chapter is to give an elementary formulation of the subject, in a way which makes the issues involved in defining a stochastic differential equation relatively straightforward to understand, rather than to make any attempt at mathematical rigour. The basic concepts in stochastic differential equations are really not very difficult to formulate in a style similar to that used for an elementary calculus course, and this is one of our principal aims. The other aim is to set the background for the formulation of quantum stochastic differential equations later in this book, where the main issues are rather similar.

4.1 Ito Stochastic Differential Equation

Let us start with a Langevin equation of a quite general form

$$\dot{x}(t) = a\big(x(t),t\big) + b\big(x(t),t\big)\xi(t), \tag{4.1.1}$$

and consider how we might solve this equation numerically.

The stochastic noise term $\xi(t)$ is defined as Gaussian and satisfying the conditions (3.2.2, 3.2.3), but the only feasible way of implementing these properties is to use the Wiener increment form in a discretized time version of the equation. We consider a time interval (t_0, t) and divide this into N equal intervals $\Delta t = (t - t_0)/N$. We also introduce N *independent* Gaussian random numbers ΔW_i such that

$$\langle \Delta W_i \rangle = 0, \tag{4.1.2}$$

$$\langle \Delta W_i \rangle = \Delta t. \tag{4.1.3}$$

We also define the times $t_i = t_0 + i\,\Delta t$ (which means that $t_N \equiv t$), and write our equation in the discretized form

$$x_{i+1} = x_i + a_i\,\Delta t + b_i\,\Delta W_i. \tag{4.1.4}$$

The concept is illustrated in Fig. 4.1. This algorithm can be shown to converge under certain mild conditions, and under these conditions we say that $x(t)$ obeys the *Ito stochastic differential equation*

$$(\mathbf{I})\; dx(t) = a[x(t),t]\,dt + b[x(t),t]\,dW(t). \tag{4.1.5}$$

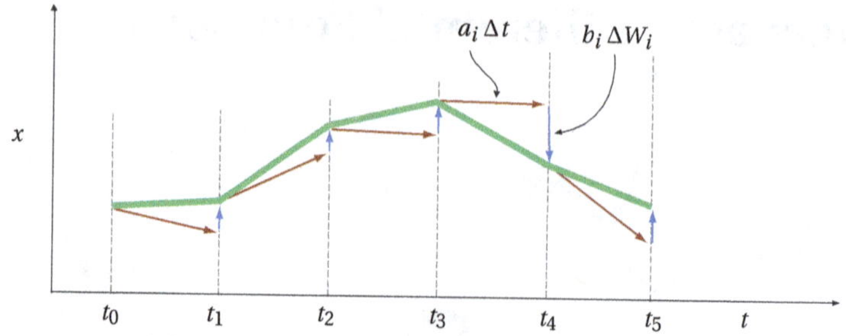

Fig. 4.1. Illustration of the Ito definition of the stochastic differential equation $dx(t) = a\big(x(t),t\big)\,dt + b\big(x(t),t\big)\,dW(t)$.

4.1.1 Calculus of Stochastic Differential Equations

The stochastic differential equation has two infinitesimals, dt and $dW(t)$ which satisfy the requirements

$$\langle dW(t)\rangle = 0, \tag{4.1.6}$$

$$\langle dW(t)^2\rangle = dt. \tag{4.1.7}$$

a) Ito Rules: This means that in developing the formalism, we have to deal with the fact that terms of second order in $dW(t)$ are of first order in dt. In fact, when used in integrals, it can be shown (see *Stochastic Methods*) that we should always keep all terms up to first order in dt by using the following *Ito rules* in all calculations:

$$\left.\begin{aligned} dW(t)^2 &= dt, \\ dW(t)\,dt &= 0, \\ dt^n &= 0, \qquad \text{for } n \ge 2, \\ dW(t)^n &= 0, \qquad \text{for } n \ge 3. \end{aligned}\right\} \tag{4.1.8}$$

The last three of these all involve terms of order $dt^{3/2}$, and therefore integrals defined in terms of them must vanish. The first rule is a little surprising, since for the increment viewed in isolation, this is only true in the mean, as in (4.1.7). The reason for this rule lies in the fact that a stochastic differential equation written in the form (4.1.5) is really a shorthand for a *stochastic integral equation*

$$x(t) = x(t_0) + \int_{t_0}^{t} a[x(t),t]\,dt + (\mathbf{I})\int_{t_0}^{t} b[x(t),t]\,dW(t). \tag{4.1.9}$$

This involves defining the stochastic integral $(\mathbf{I})\int$, and from this it can be shown that, in an integral over a finite time interval, and for an appropriate integrand $A(t)$

$$(\mathbf{I})\int_{t_0}^{t} A(t)\,dW(t)^2 = \int_{t_0}^{t} A(t)\,dt. \tag{4.1.10}$$

b) Independence of the Increment: Together with these rules, we have another useful property. Since $x(t)$ can only depend on $W(t')$ for $t' < t$, we know that $x(t)$ and $dW(t)$ must be statistically independent, hence for any function $f[x(t),t]$ we have

$$\langle f[x(t),t]\, dW(t)\rangle = \langle f[x(t),t]\rangle\langle dW(t)\rangle = 0. \tag{4.1.11}$$

4.1.2 Change of Variables: Ito's Formula

Suppose $f(x)$ is a function of $x(t)$, which obeys the stochastic differential equation (4.1.5); what equation does $f(x)$ obey? Note that

$$df(x) = f'(x)\,dx + \tfrac{1}{2}f''(x)\,dx^2 + \dots \tag{4.1.12}$$

and substitute for dx using the stochastic differential equation (4.1.5), keeping only those terms which according to (4.1.8) are non-zero; we get *Ito's formula*:

$$df[x(t)] = \left\{f'(x)a(x,t) + \tfrac{1}{2}f''(x)b(x,t)^2\right\}dt + f'(x)b(x,t)\,dW(t), \tag{4.1.13}$$

where, on the right-hand side, we have written x instead of $x(t)$ for brevity.

If we had kept only terms up to first order in either dt or $dW(t)$, we would not get the second part of coefficient of dt; this part is known as the *Ito correction.*

Example—Application of Ito's Formula: Suppose x obeys the Ito equation

$$\text{(I)}\; dx = -\alpha x\,dt + \beta\,dW(t). \tag{4.1.14}$$

Then $\langle x\rangle$ obeys the equation

$$d\langle x\rangle = -\alpha\langle x\rangle\,dt, \tag{4.1.15}$$

which is easily solved; $\langle x\rangle = \langle x_0\rangle e^{-\alpha t}$.

The second moment is found by setting $f(x) = x^2$; thus

$$d(x^2) = \left\{2x(-\alpha x) + \tfrac{1}{2}2\beta^2\right\}dt + 2x\beta\,dW(t), \tag{4.1.16}$$

and thus, using (4.1.11) to set $\langle x dW(t)\rangle$

$$d\langle x^2\rangle = \left\{-2\alpha\langle x^2\rangle + \beta^2\right\}dt. \tag{4.1.17}$$

This is also easily solved. Notice that in the stationary state $d\langle x^2\rangle_s = 0$, so

$$\langle x\rangle_s = 0, \qquad \langle x^2\rangle_s = \beta^2/2\alpha. \tag{4.1.18}$$

If we did not use the Ito correction, the equation for $\langle x^2\rangle$ would not have the β^2 term, and we would get a different and wrong answer.

We will shortly show how a different (Stratonovich) definition of the stochastic differential equation can avoid using an Ito correction, but in this case the term $\langle 2x\beta\,dW(t)\rangle$ does not vanish, and instead gives a value equal to the Ito correction.

4.2 The Fokker–Planck Equation

The solution of a stochastic differential equation is obviously a Markov process, since knowledge of the initial condition $x(t_0)$ determines $x(t_1)$—no knowledge other than the value $x(t_0)$, and the random process $W(t)$, is necessary. Hence, we can introduce a conditional probability $p(x,t\,|\,x_0,t_0)$ for the process $x(t)$ and evaluate averages using this conditional probability. The conditional probability will have a time development equation, which we can derive from the stochastic differential equation as follows.

Suppose that $f(x)$ is an arbitrary function,

$$\frac{d}{dt}\langle f[x(t)]\rangle = \int dx\, f(x)\frac{\partial}{\partial t}p(x,t\,|\,x_0,t_0). \tag{4.2.1}$$

On the other hand, Ito's formula says

$$\frac{d}{dt}\langle f[x(t)]\rangle = \frac{\langle df[x(t)]\rangle}{dt} = \left\langle a[x(t),t]\frac{\partial f}{\partial x} + \tfrac{1}{2}b[x(t),t]^2\frac{\partial^2 f}{\partial x^2}\right\rangle. \tag{4.2.2}$$

But this must be able to be written, using the conditional probability density, as

$$= \int dx\left\{a(x,t)\frac{\partial f(x)}{\partial x} + \tfrac{1}{2}b(x,t)^2\frac{\partial^2 f}{\partial x^2}\right\}p(x,t\,|\,x_0,t_0). \tag{4.2.3}$$

We can integrate by parts, discarding surface terms, getting

$$= \int dx\left\{-\frac{\partial}{\partial x}\big(a(x,t)p(x,t\,|\,x_0,t_0)\big) + \tfrac{1}{2}\frac{\partial^2}{\partial x^2}\big(b(x,t)^2p(x,t\,|\,x_0,t_0)\big)\right\}f(x,t). \tag{4.2.4}$$

However, $f(x)$ is an *arbitrary* function, so we can use equations (4.2.1) and (4.2.4), to get

$$\frac{\partial}{\partial t}p(x,t\,|\,x_0,t_0) = -\frac{\partial}{\partial x}\big(a(x,t)p(x,t\,|\,x_0,t_0)\big) + \tfrac{1}{2}\frac{\partial^2}{\partial x^2}\big(b(x,t)^2p(x,t\,|\,x_0,t_0)\big). \tag{4.2.5}$$

This is known as the *Fokker–Planck equation* for the conditional probability density. The stochastic process is determined equivalently by the Fokker–Planck equation (4.2.5) or the stochastic differential equation

$$\textbf{(I)}\; dx = a(x,t)\,dt + b(x,t)\,dW(t). \tag{4.2.6}$$

4.3 The Stratonovich Stochastic Differential Equation

The algorithm (4.1.4) is based on the Euler algorithm for solving differential equations in which the coefficients are evaluated at the initial point of the time interval.

Stratonovich introduced the alternative algorithm, based on a Runge–Kutta midpoint rule.

$$x_{i+1} = x_i + a_i\,\Delta t + \frac{b_i + b_{i+1}}{2}\,\Delta W_i, \tag{4.3.1}$$

and the solution of this algorithm is said to satisfy the *Stratonovich stochastic differential equation*

$$(\mathbf{S})\; dx(t) = a[x(t),t]\,dt + b[x(t),t]\,dW(t). \tag{4.3.2}$$

In this notation, the increment $dW(t)$ cannot be viewed as being independent of its coefficient, since in the algorithm (4.3.1), the term b_{i+1} clearly depends on ΔW_i. We might have expected the coefficient of Δt to be $(a_i + a_{i+1})/2$, but in the limit $\Delta t \to 0$ it makes no difference which form we choose for this coefficient.

4.3.1 Change of Variables for the Stratonovich Stochastic Differential Equation

The definition of the Stratonovich integral is such as to make the ordinary rules of calculus valid for change of variables. This means, that for the Stratonovich integral, Ito's formula (4.1.13) is replaced by the simple calculus rule

$$(\mathbf{S})\; df[x(t)] = f'[x(t)]\Big(a[x(t),t]\,dt + b[x(t),t]\,dW(t)\Big). \tag{4.3.3}$$

The essence of the proof can be explained by using the simple stochastic differential equation

$$(\mathbf{S})\; dx(t) = B[x(t)]\,dW(t). \tag{4.3.4}$$

In discretized form, this can be written

$$x_{i+1} = x_i + B\big[\tfrac{1}{2}(x_{i+1} + x_i)\big]\,\Delta W_i. \tag{4.3.5}$$

To find the Stratonovich stochastic differential equation for $f[x(t)]$, we need only use the Taylor series expansion of a function about a midpoint in the form

$$f(x+a) = f(x-a) + \sum_{n=0}^{\infty} \frac{f^{2n+1}(x)\,a^{2n+1}}{(2n+1)!}. \tag{4.3.6}$$

In expanding $f(x_{i+1})$ we only need to keep terms up to second order, so we drop all but the first two terms and write

$$f(x_{i+1}) = f(x_i) + f'\big[\tfrac{1}{2}(x_{i+1} + x_i)\big](x_{i+1} - x_i), \tag{4.3.7}$$

$$= f(x_i) + f'\big[\tfrac{1}{2}(x_{i+1} + x_i)\big]\,B\big[\tfrac{1}{2}(x_{i+1} + x_i)\big]\,\Delta W_i. \tag{4.3.8}$$

This means that the Stratonovich stochastic differential equation for $f[x(t)]$ is

$$(\mathbf{S})\; df[x(t)] = f'[x(t)]\,B[x(t)]\,dW(t), \tag{4.3.9}$$

which is the ordinary calculus rule. The extension to the general case is straightforward.

4.3.2 Equivalent Stratonovich and Ito Stochastic Differential Equations

If we use the Stratonovich definition of the stochastic integral, we can write the Stratonovich stochastic differential equation as

$$x(t+dt) - x(t) \equiv dx(t) = a(x,t)\,dt + \frac{b[x(t)] + b[x(t+dt)]}{2}\,dW(t), \tag{4.3.10}$$

$$\approx a(x,t)\,dt + b(x,t)\,dW(t) + b'(x,t)\frac{dx(t)}{2}\,dW(t). \tag{4.3.11}$$

We now resubstitute for $dx(t)$, and use $dW^2 = dt$, neglecting all higher terms to get the equivalent *Ito* stochastic differential equation

$$\textbf{(I)}\; dx = \left[a(x,t) + \tfrac{1}{2}b(x,t)\frac{\partial b(x,t)}{\partial x}\right]dt + b(x,t)\,dW(t). \tag{4.3.12}$$

What we have proved is that we can transform from Stratonovich to Ito and back using the following rules

Stratonovich			**Ito**		
$a(x)$	$b(x)$	$\longrightarrow$	$a(x) + \frac{1}{2}b\frac{\partial b}{\partial x}$	$b(x)$	(4.3.13)
$a(x) - \frac{1}{2}b\frac{\partial b}{\partial x}$	$b(x)$	$\longleftarrow$	$a(x)$	$b(x)$	

4.3.3 Comparison of Ito and Stratonovich Formalisms

a) Change of Variables: The Stratonovich calculus has the advantage that we use ordinary calculus in evaluating change of variables, that is, if $x(t)$ obeys the stochastic differential equation

$$\textbf{(S)}\; dx = a(x,t)\,dt + b(x,t)\,dW(t), \tag{4.3.14}$$

then if $f(x)$ is an arbitrary function of x, the corresponding stochastic differential equation is

$$\textbf{(S)}\; df(x) = \big(a(x,t)\,dt + b(x,t)\,dW(t)\big)\frac{df(x)}{dx}, \tag{4.3.15}$$

and there is no Ito correction like that in (4.1.13).

b) The White Noise Limit: If we consider a physical system of the kind

$$\dot{x} = a(x,t) + b(x,t)\alpha(t) \tag{4.3.16}$$

where $\alpha(t)$ is a genuine non-delta-correlated random noise, then changing variables in this case can cause no problems, and must obey ordinary calculus.

Thus one expects the physical situation, in which the delta-correlated noise is an approximation, to be better represented by the Stratonovich stochastic differential equation. In Chap. 8 we will consider the limit of systems driven by almost white noise, like $\alpha(t)$, and show explicitly how the Stratonovich form arises naturally in most (but not all) circumstances.

c) Advantages of the Ito Form: In spite of the unusual formula for change of variables, the Ito form is very valuable for the following reasons:

i) It is mathematically better defined than the Stratonovich form.

ii) It very often arises naturally, in spite of the argument above that the physics favours the Stratonovich interpretation.

iii) The fact that $dW(t)$ is independent of $x(t)$ makes most stochastic calculations much easier than the corresponding calculations using the Stratonovich form.

4.4 Systems with Many Variables

We need an n-variable Wiener process, consisting of n independent Wiener processes, $W_i(t)$, and in generating stochastic differential equations, we use the rules

$$\left.\begin{aligned} dW_i(t)\,dW_j(t) &= \delta_{ij}\,dt, \\ dW_i(t)\,dt &= 0, \qquad \text{as are all other products,} \end{aligned}\right\} \tag{4.4.1}$$

and we generate many-variable stochastic differential equations in the form

$$\text{(I)}\; d\boldsymbol{x}(t) = \boldsymbol{A}(\boldsymbol{x},t)\,dt + \mathsf{B}(\boldsymbol{x},t)\,d\boldsymbol{W}(t), \tag{4.4.2}$$

where $\mathsf{B}(\boldsymbol{x},t)$ is an $n\times n$ matrix.

a) Note: B need not be square—the vector $d\boldsymbol{W}(t)$ can be m-dimensional and thus $\boldsymbol{x}$ and $\boldsymbol{A}$ are n-dimensional vectors, while B is an $m\times n$ matrix.

b) Ito–Stratonovich Conversion: For many variables, the conversion which corresponds to (4.3.13) is accomplished by

$$A_i^{\mathrm{S}} = A_i - \tfrac{1}{2}\sum_{j,k} B_{kj}\partial_k B_{ij}, \qquad \mathsf{B}^{\mathrm{S}} = \mathsf{B}. \tag{4.4.3}$$

4.4.1 Fokker–Planck Equations with Many Variables

The conditional probability $p(\boldsymbol{x},t\,|\,\boldsymbol{x}_0,t_0)$ obeys the Fokker–Planck equation

$$\frac{\partial p}{\partial t} = -\sum_i \partial_i[A_i(\boldsymbol{x},t)p] + \tfrac{1}{2}\sum_{i,j}\partial_i\partial_j\left\{[\mathsf{B}(\boldsymbol{x},t)\mathsf{B}^T(\boldsymbol{x},t)]_{ij}\,p\right\}, \tag{4.4.4}$$

for which the derivation is essentially the same as that in one variable.

4.4.2 Complex Variable Systems

The kinds of stochastic differential equations and Fokker–Planck equations which arise in quantum systems very often involve complex variables. We demonstrate how these can be used in Sect. 7.3.

Exercise 4.1 Many Variables: Work out Ito's formula for the case of many variables, corresponding to (4.1.13), and hence derive the Fokker–Planck equation (4.4.4).

Exercise 4.2 "Stratonovich Form" of the Fokker–Planck Equation: Show that if (4.4.2) is taken as a Stratonovich stochastic differential equation, then the Fokker–Planck equation can be written

$$\partial_t p = -\sum_i \partial_i \{A_i p\} + \tfrac{1}{2} \sum_{ijk} \partial_i \{B_{ik} \partial_j [B_{jk} p]\}. \tag{4.4.5}$$

4.5 Numerical Simulation of Stochastic Differential Equations

The formulae (4.1.1) and (4.3.1) both provide starting points for the simulation of stochastic differential equations, but require considerable caution. The first of these is essentially an Euler algorithm, and is neither very stable nor rapidly convergent. It is not recommended for simulation. The second formula is more stable and more rapidly convergent, and requires some kind of iteration to implement. In the last chapter of *Stochastic Methods* we give a range of algorithms, and references to the relevant literature, which should be consulted before attempting any serious simulations.

5. The Fokker–Planck Equation

The conditional probability $p(x,t|x_0,t_0)$ is the central object in a Markov process, from which everything else can be deduced. The formulation of stochastic differential equations as in the previous chapter shows that the processes they describe are indeed Markov processes, and the Fokker–Planck equation is a fundamental partial differential equation for the conditional probability relevant to processes described by the stochastic differential equations of the previous chapter. Such processes are known as *diffusion processes*, since they can be regarded as generalizations of the fundamental physical process of diffusion. In this chapter we will develop the Fokker–Planck description of diffusion processes in a form oriented towards applications.

5.1 Fokker–Planck Equation in One Dimension

In Sect. 4.2 we showed that if the variable $x(t)$ satisfies the stochastic differential equation

$$dx(t) = A[x(t),t]\,dt + \sqrt{B[x(t),t]}\,dW(t), \tag{5.1.1}$$

then the conditional probability $p(x,t\,|\,x_0,t_0)$ satisfies the Fokker–Planck equation

$$\frac{\partial p(x,t\,|\,x_0,t_0)}{\partial t} = -\frac{\partial}{\partial x}\big[A(x,t)p(x,t\,|\,x_0,t_0)\big] + \frac{1}{2}\frac{\partial^2}{\partial x^2}\big[B(x,t)p(x,t\,|\,x_0,t_0)\big]. \tag{5.1.2}$$

a) Initial Conditions: The appropriate initial condition for the conditional probability is

$$p(x,t_0\,|\,x_0,t_0) = \delta(x-x_0). \tag{5.1.3}$$

However, if we know initially that x has probability distribution $p_i(x)$ at time t_0, then its probability distribution $p(x,t)$ at a later time t will satisfy the Fokker–Planck equation above with initial condition $p(x,t_0) = p_i(x)$.

b) Probability Current: The probability current is defined by

$$J(x,t) = \big[A(x,t)p(x,t\,|\,x_0,t_0)\big] + \frac{1}{2}\frac{\partial}{\partial x}\big[B(x,t)p(x,t\,|\,x_0,t_0)\big], \tag{5.1.4}$$

and in terms of this current the Fokker–Planck equation takes the form of a conservation equation for a density $\rho(x,t) \equiv p(x,t\,|\,x_0,t_0)$

$$\frac{\partial \rho}{\partial t} + \frac{\partial J}{\partial x} = 0. \tag{5.1.5}$$

Thus, we can view the Fokker–Planck equation as an equation describing the motion of a one-dimensional conserved fluid whose density is equal to the probability density.

c) Stationary Solutions: Let us consider now only the case where the coefficients are independent of time and can be written as $A(x)$, $B(x)$. The stationary solution from (5.1.4) requires that the stationary current is a constant J_s, independent of x. If the range is infinite, the only acceptable solution is usually $J_s = 0$, in which case the stationary probability distribution becomes

$$p_s(x) = \frac{\mathcal{N}}{B(x)} \exp\left(\int_0^x \frac{2A(x)\, dx}{B(x)} \right), \tag{5.1.6}$$

where $\mathcal{N}$ is a normalization factor to ensure that $\int p_s(x)\, dx = 1$.

d) The Backward Fokker–Planck Equation: The conditional probability density $p(x,t|x_0,t_0)$ also obeys the differential equation in terms of of the initial parameters x_0 and t_0, known as the *backward Fokker–Planck equation*

$$\frac{\partial p(x,t\,|\,x_0,t_0)}{\partial t_0} = A(x_0,t_0)\frac{\partial p(x,t\,|\,x_0,t_0)}{\partial x_0} + \tfrac{1}{2}B(x_0,t_0)\frac{\partial^2 p(x,t\,|\,x_0,t_0)}{\partial x_0^2}. \tag{5.1.7}$$

This will not arise very frequently in this book; the reader should consult *Stochastic Methods* for its derivation and application. When appropriate, the Fokker–Planck equation (5.1.2) is called the *forward* Fokker–Planck equation.

Exercise 5.1 General Stationary Solution: Compute the formula for the stationary solution in the case that $J_s \neq 0$.

5.1.1 Boundary Conditions

The stochastic differential equation (5.1.1) does not directly specify any boundary conditions. However, very often the range of interest is $-\infty < x < \infty$, and no boundary condition is necessary.

a) The Ornstein–Uhlenbeck Process: The Ornstein–Uhlenbeck process is the most fundamental process after the Wiener process, and it and its applications are treated thoroughly in Sect. 7.1.1. It corresponds to the Langevin equation (3.2.1), and has the stochastic differential equation and corresponding Fokker–Planck equation

$$dx = -kx\,dt + \sqrt{D}\,dW(t), \tag{5.1.8}$$

$$\frac{\partial p}{\partial t} = \left(\frac{\partial}{\partial x}kx + \tfrac{1}{2}D\frac{\partial^2}{\partial x^2}\right)p, \tag{5.1.9}$$

whose stationary solution is

$$p_s(x) = \mathcal{N}\exp(-kx^2/D). \tag{5.1.10}$$

Provided $k > 0$, this is normalizable on $(-\infty,\infty)$, and no special boundary condition is needed.

b) Reflecting Boundary Condition—Diffusion in a Gravitational Field: We can describe a strongly damped Brownian particle moving in a constant gravitational field using the stochastic differential equation

$$dx = -g\,dt + \sqrt{D}\,dW(t), \tag{5.1.11}$$

for which the Fokker–Planck equation is

$$\frac{\partial p}{\partial t} = \frac{\partial}{\partial x}(gp) + \frac{1}{2}D\frac{\partial^2 p}{\partial x^2}. \tag{5.1.12}$$

The stationary solution given by (5.1.6) becomes

$$p_s(x) = \mathcal{N}\exp(-2gx/D), \tag{5.1.13}$$

where we have absorbed constant factors into the definition of $\mathcal{N}$.

This solution is normalizable on (a, b) only if a is finite, though b may be infinite. The probability current is given by

$$J(x) = -gp - \tfrac{1}{2}D\frac{\partial p}{\partial x}, \tag{5.1.14}$$

and the two parts correspond to a gravitational drift and a diffusional drift, and the condition that the total current is zero requires their cancellation.

Mathematically this arises because we must have a lower boundary a and since no current can pass through this boundary $J(a) = 0$, which requires that the stationary current J_s be zero everywhere. The condition that the current should be zero on a boundary is a *reflecting boundary condition.*

In the general case, a reflecting boundary condition for the Fokker–Planck equation (5.1.2) at the position a, requires the probability current (5.1.14) vanish at a, that is

$$A(a,t)p(a,t) + \tfrac{1}{2}\partial_x[B(x,t)p(x,t)\big|_a = 0. \tag{5.1.15}$$

c) Absorbing Boundary Condition: If we set $p(a, t|x_0, t_0) = 0$ for some value a we have an *absorbing boundary condition.* In this case total probability is no longer conserved, and we must interpret the total probability as being appropriate to the interval either above or below a. To simulate this, one would consider an ensemble of particles each of which obeys the stochastic differential equation under consideration. The particles move in the region $x \geq a$, and whenever one reaches $x = a$, it is removed from the system.

The only stationary solution is $p_s(x) = 0$, and attention is focused more on how long the particle stays in the interval.

d) General Boundary Conditions: The description of the possible boundary conditions even for a one-dimensional Fokker–Planck equation is quite a complex task—see *Stochastic Methods* for more detail.

5.1.2 Deterministic Motion

When the coefficient of the second derivative vanishes, the corresponding stochastic differential equation becomes an ordinary differential equation, so the motion is deterministic. The resulting Fokker–Planck equation then has the form

$$\frac{\partial p}{\partial t} = \frac{\partial}{\partial x}\big(A(x)p(x,t)\big). \tag{5.1.16}$$

If $u(y,t)$ is the solution of

$$\dot{u}(t) = -A\big(u(t)\big), \tag{5.1.17}$$

with the initial condition $u(y,0) = y$, then the solution of (5.1.16) with this initial condition must be

$$p(x,t) = \delta\big(x - u(y,t)\big). \tag{5.1.18}$$

Exercise 5.2 : By direct substitution, show that (5.1.18) is the solution of the equation of motion (5.1.16).

a) Interpretation: The solution (5.1.18) represents a point which starts at $x = y$, and then moves according to the equation of motion (5.1.17).

Suppose we have an initial probability distribution

$$p(x,0) = p_0(x) \equiv \int dy\, p_0(y)\delta(x-y). \tag{5.1.19}$$

Since the equation of motion (5.1.16) is linear, the solution is

$$p(x,t) = \int dy\, p_0(y)\delta(x - u(y,t)). \tag{5.1.20}$$

This corresponds to an initial "cloud" of particles, distributed over the range of values of x according to (5.1.19), each of which then follows a trajectory given by the equation of motion (5.1.17).

Exercise 5.3 Simple Damping: Suppose $A(x) = -kx$. Both by substitution into equation (5.1.16), and by following the method of solution given above, show that the solution of the equation (5.1.16) is given by $p(x,t) = e^{kt}p(xe^{kt},0)$.

b) Differential Equation Interpretation: Formally, we can say that the equation of motion

$$\frac{\partial p}{\partial t} = \frac{\partial}{\partial x}\big\{A(x)p\big\}, \tag{5.1.21}$$

is the equation of motion for a cloud of particles whose distribution is $p(x,t)$, each point of which obeys the equation of motion

$$\frac{dx}{dt} = -A(x). \tag{5.1.22}$$

This is simply a special case of the correspondence between the Fokker–Planck equation and the stochastic differential equation.

5.1.3 The Wiener Process

The Wiener process has the Fokker–Planck equation and corresponding stochastic differential equation

$$\frac{\partial p(x,t|x_0,t_0)}{\partial t} = \frac{1}{2}\frac{\partial^2 p(x,t|x_0,t_0)}{\partial x^2}, \tag{5.1.23}$$

$$dx(t) = dW(t). \tag{5.1.24}$$

Thus we can write

$$x(t) = x_0 + \int_{t_0}^{t} dW(t') \equiv x_0 + W(t) - W(t_0), \tag{5.1.25}$$

$$\langle x(t)\rangle = x_0, \tag{5.1.26}$$

$$\langle [x(t)-x_0]^2\rangle = \int_{t_0}^{t}\int_{t_0}^{t} \langle dW(t'')\,dW(t')\rangle = \int_{t_0}^{t} dt' \int_{t_0}^{t} dt''\,\delta(t'-t'') = t - t_0. \tag{5.1.27}$$

Thus $x(t)$ is a Gaussian quantity with mean x_0 and variance $t-t_0$; this means its conditional probability density is given by

$$p(x,t\,|\,x_0,t_0) = \frac{1}{\sqrt{2\pi(t-t_0)}}\exp\left(-\frac{(x-x_0)^2}{2(t-t_0)}\right). \tag{5.1.28}$$

5.2 Eigenfunctions of the Fokker–Planck Equation

When the drift and diffusion coefficients are time-independent it is useful to express solutions of the Fokker–Planck equation in terms of eigenfunctions. To show how this is done, we will treat the case of the Fokker–Planck equation

$$\partial_t p(x,t) = -\partial_x[A(x)p(x,t)] + \tfrac{1}{2}\partial_x^2[B(x)p(x,t)], \tag{5.2.1}$$

for a process on a interval (a,b) with reflecting boundaries, and for which the stationary solution is $p_s(x)$.

5.2.1 Construction of Eigenfunctions

There are eigenfunctions for both the forward and the backward Fokker–Planck operators, and both are useful. We first introduce the function $q(x,t)$, defined by

$$p(x,t) = p_s(x)q(x,t). \tag{5.2.2}$$

Substituting in the Fokker–Planck equation (5.2.1), we find that $q(x,t)$ satisfies the backward Fokker–Planck equation

$$\partial_t q(x,t) = A(x)\partial_x q(x,t) + \tfrac{1}{2}B(x)\partial_x^2 q(x,t). \tag{5.2.3}$$

We introduce the eigenfunctions by considering solutions of the form

$$p(x,t) = P_\lambda(x)\mathrm{e}^{-\lambda t}, \tag{5.2.4}$$

$$q(x,t) = Q_\lambda(x)\mathrm{e}^{-\lambda t}. \tag{5.2.5}$$

By substitution in the Fokker–Planck equation (5.2.1), we find the eigenfunction equations

$$-\partial_x[A(x)P_\lambda(x)] + \tfrac{1}{2}\partial_x^2[B(x)P_\lambda(x)] = -\lambda P_\lambda(x), \tag{5.2.6}$$

$$A(x)\partial_x Q_\lambda(x) + \tfrac{1}{2}B(x)\partial_x^2 Q_\lambda(x) = -\lambda Q_\lambda(x). \tag{5.2.7}$$

a) Forward and Backward Eigenfunctions: The P_λ are eigenfunctions of the forward Fokker–Planck equation, while the Q_λ are eigenfunctions of the backward Fokker–Planck equation. From (5.2.2) and (5.2.3) it follows that

$$P_\lambda(x) = p_\mathrm{s}(x)Q_\lambda(x). \tag{5.2.8}$$

This simple result does not generalize completely to many dimensional situations. These are treated in *Stochastic Methods.*

b) Bi-Orthogonality of Eigenfunctions: Using partial integration we can deduce that

$$(\lambda' - \lambda)\int_a^b dx P_\lambda(x)Q_{\lambda'}(x)$$

$$= \Big[Q_{\lambda'}(x)\left\{-A(x)P_\lambda(x) + \tfrac{1}{2}\partial_x[B(x)P_\lambda(x)]\right\} - \tfrac{1}{2}B(x)P_\lambda(x)\partial_x Q_{\lambda'}(x)\Big]_a^b. \tag{5.2.9}$$

The coefficient of $Q_{\lambda'}(x)$, vanishes because of the reflecting boundary conditions (see 5.1.15) at a and b. From the definition (5.2.2) of $q(x,t)$, it follows that

$$\tfrac{1}{2}B(x)\partial_x Q_{\lambda'}(x) = -A(x)P_{\lambda'}(x) + \tfrac{1}{2}\partial_x[B(x)P_{\lambda'}(x)]. \tag{5.2.10}$$

Thus, this term also vanishes as a result of the reflecting boundary condition. This means that the forward and backward eigenfunctions form a bi-orthogonal set

$$\int_a^b dx P_\lambda(x)Q_{\lambda'}(x) = \delta_{\lambda\lambda'}. \tag{5.2.11}$$

As a consequence of the proportionality (5.2.8) of the two sets of eigenfunctions, we also have two *orthogonality* systems,

$$\int_a^b dx\, p_\mathrm{s}(x)Q_\lambda(x)Q_{\lambda'}(x) = \delta_{\lambda\lambda'}, \tag{5.2.12}$$

$$\int_a^b dx[p_\mathrm{s}(x)]^{-1}P_\lambda(x)P_{\lambda'}(x) = \delta_{\lambda\lambda'}. \tag{5.2.13}$$

If we set $\lambda = \lambda' = 0$, both of these equations are equivalent to normalization of the stationary solution $p_\mathrm{s}(x)$, because

$$P_0(x) = p_\mathrm{s}(x), \tag{5.2.14}$$

$$Q_0(x) = 1. \tag{5.2.15}$$

5.2.2 Expansion in Eigenfunctions

If we assume the eigenfunctions form a complete set, we can expand any solution in terms of eigenfunctions.

$$p(x,t) = \sum_\lambda A_\lambda P_\lambda(x) e^{-\lambda t}, \tag{5.2.16}$$

and the coefficients will be given by using bi-orthogonality

$$\int_a^b dx\, Q_\lambda(x) p(x,0) = A_\lambda. \tag{5.2.17}$$

a) Conditional Probability: We will now find an expression for the conditional probability $p(x,t\,|\,x_0,0)$ in terms of eigenfunctions. The initial condition is

$$p(x,0\,|\,x_0,0) = \delta(x-x_0), \tag{5.2.18}$$

from which it follows that

$$A_\lambda = \int_a^b dx\, Q_\lambda(x)\delta(x-x_0) = Q_\lambda(x_0). \tag{5.2.19}$$

The conditional probability therefore has the expansion

$$p(x,t\,|\,x_0,0) = \sum_\lambda P_\lambda(x) Q_\lambda(x_0) e^{-\lambda t}. \tag{5.2.20}$$

b) Autocorrelation Function: Using the expansion for the conditional probability, the autocorrelation function can be expressed as

$$\langle x(t)x(0)\rangle = \int dx \int dx_0\, x\, x_0\, p(x,t\,|\,x_0,0)\, p_s(x), \tag{5.2.21}$$

$$= \sum_\lambda \left(\int dx\, x P_\lambda(x) \right)^2 e^{-\lambda t}, \tag{5.2.22}$$

where we have substituted for $Q_\lambda(x)$ using (5.2.8).

c) Spectrum: As noted in Sect. 3.3.4, the Fourier transform of the autocorrelation function gives the spectrum—applied to (5.2.22) this gives

$$S(\omega) = \sum_\lambda \frac{\lambda \left(\int dx\, x P_\lambda(x) \right)^2}{\pi(\omega^2+\lambda^2)}. \tag{5.2.23}$$

5.2.3 Eigenfunctions for the Ornstein–Uhlenbeck Process

The Ornstein–Uhlenbeck process is one of the simplest stochastic processes, for which the Fokker–Planck equation is,

$$\partial_t p(x,t) = \partial_x [kxp(x,t)] + \tfrac{1}{2} D \partial_x^2 p(x,t). \tag{5.2.24}$$

a) The Stationary Solution: For this Fokker–Planck equation this is

$$p_s(x) = \sqrt{\frac{k}{\pi D}} \exp\left(-\frac{kx^2}{D} \right). \tag{5.2.25}$$

b) The Case of a Brownian Particle: This is in fact the Fokker–Planck equation for the velocity of a Brownian particle, as introduced in (3.1.1–3.1.3), with the replacements $D \leftrightarrow 2f = 2\gamma k_B T/m$, and $k \leftrightarrow \gamma$. For a Brownian particle, the stationary solution for the distribution of the velocity u becomes, on replacing $x \to u$,

$$p_s(u) = \sqrt{\frac{m}{2\pi k_B T}} \exp\left(-\frac{mu^2}{2k_B T}\right), \tag{5.2.26}$$

as expected from statistical mechanics.

c) Eigenfunctions: The eigenfunction equation for Q_λ is

$$d_x^2 Q_\lambda - \frac{2kx}{d} d_x Q_\lambda + \frac{2\lambda}{D} Q_\lambda = 0, \tag{5.2.27}$$

which is equivalent to the differential equation for *Hermite polynomials* $\mathrm{H}_n(y)$ [43] in the variable $y = x\sqrt{k/D}$:

$$d_y^2 Q_\lambda - 2y d_y Q_\lambda + (2\lambda/k) Q_\lambda = 0. \tag{5.2.28}$$

Hence the eigenfunctions can be written

$$Q_\lambda = (2^n n!)^{-1/2} \mathrm{H}_n\left(x\sqrt{k/D}\right), \tag{5.2.29}$$

where the eigenvalues are

$$\lambda_n = nk, \tag{5.2.30}$$

and these solutions are normalized as in (5.2.11–5.2.13).

The time scale of relaxation of any solution to the stationary state is given by the these eigenvalues, and since k is the rate constant for deterministic relaxation, it thus determines the slowest time in the relaxation process.

5.3 Many-Variable Fokker–Planck Equations

Many-variable Fokker–Planck equations are very much more complex than those in one variable only. Boundaries are curves or surfaces, rather than isolated end points, and a non-zero probability current in the stationary state can exist, even when the boundary conditions are reflecting.

5.3.1 Boundary Conditions

Here we shall only consider two principal kinds of boundary conditions at a surface $\boldsymbol{S}$, namely,

i) *Reflecting Barrier Boundary Condition*:

$$\boldsymbol{n} \cdot \boldsymbol{J} = 0 \quad \text{for } \boldsymbol{x} \in \boldsymbol{S}, \tag{5.3.1}$$

where $\boldsymbol{n}$ is the normal to the surface and

$$J_i(\boldsymbol{x}, t) = A_i(\boldsymbol{x}, t) p(\boldsymbol{x}, t) - \tfrac{1}{2} \sum_j \frac{\partial}{\partial x_j} B_{ij}(\boldsymbol{x}, t) p(\boldsymbol{x}, t). \tag{5.3.2}$$

ii) *Absorbing Barrier Boundary Condition*:

$$p(\boldsymbol{x}, t) = 0 \quad \text{for } \boldsymbol{x} \in \boldsymbol{S}. \tag{5.3.3}$$

5.3.2 Stationary Solutions and Potential Conditions

In physical situations it is very often the case that the probability current vanishes in the stationary state. If this is the case, (5.3.2) shows that

$$\tfrac{1}{2}\sum_j B_{ij}(\boldsymbol{x})\frac{\partial p_s(\boldsymbol{x})}{\partial x_j} = p_s(\boldsymbol{x})\left(A_i(\boldsymbol{x}) - \tfrac{1}{2}\sum_j \frac{\partial}{\partial x_j}B_{ij}(\boldsymbol{x})\right). \tag{5.3.4}$$

If the matrix $B_{ij}(\boldsymbol{x})$ always has an inverse we can write (5.3.4) in the form

$$\frac{\partial}{\partial x_i}\log[p_s(\boldsymbol{x})] = \sum_k B^{-1}_{ik}(\boldsymbol{x})\left(2A_k(\boldsymbol{x}) - \sum_j \frac{\partial}{\partial x_j}\boldsymbol{B}_{kj}(\boldsymbol{x})\right), \tag{5.3.5}$$

$$\equiv Z_i(A, B, \boldsymbol{x}). \tag{5.3.6}$$

Because the left-hand side is a gradient, this imposes the condition

$$\frac{\partial Z_i}{\partial x_j} = \frac{\partial Z_j}{\partial x_i}. \tag{5.3.7}$$

When this condition is satisfied, the stationary solution is given by integrating (5.3.5):

$$p_s(\boldsymbol{x}) = \exp\left(\int^{\boldsymbol{x}} d\boldsymbol{x}' \cdot \boldsymbol{Z}(A, B, \boldsymbol{x}')\right). \tag{5.3.8}$$

The conditions (5.3.7) are known as *potential conditions* since we derive the quantities Z_i from derivatives of log $[p_s(\boldsymbol{x})]$, which, therefore, is often thought of as a potential $-\phi(\boldsymbol{x})$.

Exercise 5.4 Stationary Solution Using Potential Conditions: Show that the two-variable Fokker–Planck equation

$$\frac{\partial P}{\partial t} = \left\{-\frac{\partial}{\partial x}\gamma x(1 - x^2 - y^2) - \frac{\partial}{\partial y}\gamma y(1 - x^2 - y^2) + \tfrac{1}{2}g\left(\frac{\partial^2}{\partial x^2} + \frac{\partial^2}{\partial y^2}\right)\right\}P, \tag{5.3.9}$$

satisfies the potential conditions (5.3.7), and that the stationary solution can be written

$$p_s(x, y) = \mathcal{N}\exp\left(\frac{\gamma}{2g}\left(2(x^2 + y^2) - (x^2 + y^2)^2\right)\right). \tag{5.3.10}$$

Exercise 5.5 Detailed Balance: Show that the Fokker–Planck equation (7.3.32) considered in Sect. 7.3.4 can be put in a similar form to the previous Fokker–Planck equation by writing $\alpha = x + \mathrm{i}y$, yielding

$$\frac{\partial P}{\partial t} = \left\{-\frac{\partial}{\partial x}\gamma x(1 - x^2 - y^2) - \frac{\partial}{\partial y}\gamma y(1 - x^2 - y^2) + \frac{\partial}{\partial x}\Omega y - \frac{\partial}{\partial y}\Omega x - \tfrac{1}{2}g\left(\frac{\partial^2}{\partial x^2} + \frac{\partial^2}{\partial y^2}\right)\right\}P. \tag{5.3.11}$$

This equation has the same stationary solution (5.3.10), but does not satisfy the potential conditions. In fact, in *Stochastic Methods* we show that there are *detailed balance conditions*, which are more general than the potential conditions. This equation satisfies these detailed balance conditions, a fact which is related to this equation's symmetry in the exchange $x \leftrightarrow -y$.

What is the current in the stationary state?

Exercise 5.6 Fokker–Planck Equation for Brownian Motion: Show that the Fokker–Planck equation corresponding to the Brownian motion stochastic differential equations (3.1.1, 3.1.2) is

$$\frac{\partial p}{\partial t} = \left\{ -\frac{\partial}{\partial x} u - \gamma \frac{\partial}{\partial u} u + \frac{\gamma k_B T}{m} \frac{\partial^2}{\partial u^2} \right\} p. \tag{5.3.12}$$

Show that the stationary solution to this equation is independent of x, and explicitly is

$$p_s(x, u) \propto \exp\left(-\frac{mu^2}{2k_B T} \right). \tag{5.3.13}$$

Suppose a potential is included by adding a term $-dV(x)/dx$ to the right-hand side of (3.1.2). Show that the Fokker–Planck equation becomes

$$\frac{\partial p}{\partial t} = \left\{ -\frac{\partial}{\partial x} u + \frac{\partial}{\partial u} \left(\frac{dV(x)}{dx} - \gamma u \right) + \frac{\gamma k_B T}{m} \frac{\partial^2}{\partial u^2} \right\} p, \tag{5.3.14}$$

and the stationary solution becomes

$$p_s(x, u) \propto \exp\left(-\frac{V(x) + \frac{1}{2} mu^2}{k_B T} \right). \tag{5.3.15}$$

Find the current in the stationary state, and explain why it is independent of γ.

6. Master Equations and Jump Processes

The defining property of a Markov process is that it is only the present value of a stochastic variable that is required to determine the probabilistic future, leading to the introduction of a *conditional probability density* $p(x_2,t_2|x_1,t_1)$. Thus, the probability that the random variable $X(t)$ has a value in the range (x_2, x_2+dx_2) at time t_2, under the condition that it had the value x_1 at time $t_1 < t_2$, is $p(x_2,t_2|x_1,t_1)\,dx_2$. Consistency then requires the conditional probability to satisfy the Chapman–Kolmogorov equation, (3.1.7), which we have already met:

$$\int dx_2\, p(x_3,t_3\,|\,x_2,t_2)\,p(x_2,t_2\,|\,x_1,t_1) = p(x_3,t_3\,|\,x_1,t_1).$$

This equation expresses a fundamental classical property, that the variable $X(t)$ must have a value at time t_2, and that the probability of having the value at time t_3 can be expressed as the sum over the probabilities of passing through the intervals (x_2, x_2+dx_2) at any intermediate time t_2.

The essential difference in quantum mechanics, is that the world is governed by complex *amplitudes* $\psi(x_2,t_2|x_1,t_1)$, the squares of whose moduli give the probability densities, and instead of the Chapman–Kolmogorov equation, we have the *Feynman* path integral formula [44]

$$\int dx_2\, \psi(x_3,t_3\,|\,x_2,t_2)\,\psi(x_2,t_2\,|\,x_1,t_1) = \psi(x_3,t_3\,|\,x_1,t_1).$$

We will not follow this path in developing the theory of quantum Markov processes, but it nevertheless is the root of the difference between quantum and classical mechanics.

6.1 The Master Equation

Let us choose $t_1 = t_2 - \Delta t$, and assume that the conditional probability is differentiable in the sense

$$p(x_2,t_1+\Delta t|x_1,t_1) = W(x_2|x_1,t_1)\,\Delta t + \delta(x_2-x_1)\left(1-\int dx_2 W(x_2|x_1,t)\,\Delta t\right). \tag{6.1.1}$$

This states that in a time Δt, we may either:

i) Stay at the point x_1, with probability $1-\int dx_2 W(x_2|x_1,t)\,\Delta t$.

ii) Jump to x_2, with probability density $W(x_2|x_1,t_1)\,\Delta t$. The jump probability thus increases with Δt.

Now assume that $p(x_1, t_2 + \Delta t | x_3, t_3)$ is differentiable with respect to t; we get the *master equation*

$$\frac{\partial p(x,t|x_0,t_0)}{\partial t} = \int dz \Big\{ W(x|z,t)\, p(z,t|x_0,t_0) - W(z|x,t)\, p(x,t|x_0,t_0) \Big\}. \qquad (6.1.2)$$

a) Interpretations: This equation, also known as Kolmogorov's equation, can be interpreted as a rate equation for the increase and decrease of probability. The probability of being at x changes in two ways:

i) It increases by transitions *from* all other points z.

ii) It decreases by transitions *to* all other points z.

b) Discrete Version of the Master Equation: If $X(t) \to N(t)$, a random variable which takes on *integral* values, (or a discrete range of variables) then the Chapman–Kolmogorov equation becomes

$$\sum_{n_2} P(n_3, t_3 | n_2, t_2) P(n_2, t_2 | n_1, t_1) = P(n_3, t_3 | n_1, t_1). \qquad (6.1.3)$$

In this case the master equation takes the form

$$\frac{\partial P(n,t|n_0,t_0)}{\partial t} = \sum_n W(n|m,t) P(m,t|n_0,t_0) - W(m|n,t) P(n,t|n_0,t_0). \qquad (6.1.4)$$

c) Time Between Jumps: The probability $Q(y,t,t_0)$ that, given that we start from point y at time t_0, we are still at point y at time t is given for infinitesimal Δt by

$$Q(y, t_0 + \Delta t, t_0) = 1 - \int dx\, W(x\,|\,y, t_0)\, \Delta t. \qquad (6.1.5)$$

Clearly this means that

$$\frac{\partial Q(y,t,t_0)}{\partial t} = -\int dx\, W(x\,|\,y,t)\, Q(y,t,t_0). \qquad (6.1.6)$$

If the jump probability is independent of t, then this has the simple solution

$$Q(y,t,t_0) = \mathrm{e}^{-\lambda(y)(t-t_0)}, \qquad (6.1.7)$$

$$\text{with } \lambda(y) = \int dx\, W(x\,|\,y). \qquad (6.1.8)$$

Thus, the jump times are exponentially distributed, and can be simulated very simply.

d) Simulating a Jump Process: To simulate, one first chooses a jump time according to the probability law (6.1.7), and then chooses the value of x to which the jump was made according to a probability law

$$w(x\,|\,y) = W(x\,|\,y)/\lambda(y). \qquad (6.1.9)$$

If the jump probability is not independent of t, the same procedure is in principle possible, but instead of using the exponential form (6.1.7), one must solve the differential equation (6.1.6).

In Sect. 6.1.4 we give an explicit example of this kind of simulation.

6.1.1 The Continuous Time Random Walk

We consider the model of a person who takes a step to the left or to the right at random, with a probability per unit time d for either direction. Thus the probability of jumping left or right can, for small Δt, be written

$$t^{\pm}(n) = d\,\Delta t. \tag{6.1.10}$$

We obtain an equation of motion (a genuine master equation)

$$\frac{\partial P(n,t)}{\partial t} = d\{P(n+1,t) + P(n-1,t) - 2P(n,t)\}. \tag{6.1.11}$$

We can solve this by a generating function method. We write

$$G(s,t) = \sum_{n=-\infty}^{\infty} P(n,t)\mathrm{e}^{ins}, \tag{6.1.12}$$

which gives the equation of motion

$$\frac{\partial G(s,t)}{\partial t} = d(\mathrm{e}^{\mathrm{i}s} + \mathrm{e}^{-\mathrm{i}s} - 2)G(s,t), \tag{6.1.13}$$

with an obvious solution

$$G(s,t) = \exp\left[2dt(\cos s - 1)\right], \tag{6.1.14}$$

assuming

$$P(n,0) = \delta_{n,0}. \tag{6.1.15}$$

The analytic solution (6.1.14) can then be re-expanded in powers of $\mathrm{e}^{\mathrm{i}s}$ to give

$$P(n,t) = \mathrm{e}^{-2td} I_n(2td), \tag{6.1.16}$$

where $I_n(z)$ is a modified Bessel function.

Exercise 6.1 Generating Function Solution: Use the generating function [45] for the modified Bessel function $I_n(x)$

$$\exp(x\cos s) = I_0(x) + 2\sum_{n=1}^{\infty} I_n(x)\cos(ns), \tag{6.1.17}$$

to prove (6.1.16).

Exercise 6.2 Mean and Variance: Show that the mean of n and the variance of n are

$$\langle n\rangle = \sum_{-\infty}^{\infty} nP(n,t) = -\mathrm{i}\frac{\partial}{\partial s}G(s,t)\Big|_{s=0} = 0, \tag{6.1.18}$$

$$\mathrm{var}[n] = \langle n^2\rangle - \langle n\rangle^2 = -\frac{\partial^2 G(s,t)}{\partial s^2}\Big|_{s=0} - \langle n\rangle^2 = 2td. \tag{6.1.19}$$

Notice that the mean *square distance* travelled is proportional to t—this is a characteristic of diffusion. Uniform velocity gives a mean *distance* proportional to t.

Exercise 6.3 Three Dimensions: Formulate the continuous time random walk in three dimensions, and show that the same kind of conclusions can be deduced.

a) Approximate Fokker–Planck Equation—the Kramers–Moyal Expansion: We would like the continuous *space* limit of a random walk. Let us write $x = nl$, where we understand that l is the distance between the points the walker can be at. Then (6.1.11) becomes

$$\frac{\partial p(x,t)}{\partial t} = d\{p(x+l,t) + p(x-l,t) - 2p(x,t)\}, \tag{6.1.20}$$

$$\approx l^2 d\frac{\partial^2 p}{\partial x^2}. \tag{6.1.21}$$

We have dropped higher-order derivatives, which will be justifiable if the function $p(x,t)$ is sufficiently smooth in x. The resultant differential equation is the *diffusion* equation, and is a special case of a *Fokker–Planck equation.*

Solving it is simple:

i) Again we define a generating function, this time as a Fourier transform

$$g(q,t) = \frac{1}{\sqrt{2\pi}}\int dx\, p(x,t)\mathrm{e}^{\mathrm{i}qx}, \tag{6.1.22}$$

and it is easy to see that (6.1.21) requires

$$\frac{\partial g(q,t)}{\partial t} = -l^2 dq^2 g(q,t), \tag{6.1.23}$$

so that

$$g(q,t) = \exp\left(-q^2 l^2 dt\right) g(q,0). \tag{6.1.24}$$

ii) Now consider the case where $p(x,0) = \delta(x)$ corresponding to a particle localised at $x = 0$, so that $g(q,0) = 1/\sqrt{2\pi}$. This means that we have $g(q,t)$, and that $p(x,t)$ is given by the Fourier inversion formula.

$$p(x,t) = \frac{1}{2\pi}\int dq\, \exp\left(-q^2 l^2 dt\right)\mathrm{e}^{-\mathrm{i}qx}, \tag{6.1.25}$$

$$= \frac{1}{2l\sqrt{\pi dt}}\exp\left(-\frac{x^2}{4l^2 dt}\right). \tag{6.1.26}$$

b) Comments: The results of the Fokker–Planck approximation are a simplified version of the accurate treatment:

i) The result is a Gaussian distribution, which can be seen as a result of the central limit theorem, in that the jumps are very small, and there are very many of them.

ii) The mean and variance of the Gaussian are given by

$$\langle x(t)\rangle = 0, \qquad \langle x(t)^2\rangle = 2l^2 dt. \tag{6.1.27}$$

iii) We see that the mean *square* distance travelled per unit time is $2l^2 d$, equivalent to the exact result (6.1.19).

6.1.2 The Poisson Process

The Poisson process models *independent arrivals*, such as electrons arriving at an anode, or customers arriving at a shop, with a well-defined probability per unit time λ of arriving. This is governed by the master equation for which only the jump $n \to n+1$ occurs. Thus

$$W(n+1\,|\,n,t) = \lambda, \tag{6.1.28}$$

$$W(n\,|\,m,t) = 0, \qquad m \neq n+1. \tag{6.1.29}$$

This master equation becomes

$$\partial_t P(n,t\,|\,n',t') = \lambda[P(n-1,t\,|\,n',t') - P(n,t\,|\,n',t')]. \tag{6.1.30}$$

If we compare this to (6.1.11), we can see that this amounts to a "one-sided" random walk, in which the walker makes steps only to the right, with step probability per unit time equal to λ. A numerical simulation is illustrated in Fig. 6.1

We can solve this problem using the characteristic function, whose evaluation is similar to the method used in Sect. 6.1.1. Doing so, we find

$$\partial_t G(s,t) = \lambda\big(\exp(\mathrm{i}s) - 1\big)G(s,t). \tag{6.1.31}$$

Let us set the initial condition as $n = 0$ at time $t = 0$. This means that $G(s,0) = 1$, and the solution of this equation is

$$G(s,t) = \exp\big[\lambda t\big(\exp(\mathrm{i}s) - 1\big)\big]. \tag{6.1.32}$$

The coefficients of $\mathrm{e}^{\mathrm{i}ns}$ then yield the probabilities, which are

$$P(n,t\,|\,0,0) = \frac{\exp(-\lambda t)(\lambda t)^n}{n!}. \tag{6.1.33}$$

This is a Poisson distribution whose mean is given by

$$\langle N(t)\rangle = \lambda t. \tag{6.1.34}$$

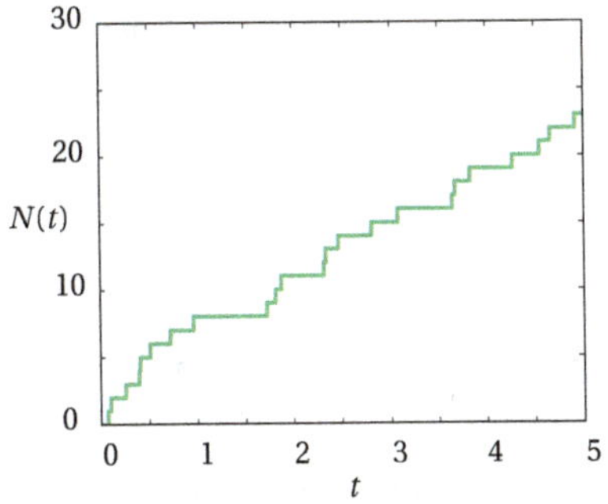

Fig. 6.1. Simulation of a Poisson process with intensity parameter $\lambda = 5$.

6.1.3 The Random Telegraph Process

The random telegraph process is a model of a random Morse signal, that is, of a signal $X(t)$ which can have either of two values a and b. The signal is taken to switch between the two values with certain probabilities per unit time. A master equation which models this can be written

$$\left.\begin{aligned} \partial_t P(a,t\,|\,x,t_0) &= -\lambda P(a,t\,|\,x,t_0)+\mu P(b,t\,|\,x,t_0)\,, \\ \partial_t P(b,t\,|\,x,t_0) &= \lambda P(a,t\,|\,x,t_0)-\mu P(b,t\,|\,x,t_0)\,. \end{aligned}\right\} \tag{6.1.35}$$

In Fig. 6.2 we show a numerical simulation of a typical example of the random telegraph process.

a) Time-Dependent Solutions: These two coupled differential equations can be simplified to a single equation by using the condition

$$P(a,t\,|\,x,t_0)+P(b,t\,|\,x,t_0)=1\,. \tag{6.1.36}$$

The initial condition for the conditional probability is

$$P(x',t_0\,|\,x,t_0)=\delta_{x,x'}\,. \tag{6.1.37}$$

We can straightforwardly derive an equation for $\lambda P(a,t\,|\,x,t_0)-\mu P(b,t\,|\,x,t_0)$, for which the solution becomes

$$\lambda P(a,t\,|\,x,t_0)-\mu P(b,t\,|\,x,t_0)=\mathrm{e}^{-(\lambda+\mu)(t-t_0)}\,(\lambda\delta_{a,x}-\mu\delta_{bx})\,. \tag{6.1.38}$$

Using (6.1.38) we can now construct the solution for the conditional probabilities in the form

$$\left.\begin{aligned} P(a,t\,|\,x,t_0) &= \frac{\mu}{\lambda+\mu}+\mathrm{e}^{-(\lambda+\mu)(t-t_0)}\left(\frac{\lambda}{\lambda+\mu}\delta_{a,x}-\frac{\mu}{\lambda+\mu}\delta_{b,x}\right), \\ P(b,t\,|\,x,t_0) &= \frac{\lambda}{\lambda+\mu}-\mathrm{e}^{-(\lambda+\mu)(t-t_0)}\left(\frac{\lambda}{\lambda+\mu}\delta_{a,x}-\frac{\mu}{\lambda+\mu}\delta_{b,x}\right). \end{aligned}\right\} \tag{6.1.39}$$

Corresponding to this solution, the mean of $X(t)$ is

$$\begin{aligned} \langle X(t)\,|\,[x_0,t_0]\rangle &= \sum xP(x,t\,|\,x_0,t_0)\,, \\ &= \frac{a\mu+b\lambda}{\mu+\lambda}+\mathrm{e}^{-(\lambda+\mu)(t-t_0)}\left(x_0-\frac{a\mu+b\lambda}{\mu+\lambda}\right). \end{aligned} \tag{6.1.40}$$

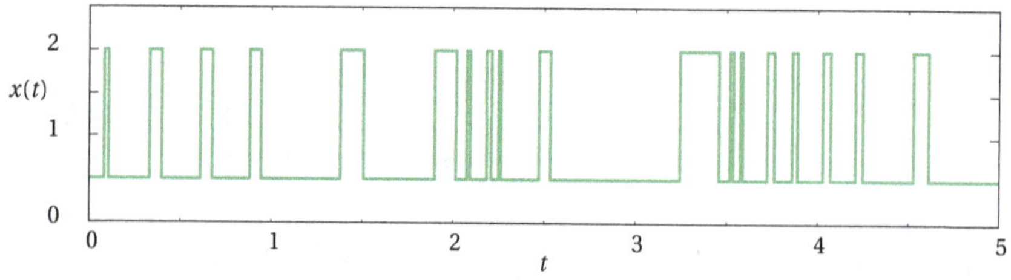

Fig. 6.2. Simulation of a random telegraph process with $a=0.5$, $b=2$, $\lambda=3$ and $\mu=3$.

Exercise 6.4 Mean and Variance of $X(t)$: Show that the mean is given by (6.1.40), and find an expression for the variance of $X(t)$.

b) Stationary Solutions: If we let $t_0 \to -\infty$ we find the stationary solution

$$P_s(a) = \frac{\mu}{\lambda+\mu}, \qquad P_s(b) = \frac{\lambda}{\lambda+\mu}. \tag{6.1.41}$$

We could also have got this result by directly using the master equation (6.1.35).

The mean and variance in the stationary state are

$$\langle X\rangle_s = \frac{a\mu+b\lambda}{\mu+\lambda}, \qquad \mathrm{var}\,[X]_s = \frac{(a-b)^2\mu\lambda}{(\lambda+\mu)^2}. \tag{6.1.42}$$

c) Stationary Correlation Functions: The stationary correlation function for two times $t \geqslant s$ is calculated by writing

$$\langle X(t)X(s)\rangle_s = \sum_{xx'} xx' P(x,t\,|\,x',s)P_s(x'), \tag{6.1.43}$$

$$= \sum_{x'} x'\langle X(t)\,|\,[x',s]\rangle P_s(x'). \tag{6.1.44}$$

We can then use the explicit solutions (6.1.40–6.1.42) to find

$$\langle X(t)X(s)\rangle_s = \langle X\rangle_s^2 + \exp[-(\lambda+\mu)(t-s)](\langle X^2\rangle_s - \langle X\rangle_s^2), \tag{6.1.45}$$

$$= \left(\frac{a\mu+b\lambda}{\mu+\lambda}\right)^2 + \exp[-(\lambda+\mu)(t-s)]\frac{(a-b)^2\mu\lambda}{(\lambda+\mu)^2}. \tag{6.1.46}$$

Thus, it follows that

$$\langle X(t), X(s)\rangle_s = \langle X(t)X(s)\rangle_s - \langle X\rangle_s^2 = \frac{(a-b)^2\mu\lambda}{(\lambda+\mu)^2}\,\mathrm{e}^{-(\lambda+\mu)|t-s|}. \tag{6.1.47}$$

This time correlation function is of exactly the same form as that of the Ornstein–Uhlenbeck process. Of course, the higher-order correlation functions are different from those of the Ornstein–Uhlenbeck process, but these are not always important. Because of this simple correlation function, and because the two-state process is very simple to implement, the random telegraph is widely used in model building.

6.1.4 Example—Simulating Jumps in a Two-Level Atom

In Sect. 14.1.3 we develop a model of a two-level atom in which the occupation probabilities for the excited and ground states (labelled e and g) obey the random telegraph master equation

$$\dot{p}(e) = -\gamma(\bar{N}+1)p(e) + \gamma\bar{N}p(g), \tag{6.1.48}$$

$$\dot{p}(g) = \gamma(\bar{N}+1)p(e) - \gamma\bar{N}p(g). \tag{6.1.49}$$

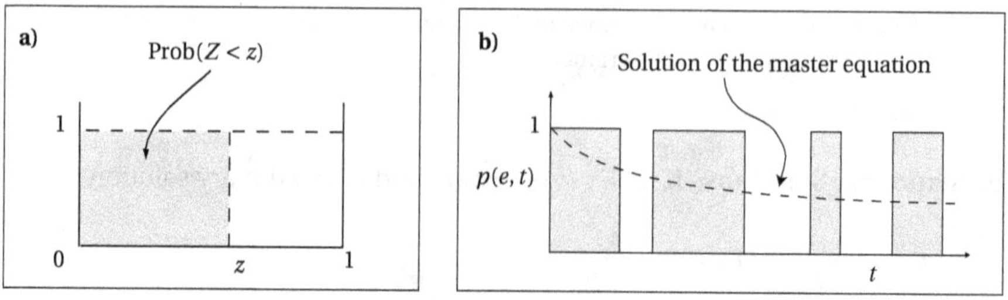

Fig. 6.3. a) We choose the value z of the random variable Z from a uniform distribution of random numbers on $(0,1)$; **b)** The average over occupations of 100% and 0% gives the exponential decay to the average value, as indicated by the dashed line.

This can be simulated using the method of Sect. 6.1d. Here there are only two levels, and the levels are discrete so we have

$$\begin{aligned} \lambda(e) &= \textstyle\sum_f W(f\,|\,e) &= \gamma(N+1) \\ \lambda(g) &= \textstyle\sum_f W(f\,|\,g) &= \gamma N, \\ Q(e,t,t_0) &= \mathrm{e}^{-\gamma(N+1)(t-t_0)}, \\ Q(g,t,t_0) &= \mathrm{e}^{-\gamma N(t-t_0)}. \end{aligned} \tag{6.1.50}$$

Thus, starting initially with the atom in the upper level "e", if T is the (random) time at which the atom jumps to the lower state, what we can say is

$$\mathrm{Prob}(T > t) = Q(e,t,t_0) = \exp\big(-\gamma(\bar N+1)t\big). \tag{6.1.51}$$

Let us define

$$z = \exp\big(-\gamma(\bar N+1)t\big)\,, \qquad Z = \exp\big(-\gamma(\bar N+1)T\big)\,. \tag{6.1.52}$$

Then from (6.1.51) we have

$$\mathrm{Prob}(Z < z) = z. \tag{6.1.53}$$

The range of both z and Z is $(0,1)$, so that (6.1.53) simply says that Z is a random variable uniformly distributed on the range $(0,1)$.

The simulation method, which is illustrated in Fig. 6.3, proceeds as follows:

i) Using a random number generator, choose a random number Z in the range $(0,1)$.

ii) Inverting (6.1.52), let

$$T = -\log Z/\gamma(\bar N+1). \tag{6.1.54}$$

iii) The atom then starts in level "e", and stays there until the time T, when it makes a jump to "g".

iv) The atom is now in the ground state "g", and we now must calculate how long it will remain there. From (6.1.48) we can see that the probability of leaving "g" in a time interval dt is $\gamma\bar N dt$, and then the argument follows in exactly the same way, with $\bar N+1$ replaced by $\bar N$.

v) Therefore, choose Z' randomly in $(0, 1)$, and let

$$T' - T = -\log Z'/\gamma\bar{N} \tag{6.1.55}$$

determine the time of the jump from "g" back to "e".

vi) The process continues indefinitely, giving a graph of $p(e)$ which is a series of random jumps.

The time-dependent probability $p(e, t)$ is obtained by repeating this process many times, with different random number sequences, and this yields the result obtained by solving the master equation as in (6.1.39).

7. Applications of Random Processes

The process of modelling physical systems is dominated by the aim of simplicity, and this chapter concentrates on providing the descriptions of the most common models, and some of their simple applications. Classically, there is considerable freedom available in choosing models, a situation which does not pertain in quantum mechanics. Nevertheless, it is often possible to develop representations of noisy quantum systems, sometimes exact, sometimes approximate, which take the form of classical stochastic equations. In particular, Bosonic systems can be represented in various ways in terms of a complex oscillator variable $\alpha(t)$ which obeys a classical stochastic differential equation.

7.1 The Ornstein–Uhlenbeck Process

The Ornstein–Uhlenbeck process represents the simplest kind of stochastic differential equation in which non-trivial dynamics exists. It is the abstraction of the Brownian motion model considered in Sect. 3.2, and we have already seen its Fokker–Planck equation in Sect. 5.1.1a, when we were considering boundary conditions for the Fokker–Planck equation. It is one of the most basic processes one uses in modelling a noise with a finite correlation time.

7.1.1 Ornstein–Uhlenbeck Process in One Dimension

The stochastic differential equation is

$$dx = -kx\,dt + \sqrt{D}\,dW(t), \tag{7.1.1}$$

and its solution is

$$x(t) = x(0)\,\mathrm{e}^{-kt} + \sqrt{D}\int_0^t \mathrm{e}^{-k(t-t')}\,dW(t'). \tag{7.1.2}$$

If the initial condition is deterministic or Gaussian distributed, then $x(t)$ is clearly Gaussian.

a) Mean and Variance: These can be computed directly from the solution (7.1.2), and are

$$\langle x(t)\rangle = \langle x(0)\rangle\, \mathrm{e}^{-kt}, \tag{7.1.3}$$

$$\mathrm{var}\,[x(t)] = \left\langle \left\{ [x(0) - \langle x(0)\rangle]\,\mathrm{e}^{-kt} + \sqrt{D}\int_0^t \mathrm{e}^{-k(t-t')}\, dW(t') \right\}^2 \right\rangle. \tag{7.1.4}$$

The initial condition is independent of $dW(t)$ for $t > 0$, so we can write

$$\begin{aligned}\mathrm{var}\,[x(t)] &= \mathrm{var}\,[x(0)]\,\mathrm{e}^{-2kt} + D\int_0^t \mathrm{e}^{-2k(t-t')}\, dt', \\ &= \left(\mathrm{var}\,[x(0)] - D/2k\right)\mathrm{e}^{-2kt} + D/2k.\end{aligned} \tag{7.1.5}$$

b) Time Correlation Function: This can also be calculated directly and is,

$$\begin{aligned}\langle x(t), x(s)\rangle &= \mathrm{var}\,[x(0)]\,\mathrm{e}^{-k(t+s)} + D\left\langle \int_0^t \mathrm{e}^{-k(t-t')}\, dW(t') \int_0^s \mathrm{e}^{-k(s-s')}\, dW(s') \right\rangle, \\ &= \mathrm{var}\{x(0)\}\,\mathrm{e}^{-k(t+s)} + D \int_0^{\min(t,s)} \mathrm{e}^{-k(t+s-2t')}\, dt', \\ &= \left[\mathrm{var}\{x(0)\} - \frac{D}{2k}\right]\mathrm{e}^{-k(t+s)} + \frac{D}{2k}\,\mathrm{e}^{-k|t-s|}.\end{aligned} \tag{7.1.6}$$

Notice that if $k > 0$, as $t, s \to \infty$ with finite $|t - s|$, the correlation function becomes stationary and of the form

$$\langle x(t), x(s)\rangle_{\mathrm{s}} = \frac{D}{2k}\,\mathrm{e}^{-k|t-s|}. \tag{7.1.7}$$

c) Conditional Probability: The conditional probability density $p(x, t|x_0, 0)$ is the probability distribution of $x(t)$ when the initial condition is the definite number x_0. Because $x(t)$ is Gaussian, this distribution is given by the knowledge of the mean and variance, which we have computed above, and is therefore given by the usual Gaussian formula as

$$p(x, t|x_0, 0) = \frac{\sqrt{k}}{\sqrt{\pi D\big(1 - \exp(-2kt)\big)}}\exp\left(-\frac{k\big(x - x_0\exp(-kt)\big)^2}{D\big(1 - \exp(-2kt)\big)}\right). \tag{7.1.8}$$

7.1.2 Many-Variable Ornstein–Uhlenbeck Process

The generalization to many dimensions can be formulated as an n-variable linear system of Langevin equations, with variables and noise given by

$$\boldsymbol{u} = \begin{bmatrix} u_1 \\ u_2 \\ \vdots \\ u_n \end{bmatrix}, \qquad \boldsymbol{\xi}(t) = \begin{bmatrix} \xi_1(t) \\ \xi_2(t) \\ \vdots \\ \xi_n(t) \end{bmatrix}, \tag{7.1.9}$$

so that an equation of motion can be written

$$\frac{d\boldsymbol{u}}{dt} = -A\boldsymbol{u} + \boldsymbol{\xi}(t), \tag{7.1.10}$$

where A is some matrix. Here the correlation functions would take the form

$$\langle \xi_i(t)\xi_j(t')\rangle = B_{ij}\delta(t-t'), \tag{7.1.11}$$

which can be written in matrix form as

$$\langle \boldsymbol{\xi}(t)\boldsymbol{\xi}^{\mathrm{T}}(t')\rangle = B\delta(t-t'), \tag{7.1.12}$$

where B is a symmetric positive semidefinite matrix.

a) Equivalent Stochastic Differential Equation: This would take the form

$$d\boldsymbol{u} = -A\boldsymbol{u}\,dt + \mathsf{B}\,d\boldsymbol{W}(t), \tag{7.1.13}$$

in which the correlation function (7.1.12) requires

$$\boldsymbol{\xi}(t)\,dt = \mathsf{B}\,d\boldsymbol{W}(t), \tag{7.1.14}$$

$$\mathsf{B}\mathsf{B}^{\mathrm{T}} = B. \tag{7.1.15}$$

b) Solution of the Equation: It is slightly simpler to keep the Langevin notation—thus we can now solve (7.1.10) to obtain the exact solution

$$\boldsymbol{u}(t) = \mathrm{e}^{-At}\boldsymbol{u}(0) + \int_0^t dt'\,\mathrm{e}^{-A(t-t')}\boldsymbol{\xi}(t'). \tag{7.1.16}$$

c) Solution for the Mean: Since the mean of the noise is zero, the mean $\langle \boldsymbol{u}(t)\rangle$ is given by

$$\langle \boldsymbol{u}(t)\rangle = \mathrm{e}^{-At}\boldsymbol{u}(0). \tag{7.1.17}$$

d) Correlation Matrix: This is defined by

$$\Big\langle \big[\boldsymbol{u}(t) - \mathrm{e}^{-At}\boldsymbol{u}(0)\big]\big[\boldsymbol{u}(t) - \mathrm{e}^{-At}\boldsymbol{u}(0)\big]^{\mathrm{T}}\Big\rangle = \int_0^t dt' \int_0^t dt''\,\mathrm{e}^{-A(t-t')}\langle \boldsymbol{\xi}(t')\boldsymbol{\xi}^{\mathrm{T}}(t'')\rangle \mathrm{e}^{-A^{\mathrm{T}}(t-t'')}, \tag{7.1.18}$$

so that

$$\langle \boldsymbol{u}(t)\boldsymbol{u}^{\mathrm{T}}(t)\rangle = \mathrm{e}^{-At}\boldsymbol{u}(0)\boldsymbol{u}^{\mathrm{T}}(0)\mathrm{e}^{-A^{\mathrm{T}}t} + \int_0^t dt'\,\mathrm{e}^{-A(t-t')}B\mathrm{e}^{-A^{\mathrm{T}}(t-t')}. \tag{7.1.19}$$

e) Stationary Solutions: If the system is stable, we can set $e^{-At} \to 0$ as $t \to \infty$. In this case there is a stationary state, and we will have

$$\langle \boldsymbol{u} \rangle_s = 0, \tag{7.1.20}$$

$$\langle \boldsymbol{u}\boldsymbol{u}^T \rangle_s \equiv \sigma = \int_0^\infty dt\, e^{-At} B e^{-A^T t}. \tag{7.1.21}$$

Here σ is the stationary covariance matrix, and is in principle measurable. We can derive the equation

$$A\sigma + \sigma A^T = -\int_0^\infty dt \frac{d}{dt}\left\{e^{-At} B e^{-A^T t}\right\}, \tag{7.1.22}$$

$$= -e^{-At} B e^{-A^T t}\Big|_0^\infty. \tag{7.1.23}$$

As $t \to \infty$ this vanishes, so we get the simple result

$$A\sigma + \sigma A^T = B. \tag{7.1.24}$$

These equations are sufficient to determine σ when A and B are known.

Exercise 7.1 Brownian Motion of a Particle in a Harmonic Potential: Generalizing the formulation of Sect. 3.1, Sect. 3.2, show that the appropriate equations will take the form

$$\frac{dx}{dt} = u, \tag{7.1.25}$$

$$\frac{du}{dt} = -\gamma u - \omega^2 x + f\xi(t). \tag{7.1.26}$$

Write these as a matrix Langevin equation, and solve by the methods above. Show that the solution satisfies the principle of equipartition of energy.

7.1.3 Stationary Correlation Functions and Spectrum

a) Stationary Time Correlation Matrix: We can use the explicit solution (7.1.16) of the Langevin equation (7.1.10) in a similar way to show that for $t > s$,

$$\langle \boldsymbol{u}_s(t), \boldsymbol{u}_s^T(s)\rangle = \exp[-A(t-s)] \int_{-\infty}^{s} \exp[-A(s-t')]BB^T \exp[-A^T(s-t')]\,dt',$$

$$= \exp[-A(t-s)]\sigma, \qquad t > s, \tag{7.1.27a}$$

and similarly,

$$\langle \boldsymbol{u}_s(t), \boldsymbol{u}_s^T(s)\rangle = \sigma \exp[-A^T(s-t)], \quad t < s. \tag{7.1.27b}$$

As expected for a stationary solution, these depend only on $s - t$. If we define

$$G_s(t-s) = \langle \boldsymbol{u}_s(t), \boldsymbol{u}_s^T(s)\rangle, \tag{7.1.28}$$

it follows that (since $\sigma = \sigma^T$) that

$$G_s(t-s) = [G_s(s-t)]^T. \tag{7.1.29}$$

b) Spectrum Matrix: A very elegant formula can be derived for the spectrum matrix. As in Sect. 3.3.4, we define

$$S(\omega) = \frac{1}{2\pi}\int_{-\infty}^{\infty} e^{-i\omega\tau} G_s(\tau)d\tau\,, \tag{7.1.30}$$

$$= \frac{1}{2\pi}\left(\int_0^{\infty} \exp[-(i\omega + A)\tau]\sigma\, d\tau + \int_{-\infty}^{0} \sigma \exp[(-i\omega + A^{\mathrm{T}})\tau]\, d\tau\right), \tag{7.1.31}$$

$$= \frac{1}{2\pi}\left((A+i\omega)^{-1}\sigma + \sigma(A^{\mathrm{T}} - i\omega)^{-1}\right). \tag{7.1.32}$$

Therefore it follows that

$$(A+i\omega)S(\omega)(A^T - i\omega) = \frac{1}{2\pi}(\sigma A^{\mathrm{T}} + A\sigma)\,, \tag{7.1.33}$$

and now using (7.1.24), we get

$$\boxed{S(\omega) = \frac{1}{2\pi}(A+i\omega)^{-1}BB^{\mathrm{T}}(A^{\mathrm{T}} - i\omega)^{-1}\,.} \tag{7.1.34}$$

7.2 Johnson Noise

Classical noise is always associated with dissipation, and Johnson noise is the noise in an electrical circuit associated with the dissipation in a resistor. Let us write a series *LRC* circuit as in Fig. 7.1, and assume that there is a fluctuating voltage $v(t) = \sqrt{f}\,\xi(t)$ which comes from the resistor. We will show that the requirement that the system comes to thermal equilibrium determines that the noise from the resistor is proportional to the absolute temperature.

The equations of motion follow from ordinary circuit theory, and are

$$\frac{dQ}{dt} = i, \tag{7.2.1}$$

$$L\frac{di}{dt} + iR + \frac{Q}{c} = v(t). \tag{7.2.2}$$

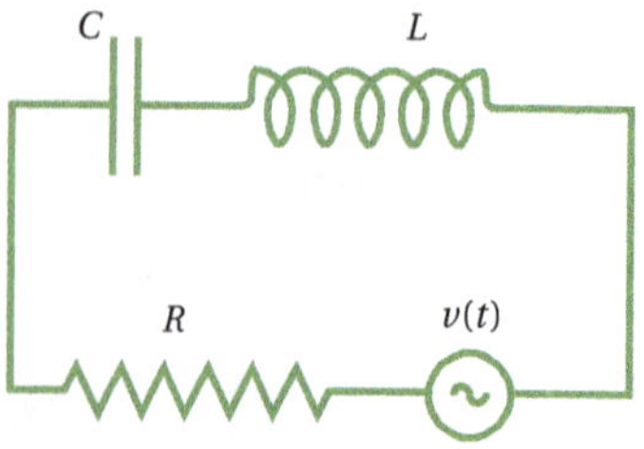

Fig. 7.1. *LRC* circuit with voltage noise arising from the resistor.

We can rewrite these as a two-variable Ornstein–Uhlenbeck process, by defining the matrix

$$\boldsymbol{u} = \begin{bmatrix} Q \\ i \end{bmatrix}. \tag{7.2.3}$$

This is of the Ornstein–Uhlenbeck form (7.1.10) if we set

$$A = \begin{bmatrix} 0 & -1 \\ 1/LC & R/L \end{bmatrix}, \qquad B = \begin{bmatrix} 0 & 0 \\ 0 & f/L^2 \end{bmatrix}. \tag{7.2.4}$$

Exercise 7.2 Covariance Matrix: Solve (7.1.24) for the stationary covariance matrix σ, using the above expressions for A and B, showing that

$$\sigma = \begin{bmatrix} fC/2R & 0 \\ 0 & f/2LR \end{bmatrix}. \tag{7.2.5}$$

Now we also note that the average values $\langle i \rangle = 0$ and $\langle Q \rangle = 0$, so we can therefore deduce that

$$\langle i^2 \rangle = \frac{f}{2LR}, \qquad \langle Q^2 \rangle = \frac{fC}{2R}. \tag{7.2.6}$$

Notice however that statistical mechanics will require that the mean energy in each degree of freedom must be $k_B T/2$. Therefore the mean energy in the capacitor is given by $\langle Q^2 \rangle/2C = k_B T/2$, and the mean energy in the inductor is $L\langle i^2 \rangle/2 = k_B T/2$, and using our solutions (7.2.6) both of these require

$$f = 2Rk_B T. \tag{7.2.7}$$

Thus, we deduce the formula of *Johnson noise*, where the fluctuating voltage arising from a resistor is given by

$$v(t) = \sqrt{2Rk_B T}\,\xi(t), \tag{7.2.8}$$

$$\langle \xi(t)\xi(t') \rangle = \delta(t-t'). \tag{7.2.9}$$

The results of this section are essentially a transcription of those of Ex. 7.1

7.3 Complex Variable Oscillator Processes

In this section, we will consider a number of models in which the amplitude and phase of an oscillator are represented as a complex variable $\alpha(t)$. This leads to the formulation of the noise in terms of complex Ito increments, with consequent slight modifications of the Ito rules.

7.3.1 Line Broadening in a Random Frequency Oscillator

This is a simple model of line broadening of a specific type. Our physical model would be an amplitude $\alpha(t)$ with a time-dependent frequency $\omega(t) \equiv \omega_0 + \mathcal{E}(t)$, where ω_0 is constant and $\mathcal{E}(t)$ is a small time-dependent fluctuation. This is best modelled therefore by a Stratonovich equation obtained by the replacement $\mathcal{E}(t)\,dt \to \sqrt{2\gamma}\,dW(t)$, leading to

$$(\mathbf{S})\; d\alpha = -\mathrm{i}\big(\omega_0\,dt + \sqrt{2\gamma}\,dW(t)\big)\alpha, \tag{7.3.1}$$

and, using the rule (4.3.13), we can transform this to an Ito equation (noting that here $b(\alpha) = \mathrm{i}\sqrt{2\gamma}\,\alpha$, so that $b(\alpha)db(\alpha)/d\alpha = -2\gamma\alpha$)

$$(\mathbf{I})\; d\alpha = -\Big[(\mathrm{i}\omega_0 + \gamma)dt - \mathrm{i}\sqrt{2\gamma}\,dW(t)\Big]\alpha. \tag{7.3.2}$$

We can immediately see that by taking the mean value, we get

$$\frac{d\langle\alpha\rangle}{dt} = -(\mathrm{i}\omega_0 + \gamma)\langle\alpha\rangle, \tag{7.3.3}$$

whose solution is the *damped* oscillatory motion

$$\langle\alpha(t)\rangle = \langle\alpha(0)\rangle \exp\{-(\mathrm{i}\omega_0 + \gamma)t\}. \tag{7.3.4}$$

The full solution to (7.3.1) can be obtained using the fact that the Stratonovich integral can be manipulated using ordinary calculus. Thus, setting $y = \log(\alpha)$ so that $dy = d\alpha/\alpha$, we get

$$(\mathbf{S})\; dy = -\mathrm{i}\big(\omega_0\,dt + \sqrt{2\gamma}\,dW(t)\big), \tag{7.3.5}$$

and thus

$$y(t) = y(0) - \mathrm{i}\big(\omega_0 t + \sqrt{2\gamma}\,W(t)\big), \tag{7.3.6}$$

leading to the solution

$$\alpha(t) = \alpha(0)\exp\Big\{-\mathrm{i}\Big(\omega_0 t + \sqrt{2\gamma}\,W(t)\Big)\Big\}. \tag{7.3.7}$$

a) Interpretation: The full solution gives an oscillation at frequency ω_0, as well as a *phase drift*, which is random, and given by the Wiener process. We can see that $|\alpha(t)|^2 = |\alpha(0)|^2$, and that the decay of the mean as given by (7.3.4) arises because the phase becomes random. If we take an ensemble of oscillators, all with the same initial conditions, but with different samples of the Wiener process $W(t)$, we would see the dephasing as in Fig. 7.2.

Exercise 7.3 A Theorem—Mean of the Exponential of a Gaussian Variable: Show that if u is a Gaussian variable with mean $\langle u\rangle = 0$ and variance $\langle u^2\rangle = \sigma^2$, then

$$\big\langle \exp(\alpha u)\big\rangle = \exp\left(\frac{\alpha^2\sigma^2}{2}\right). \tag{7.3.8}$$

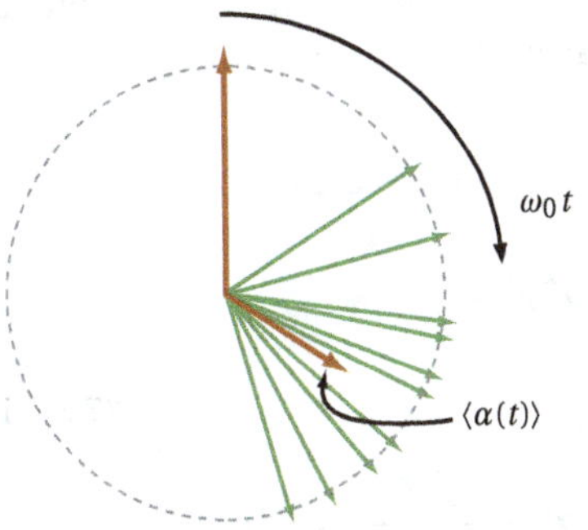

Fig. 7.2. Illustration of the decay of the mean amplitude of a complex oscillator as a result of dephasing. An ensemble of vectors evolves, all members of which share the same initial condition, represented by the vertical brown arrow. As evolution takes place, the members of the ensemble maintain the same length, but dephase with respect to each other. Their average is represented by the smaller brown arrow.

Using this result, we see that in (7.3.7) $W(t)$ is Gaussian with variance t, so we can explicitly work out the mean to get the decaying oscillation (7.3.4).

b) Correlation Function and Spectrum: Using the solution (7.3.7), we can see that

$$\langle\alpha^*(t)\alpha(t')\rangle = |\alpha(0)|^2 \left\langle \exp\left\{ i\omega_0(t-t') - i\sqrt{2\gamma}[W(t')-W(t)]\right\}\right\rangle. \tag{7.3.9}$$

Now we know that $W(t')-W(t)$ is Gaussian with mean zero and variance $|t-t'|$, hence we can say

$$\langle\alpha^*(t)\alpha(t')\rangle = |\alpha(0)|^2 \exp\{-\gamma|t-t'| + i\omega_0(t-t')\}. \tag{7.3.10}$$

Exercise 7.4 Spectrum: The spectrum is found from Khinchin's theorem to be the Fourier transform of the correlation function (7.3.10). Show it is

$$S(\omega) = \frac{I}{\pi}\frac{\gamma}{(\omega-\omega_0)^2+\gamma^2}, \tag{7.3.11}$$

where $I = |\alpha(0)|^2$. This is a standard *Lorentzian lineshape.*

7.3.2 The Thermalized Oscillator

In quantum systems we often find that thermal noise can be represented by *complex noises*

$$dw(t) = \frac{dW_1(t) + i dW_2(t)}{\sqrt{2}}, \tag{7.3.12}$$

where $dW_1(t)$ and $dW_2(t)$ are independent noises, so that

$$dw(t)^2 = dw^*(t)^2 = 0, \qquad dw^*(t)\,dw(t) = dt. \tag{7.3.13}$$

In practice, the elementary model which arises most commonly in quantum systems is a complex Ornstein–Uhlenbeck process of the form

$$d\alpha = -(i\omega + \tfrac{1}{2}\gamma)\alpha\,dt + \sqrt{\gamma\bar{N}}\,dw(t). \tag{7.3.14}$$

In contrast to the model of the previous section, the noise is in this case additive, and the equation can be solved in the same way as the Ornstein–Uhlenbeck process in Sect. 7.1.1.

a) Ito's Formula: For this kind of noise, Ito's formula takes the form

$$\begin{aligned} df(\alpha,\alpha^*) &= \frac{\partial f}{\partial \alpha} d\alpha + \frac{\partial f}{\partial \alpha^*} d\alpha^* + \frac{\partial^2 f}{\partial\alpha\partial\alpha^*} d\alpha\, d\alpha^*, \\ &\quad \tfrac{1}{2}\frac{\partial^2 f}{\partial\alpha^2} d\alpha^2 + \tfrac{1}{2}\frac{\partial^2 f}{\partial\alpha^{*2}} d\alpha^{*2} \end{aligned} \tag{7.3.15}$$

$$\begin{aligned} &= \left(\frac{\partial f}{\partial\alpha}(-\mathrm{i}\omega - \tfrac{1}{2}\gamma)\alpha + \frac{\partial f}{\partial\alpha^*}(\mathrm{i}\omega - \tfrac{1}{2}\gamma)\alpha^* + \frac{\partial^2 f}{\partial\alpha\partial\alpha^*}\gamma\bar{N}\right) dt \\ &\quad + \frac{\partial f}{\partial\alpha}\sqrt{\gamma\bar{N}}\, dw(t) + \frac{\partial f}{\partial\alpha^*}\sqrt{\gamma\bar{N}}\, dw^*(t). \end{aligned} \tag{7.3.16}$$

b) Fokker–Planck Equation: We can use the same reasoning as in Sect. 4.1.1 to show that Fokker–Planck equation is

$$\frac{\partial P}{\partial t} = \left(\frac{\partial}{\partial\alpha}(\mathrm{i}\omega + \tfrac{1}{2}\gamma)\alpha + \frac{\partial}{\partial\alpha^*}(-\mathrm{i}\omega + \tfrac{1}{2}\gamma)\alpha^* + \gamma\bar{N}\frac{\partial^2}{\partial\alpha\partial\alpha^*}\right)P. \tag{7.3.17}$$

c) Stationary Solution: The Fokker–Planck equation (7.3.17) satisfies the potential conditions of Sect. 5.3.2. Using these, the stationary solution takes the form

$$P_s(\alpha,\alpha^*) = \frac{1}{\pi\bar{N}}\exp\left(-\frac{|\alpha|^2}{\bar{N}}\right). \tag{7.3.18}$$

7.3.3 Equations for Phase and Amplitude

It is often useful to write equations like those in the previous section in terms of phase and amplitude, and this provides a nice demonstration of the power of Ito calculus.

To convert the stochastic differential equation (7.3.14) we write

$$\mu + \mathrm{i}\phi = \log\alpha, \tag{7.3.19}$$

so that using Ito rules

$$d\log\alpha = \frac{d\alpha}{\alpha} + \frac{d\alpha^2}{2\alpha^2}. \tag{7.3.20}$$

Since the noise term in (7.3.14) is proportional to $dw(t)$, whose square according to (7.3.13) is zero, the second term vanishes. Therefore we get the stochastic differential equation

$$d\mu + \mathrm{i}d\phi = \left(-\mathrm{i}\omega - \tfrac{1}{2}\gamma\right)dt + \sqrt{\gamma\bar{N}}\,\mathrm{e}^{-\mathrm{i}\phi}\frac{dW_1(t) + \mathrm{i}dW_2(t)}{\sqrt{2}|\alpha|}. \tag{7.3.21}$$

Let us define

$$\left.\begin{aligned} dW_a(t) &\equiv \quad \cos\phi \; dW_1(t) + \sin\phi \; dW_2(t), \\ dW_\phi(t) &\equiv -\sin\phi \; dW_1(t) + \cos\phi \; dW_2(t). \end{aligned}\right\} \tag{7.3.22}$$

These are *polar Ito increments.* They satisfy the Ito rules for a pair of independent Ito increments, and because $dW_1(t)$ and $dW_2(t)$ are independent of $\phi(t)$, have distribution functions independent of $\phi(t)$. Thus, since a randomly chosen pair $dW_1(t), dW_2(t)$ always gives a random pair $dW_a(t), dW_\phi(t)$, the apparent dependence on ϕ is illusory.

We can therefore write the two coupled stochastic differential equations. Firstly, the real part of (7.3.21) gives.

$$d\mu = -\tfrac{1}{2}\gamma\, dt + \sqrt{\frac{\gamma \bar{N}}{2}}\,\frac{dW_a(t)}{|\alpha|}\,. \tag{7.3.23}$$

Now define the amplitude by $|\alpha| = a \equiv \exp\mu$, and use the Ito rules to get the amplitude equation. This, together with the phase equation which follows from the imaginary part of (7.3.21) gives the pair of *phase-amplitude stochastic differential equations*

$$da = \left(-\tfrac{1}{2}\gamma a + \frac{\gamma \bar{N}}{4a}\right) dt + \sqrt{\frac{\gamma \bar{N}}{2}}\, dW_a(t), \tag{7.3.24}$$

$$d\phi = -\omega\, dt + \sqrt{\frac{\gamma \bar{N}}{2}}\,\frac{dW_\phi(t)}{a}\,. \tag{7.3.25}$$

7.3.4 The van der Pol Laser Equation

The van der Pol laser equation is a model which exhibits the basic properties of a laser, for which a full formulation is given in Sect. 17.3. The relevant stochastic differential equation can be written as

$$d\alpha = \left\{-\mathrm{i}\omega - \tfrac{1}{2}\frac{\kappa}{C}\left((1-C) + \frac{|\alpha|^2}{n_0}\right)\right\}\alpha\, dt + \sqrt{D}\, dw(t), \tag{7.3.26}$$

$$D \equiv \tfrac{1}{2}(g^2 N + \kappa). \tag{7.3.27}$$

Here α represents the electromagnetic field inside the laser cavity, and C, known as the cooperativity parameter, determines whether the device actually lases or not, depending on whether C is greater than or less than 1. The other parameters are explained in Sect. 17.3, but their precise nature is not important for the treatment in this section.

a) Deterministic Analysis: If the noise term is dropped, we can set $I = |\alpha|^2$, and develop the differential equation

$$\frac{dI}{dt} = -\frac{\kappa}{C}\left((1-C) + \frac{I}{n_0}\right) I, \tag{7.3.28}$$

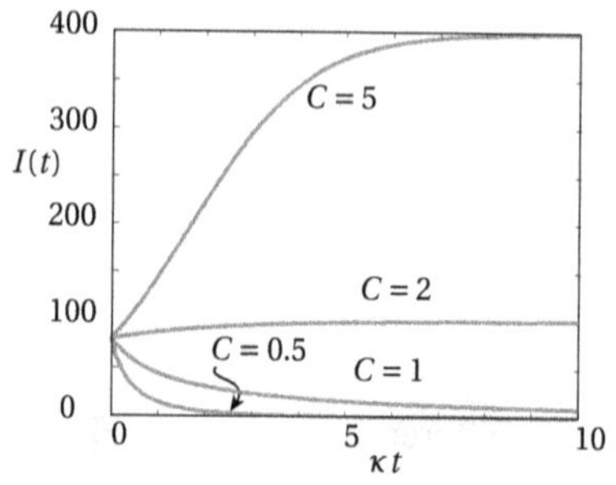

Fig. 7.3. Intensity $I(t)$ of the laser field for a range of values of the cooperativity parameter C, with $n_0 = 80$ and $I(0) = 10$. Growth occurs for $C > 1$, and decay occurs for $C \le 1$.

whose solution is

$$I(t) = \frac{n_0(C-1)}{1 - R\exp\big(\kappa(1/C-1)t\big)}, \tag{7.3.29}$$

where we define

$$R \equiv 1 + \frac{n_0(1-C)}{I(0)}. \tag{7.3.30}$$

The time-dependent nature of the deterministic solutions is illustrated in Fig. 7.3, which shows that:

i) If $C < 1$, the exponential in the denominator grows with time, so that we find $I(t \to \infty) \to 0$, and thus deterministically the laser field approaches zero.

ii) When $C > 1$ *lasing* occurs, that is, there is a non-zero field as $t \to \infty$. The exponential decreases with time, so that $I(t \to \infty) \to n_0(C-1)$.

iii) For the case $C = 1$ the solution (7.3.29) is not valid. Direct solution of the equation (7.3.28) in this case yields

$$I(t) = \frac{I(0)}{1 + \kappa t I(0)/n_0} \longrightarrow 0 \quad \text{as } t \to \infty. \tag{7.3.31}$$

This is a *very* slow decay to zero, but nevertheless yields no laser field as $t \to \infty$, as seen in Fig. 7.3.

b) Fokker–Planck Equation: The equivalent Fokker–Planck equation is

$$\frac{\partial P}{\partial t} = \left(\frac{\partial}{\partial \alpha}\left(\mathrm{i}\omega + \frac{\kappa}{2C}\left[1 - C + \frac{|\alpha|^2}{n_0}\right]\right)\alpha + \frac{\partial}{\partial \alpha^*}\left(-\mathrm{i}\omega + \frac{\kappa}{2C}\left[1 - C + \frac{|\alpha|^2}{n_0}\right]\right)\alpha^* + D\frac{\partial^2}{\partial\alpha\partial\alpha^*}\right)P. \tag{7.3.32}$$

As noted in Ex. 5.5, this equation has a stationary solution (which can be verified by inspection) given by

$$\frac{\partial}{\partial \alpha}\alpha P - \frac{\partial}{\partial \alpha^*}\alpha^* P = 0, \tag{7.3.33}$$

$$\frac{\kappa}{C}\left(1 - C + \frac{|\alpha|^2}{n_0}\right)\alpha P = -D\frac{\partial}{\partial \alpha^*}P. \tag{7.3.34}$$

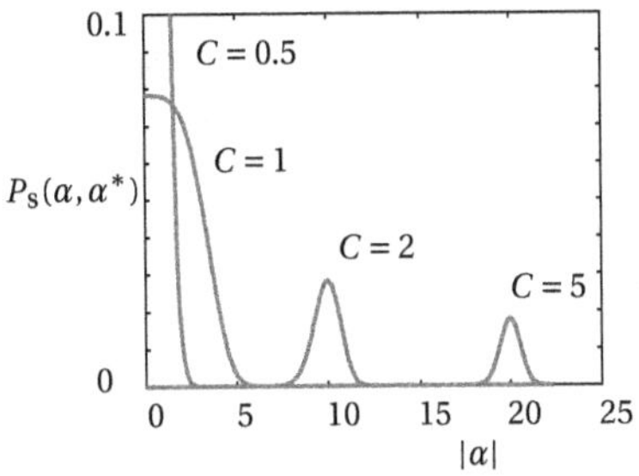

Fig. 7.4. Probability distribution of the intensity $P_s(\alpha,\alpha^*)$, for a range of values of C, as given by (7.3.35). The distribution becomes sharper as the peak number increases.

The solution to these equations is

$$P_s(\alpha,\alpha^*) = \exp\left\{-\frac{\kappa n_0}{DC}\left((1-C)\frac{|\alpha|^2}{n_0} + \frac{|\alpha|^4}{2n_0^2}\right)\right\}. \tag{7.3.35}$$

c) **The Lasing Threshold:** The value $C = 1$ is called the lasing threshold. For $C \le 1$, we have seen that deterministically the intensity $I(t)$ and hence the field $\alpha(t)$ decay to zero as $t \to \infty$. The stochastic stationary solution, as displayed in Fig. 7.4, shows that that the field is not actually zero, but fluctuates about zero.

When $C > 1$, the deterministic non-zero value of the intensity as $t \to \infty$ is mirrored by a stochastic probability distribution peaked close to the deterministic value. Since the stationary probability distribution depends only on $|\alpha|^2$, the *phase* of α is indeterminate.

A more detailed analysis of the behaviour of the solutions of the laser equation will be given in Sect. 17.3.

Exercise 7.5 The van der Pol Oscillator in Phase-Amplitude Form: Use the methods of Sect. 7.3.3 to write the van der Pol oscillator equations in phase-amplitude form like that of (7.3.24, 7.3.25).

8. The Markov Limit

Throughout the field of theoretical physics, we meet systems coupled to each other in which very different time scales occur, and in general this gives rise to interesting phenomena in which the behaviour on a fast time scale gives rise to modifications of the slow time scale behaviour. The kind of formalism required to describe this behaviour is essentially the same for both classical and quantum stochastic systems. The feature in common with all such systems is that, in the limit that one set of variables becomes very much faster than the others, a limiting Markov process arises, in which only a few parameters of the fast part of the system are relevant.

Probably the simplest example of the technique required arises when we consider a system driven by a non-white noise source, and ask how the limiting white noise behaviour is achieved as the time scale of the white noise becomes faster. This is how we will introduce the method, and after that, we will move on to more general systems in which the separation of the fast variable from the slow variables is not immediately obvious.

8.1 The White Noise Limit

As an example, let us consider a simple Langevin equation

$$\dot{u} = -\beta u + \sqrt{f}\xi(t), \tag{8.1.1}$$

but we will not explicitly require that $\xi(t)$ be white noise. Instead, let us first introduce a stationary stochastic variable $\alpha(t)$, which has zero mean, and whose correlation function is

$$\langle \alpha(t)\alpha(t')\rangle = g(|t-t'|), \tag{8.1.2}$$

and is normalized so that

$$\int_{-\infty}^{\infty} g(s)\,ds = 1. \tag{8.1.3}$$

We can now define $\xi(t)$ in terms of $\alpha(t)$ by writing

$$\xi(t) \equiv \gamma\alpha(\gamma^2 t). \tag{8.1.4}$$

The correlation function of $\xi(t)$ is then

$$\langle \xi(t)\xi(t')\rangle = \gamma^2 g(\gamma^2|t-t'|), \tag{8.1.5}$$

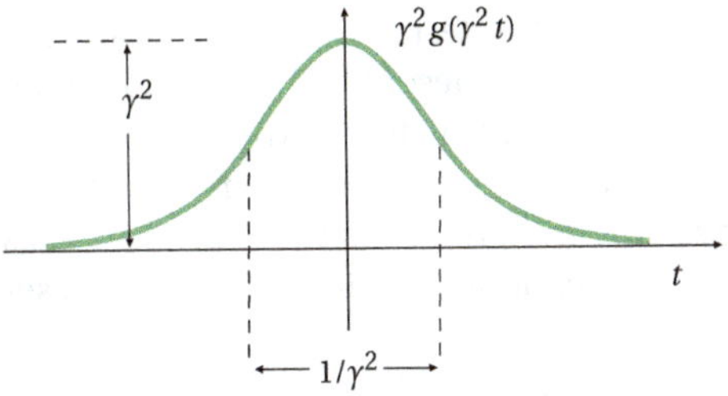

Fig. 8.1. The correlation function illustrated becomes a delta function in the limit $\gamma \to \infty$.

which approaches a delta function as $\gamma \to \infty$—the width of the correlation function on the right-hand side of (8.1.5) is $1/\gamma^2$ and as $\gamma^2 \to \infty$ this approaches zero, while the total area is clearly 1, as can be seen in Fig. 8.1.

8.1.1 Models for Non-White Noise

We will assume that $\alpha(t)$ can be modelled as a Markov process, and will show that in fact the specific kind of Markov process is unimportant as long as the mean is zero, and the correlation function is normalizable as in (8.1.3).

For clarity, however, let us first consider a model for $\alpha(t)$ as an Ornstein–Uhlenbeck process, constructed as the solution of the white noise equation,

$$d\alpha = -\alpha\, dt + dW(t). \tag{8.1.6}$$

We now take the Langevin equation (8.1.1) together with (8.1.6), and write the pair of equivalent coupled stochastic differential equations

$$du = -\beta u\, dt + \gamma\sqrt{f}\,\xi\, dt, \tag{8.1.7}$$

$$d\xi = -\gamma^2 \xi\, dt + \gamma dW(t). \tag{8.1.8}$$

8.1.2 The Projector Formalism

a) Fokker–Planck Equations: From (4.4.4), the pair of equations (8.1.7–8.1.8) is equivalent to the two-variable Fokker–Planck equation

$$\boxed{\frac{\partial P}{\partial t} = (\gamma^2 L_1 + \gamma L_2 + L_3)P,} \tag{8.1.9}$$

in which L_1, L_2, L_3 are the differential operators

$$L_1 = \frac{\partial}{\partial \xi}\xi + \tfrac{1}{2}\frac{\partial^2}{\partial \xi^2}, \tag{8.1.10}$$

$$L_2 = \sqrt{f}\frac{\partial}{\partial u}\xi, \tag{8.1.11}$$

$$L_3 = \frac{\partial}{\partial u}\beta u. \tag{8.1.12}$$

The structure of this evolution equation—with the terms L_1 and L_3 describing evolution in two different spaces, and the term L_2 describing the coupling between the two spaces—is a fundamental structure, which underpins a very wide range of physical systems. Hence the procedure we will develop in this section for finding a description as $\gamma \to \infty$, involving the projectors with the properties (8.1.18a–8.1.18d) has a very wide range of applications, both in this book and elsewhere.

b) The Large γ Limit: The white noise limit is the limit as $\gamma \to \infty$. Heuristically, we can see that the term $\gamma^2 L_1$ is then most important, and this rapidly drives the variable $\xi(t)$ to a stationary process.

c) Laplace Transform: Equations of the general form (8.1.9) will occur very frequently in the study of both quantum and classical systems, and are most easily treated using Laplace transform methods.

The Laplace transform of a function $v(t)$ is

$$\tilde{v}(s) = \int_0^\infty e^{-st} v(t)dt. \tag{8.1.13}$$

The Laplace transform of the derivative, $dv(t)/dt$, is shown by partial integration to be

$$\int_0^\infty e^{-st} v'(t)dt = s\tilde{v}(s) - v(0). \tag{8.1.14}$$

Here $v(0)$ is the initial value of $v(t)$. The Laplace transform of (8.1.9) is

$$s\tilde{P}(s) - \tilde{P}(0) = (\gamma^2 L_1 + \gamma L_2 + L_3)\tilde{P}(s). \tag{8.1.15}$$

d) Projectors: We now introduce a projection operator $\mathcal{P}$, defined in terms of the stationary solution $p_s(\xi)$, which satisfies

$$0 = L_1 p_s(\xi) = \left(\frac{\partial}{\partial \xi}\xi + \frac{1}{2}\frac{\partial^2}{\partial \xi^2}\right) p_s(\xi). \tag{8.1.16}$$

For this particular choice of L_1, the stationary solution is $p_s(\xi) \propto e^{-\xi^2}$.

If $P(\xi, u)$ is any function of ξ and u, then the projector $\mathcal{P}$ is defined by

$$(\mathcal{P}P)(\xi, u) = p_s(\xi)\int d\xi\, P(\xi, u). \tag{8.1.17}$$

The projector $\mathcal{P}$ is a projector into the space of stationary solutions of the equation $\partial f/\partial t = L_1 f$, and has the properties

$$\mathcal{P}^2 = \mathcal{P}, \tag{8.1.18a}$$

$$\mathcal{P}L_1 = L_1\mathcal{P} = 0, \tag{8.1.18b}$$

$$\mathcal{P}L_2\mathcal{P} = 0, \tag{8.1.18c}$$

$$\mathcal{P}L_3 = L_3\mathcal{P}. \tag{8.1.18d}$$

We also define the complementary projector

$$\mathcal{Q} = 1 - \mathcal{P}, \tag{8.1.19}$$

and this satisfies $\mathcal{Q}^2 = \mathcal{Q}$.

Exercise 8.1 Projector Properties: All of (8.1.18a–8.1.18d) are relatively straightforward conditions.

a) Shows that it is a projector.

b) Shows that this projector is into the space of stationary solutions of $\dot{p}(\xi) = L_1 p(\xi)$.

c) Shows that $L_2 p_s(\xi)$ is not a stationary solution, and has no component of stationary solution—it is equivalent to $\langle\xi\rangle_s = 0$.

d) Shows that L_3 is completely in another space from L_1.

Prove all of these properties.

e) Solution in the Large γ Limit Using the Projector: Let us set

$$\tilde{v}(s) = \mathcal{P}\tilde{P}(s), \qquad \tilde{w}(s) = \mathcal{Q}\tilde{P}(s). \tag{8.1.20}$$

The quantity

$$\tilde{v}(s) = p_s(\xi)\int d\xi\,\tilde{P}(\xi, u) \tag{8.1.21}$$

is proportional to the Laplace transform of the probability for u *alone*, thus we want to derive an equation for $\tilde{v}(s)$. We can straightforwardly derive from (8.1.15)

$$s\tilde{v}(s) - v(0) = L_3\tilde{v}(s) + \gamma\mathcal{P}L_2\tilde{w}(s), \tag{8.1.22}$$

$$s\tilde{w}(s) - w(0) = (L_3 + \gamma^2 L_1 + \gamma\mathcal{Q}L_2)\,\tilde{w}(s) + \gamma\mathcal{Q}L_2\,\tilde{v}(s). \tag{8.1.23}$$

We can solve for $\tilde{w}(s)$, using the second equation, and substitute into the first equation, to get

$$\begin{aligned} &s\tilde{v}(s) - \left\{v(0) + \gamma\mathcal{P}L_2[s - L_3 - \gamma^2 L_1 - \gamma\mathcal{Q}L_2]^{-1} w(0)\right\} \\ &= \left\{L_3 + \gamma^2\mathcal{P}L_2[s - L_3 - \gamma^2 L_1 - \gamma\mathcal{Q}L_2]^{-1}\mathcal{Q}L_2\right\}\tilde{v}(s). \end{aligned} \tag{8.1.24}$$

Now consider the limit $\gamma \to \infty$; we can simplify the final result by using

$$\mathcal{Q}L_2\,\tilde{v}(s) = (1 - \mathcal{P})L_2\mathcal{P}\tilde{P}(s) = L_2\mathcal{P}\tilde{P}(s) = L_2\,\tilde{v}(s), \tag{8.1.25}$$

which follows from (8.1.18c). Then, taking the limit $\gamma \to \infty$ we find the rather simple equation

$$s\tilde{v}(s) - v(0) = (L_3 - \mathcal{P}L_2 L_1^{-1} L_2)\,\tilde{v}(s). \tag{8.1.26}$$

This is the fundamental result which can be generalized in many useful ways in a wide variety of contexts.

8.1.3 The Limiting Fokker–Planck Equation

It is not difficult to see that (8.1.26) leads to a Fokker–Planck equation. Explicitly

$$\mathcal{P}L_2L_1^{-1}L_2\tilde{v}(s) = p_s(\xi)\int d\xi'\,\frac{\partial}{\partial u}\left(\sqrt{f}\right)\xi' L_1^{-1}\frac{\partial}{\partial u}\left(\sqrt{f}\right)\xi' p_s(\xi')\int d\xi''\,\tilde{P}(\xi,u,s)$$
$$= p_s(\xi) f\frac{\partial^2}{\partial u^2}\left\{\int d\xi'\,\xi' L_1^{-1}\xi' p_s(\xi')\right\}\int d\xi'\,\tilde{P}(\xi,u,s)$$
$$= fD\frac{\partial^2}{\partial u^2}\tilde{v}(\xi,u,s), \tag{8.1.27}$$

where

$$D \equiv \int d\xi'\,\xi' L_1^{-1}\xi' p_s(\xi'). \tag{8.1.28}$$

Leaving aside the evaluation of D; the limiting equation is

$$s\tilde{v}(s) - v(0) = \left(L_3 + Df\frac{\partial^2}{\partial u^2}\right)\tilde{v}(\xi,u,s). \tag{8.1.29}$$

We can invert the Laplace transform, and cancel a common factor of $p_s(\xi)$, to get

$$\frac{\partial p}{\partial t} = \left\{\frac{\partial}{\partial u}\beta u + Df\frac{\partial^2}{\partial u^2}\right\}p, \tag{8.1.30}$$

where $$p(u,t) = \int d\xi\,P(\xi,u,t), \tag{8.1.31}$$

is the probability density function for u.

8.1.4 Evaluation of the Diffusion Coefficient in Terms of Correlation Functions

The details of the dynamics of the noise have now been reduced to the evaluation of the single number D, and in this section we will show that D is a measure of the strength of the fluctuations in $\alpha(t)$, and indeed can be expressed as the time integral of the correlation function as $D = \int_0^\infty dt\,\langle\xi(t)\xi(0)\rangle_s$.

a) Preliminary Results: As defined in (8.1.27), D represents the size of the diffusion term in the Fokker–Planck equation; it is defined as

$$D = \int d\xi'\,\xi' L_1^{-1}\xi' p_s(\xi'). \tag{8.1.32}$$

i) Let us note that we can write the solution of the equation

$$\frac{\partial p}{\partial t} = L_1 p, \tag{8.1.33}$$

as

$$p(\xi,t) = \exp(L_1 t)p(\xi,0), \tag{8.1.34}$$

and the fact that, as $t \to \infty$

$$p(\xi,t) \to p_s(\xi) \equiv \mathcal{P}p(\xi,0), \tag{8.1.35}$$

means that

$$\lim_{t\to\infty} \exp(L_1 t) = \mathcal{P}. \tag{8.1.36}$$

ii) We can then also deduce that

$$\int_0^\infty dt \exp(L_1 t) = \lim_{t\to\infty} \{L_1^{-1}\{\exp(L_1 t) - 1\}\} = L_1^{-1}\{\mathcal{P} - 1\}. \tag{8.1.37}$$

iii) Notice also that requirement (8.1.18c) means that $\langle\xi\rangle_s = 0$, so that

$$\mathcal{P}\left(\xi' p_s(\xi')\right) = p_s(\xi') \int d\xi''\,\xi'' p_s(\xi'') = p_s(\xi')\langle\xi\rangle_s = 0, \tag{8.1.38}$$

so that we can use (8.1.37) to write

$$D = \int d\xi\,\xi L^{-1}\xi\, p_s(\xi), \tag{8.1.39}$$

$$= \int d\xi\,\xi L^{-1}\{1 - \mathcal{P}\}\xi\, p_s(\xi), \qquad \text{using (8.1.38)} \tag{8.1.40}$$

$$= -\int_0^\infty dt \int d\xi\,\xi \exp(L_1 t)\xi\, p_s(\xi). \tag{8.1.41}$$

iv) Now the solution of the Fokker–Planck equation

$$\frac{\partial p}{\partial t} = L_1 P, \tag{8.1.42}$$

with the initial condition

$$p(\xi, t_0) = \delta(\xi - \xi_0), \tag{8.1.43}$$

is the conditional probability $p(\xi, t|\xi_0, t_0)$, so that using (8.1.34)

$$p(\xi, t|\xi_0, t_0) = \exp(L_1(t - t_0))\delta(\xi - \xi_0), \tag{8.1.44}$$

and thus, in (8.1.41)

$$\exp(L_1 t)\xi p_s(\xi) = \exp(L_1 t) \int d\xi_0\,\xi_0 p_s(\xi_0)\delta(\xi - \xi_0)$$
$$= \int d\xi_0\, p(\xi, t|\xi_0, 0)\xi_0 p_s(\xi_0). \tag{8.1.45}$$

v) Now note that the joint probability density in the stationary state is

$$p_s(\xi, t; \xi_0, t_0) = p(\xi, t|\xi_0, t_0)\, p_s(\xi_0), \tag{8.1.46}$$

and this can only depend on time differences if the situation is stationary so that

$$D = \int_0^\infty dt \int d\xi \int d\xi_0 \, \xi \, \xi_0 \, p_s(\xi, t; \xi_0, 0). \tag{8.1.47}$$

b) Final Evaluation of *D*: This final result is most elegantly written in the form

$$D = \int_0^\infty dt \, \langle \xi(t)\xi(0)\rangle_s \tag{8.1.48}$$

and, in this case from (8.1.2–8.1.5) we can deduce that $D = \frac{1}{2}$.

8.1.5 Summary

We have shown that if u obeys a differential equation

$$\dot{u} = -\beta u + \sqrt{f}\xi(t), \tag{8.1.49}$$

and

$$\xi(t) = \gamma\alpha(\gamma^2 t), \tag{8.1.50}$$

with $\alpha(t)$ obeying the stochastic differential equation

$$d\alpha = -\alpha \, dt + dW(t), \tag{8.1.51}$$

then the limit $\gamma \to \infty$, in which

$$\langle \xi(t)\xi(t')\rangle = \gamma^2 g(\gamma^2 |t - t'|) \to \delta(t - t'), \tag{8.1.52}$$

u has a probability distribution $p(u,t)$ which obeys the Fokker–Planck equation

$$\frac{\partial P}{\partial t} = \frac{\partial}{\partial u}\beta u P(u) + \tfrac{1}{2} f \frac{\partial^2 P}{\partial u^2}, \tag{8.1.53}$$

so that there is a stochastic differential equation for u in the form

$$du = -\beta u \, dt + \sqrt{f} dW(t). \tag{8.1.54}$$

8.2 Interpretation and Generalizations of the White Noise Limit

The white noise result is in fact valid under much wider conditions than used above. The coefficients of the noise can depend on the variable of interest, u, they can depend on time, and even the evolution operator L_1 of the noise can depend *slowly* on time—in all cases an appropriate generalization can be proved.

8.2.1 Result for a General Stochastic Differential Equation

Suppose instead of (8.1.1), that u obeys the equation

$$\dot{u} = a(u) + b(u)\xi(t), \tag{8.2.1}$$

and $\xi(t)$ is defined as in (8.1.2–8.1.5); i.e., $\xi(t) = \gamma\alpha(\gamma^2 t)$. All of the reasoning comes through, but in this case

$$\begin{aligned} L_2 &= \frac{\partial}{\partial u} b(u)\xi \\ L_3 &= -\frac{\partial}{\partial u} a(u). \end{aligned} \tag{8.2.2}$$

Thus, instead of (8.1.27), we need to evaluate

$$\mathcal{P} L_2 L_1^{-1} L_2 \tilde{v}(s) = -\frac{\partial}{\partial u} b(u) \frac{\partial}{\partial u} b(u) D, \tag{8.2.3}$$

in which D, and therefore the ξ-dependent part, is exactly the same as before. Thus the resulting Fokker–Planck equation is

$$\boxed{\frac{\partial p}{\partial t} = -\frac{\partial}{\partial u} a(u) P + \tfrac{1}{2} \frac{\partial}{\partial u} b(u) \frac{\partial}{\partial u} b(u) P,} \tag{8.2.4}$$

which is the Fokker–Planck equation for the *Stratonovich* stochastic differential equation

$$(\mathbf{S})\, du = a(u)\, dt + b(u)\, dW(t). \tag{8.2.5}$$

8.2.2 Result for General Non-Gaussian Noises

Notice that L_1 is quite arbitrary—it is only necessary to have $\langle \xi \rangle_s = 0$, but otherwise the exact nature of L_1 is quite unimportant, and indeed has not been used in the derivation. The end result is that if u obeys the equation (8.2.1), and $\xi(t) = \gamma\alpha(\gamma^2 t)$ has zero stationary mean, but is an otherwise arbitrary Markov process, then in the $\gamma \to \infty$ limit, in which $\xi(t)$ becomes $\delta(t - t')$ correlated, u obeys the Stratonovich stochastic differential equation (8.2.5).

This means that the Gaussian nature of $dW(t)$ is the natural consequence of a speeding up of the time scale, since the noise described by L_1 need not itself be Gaussian. It is essentially a consequence of the central limit theorem, since the driving term $\xi(t)$ is added on very many times in any time interval.

8.2.3 Time Dependence of Coefficients

Consider the more general form

$$\dot{u} = a(u, t) + b(u, t)\xi(t), \tag{8.2.6}$$

with $\xi(t) = \gamma\alpha(\gamma^2 t)$ as before. Because of the explicit time dependence of the coefficients, the Laplace transform method now becomes invalid, and must be modified. The modification is not difficult, and it is shown in *Stochastic Methods* that the limiting Fokker–Planck equation is

$$\frac{\partial p}{\partial t} = -\frac{\partial}{\partial u} a(u,t)P + \tfrac{1}{2}\frac{\partial}{\partial u} b(u,t)\frac{\partial}{\partial u} b(u,t)P, \tag{8.2.7}$$

which is again the Fokker–Planck equation for the *Stratonovich* stochastic differential equation

$$(\mathbf{S})\; du = a(u,t)\,dt + b(u,t)\,dW(t). \tag{8.2.8}$$

We can interpret this result to mean that the time dependence of the coefficients is infinitely slower than that of the noise. Put more precisely, this means that for large but not infinite γ, the coefficients $a(u,t)$, $b(u,t)$ should not change significantly in a time of order of magnitude $1/\gamma^2$.

The proof of the result is made by introducing a new variable τ which obeys the equation of motion $d\tau/dt = 1$, so that the system of equations becomes

$$\dot{u} = a(u,\tau) + b(u,\tau)\,\xi(t), \tag{8.2.9}$$

$$\dot{\tau} = 1, \tag{8.2.10}$$

and is now a time homogeneous system in two variables, which permits the use of the Laplace transform.

8.2.4 Time Dependence of the Noise Equation

Using similar methods it is even possible to deal with a situation in which the noise evolution operator is $L_1(t)$, and this is also treated in *Stochastic Methods.*

Exercise 8.2 Random Telegraph Noise: Instead of the Ornstein–Uhlenbeck noise (8.1.51), use a noise based on the *random telegraph process* $\beta(t)$, as defined in Sect. 6.1.3. Formulate the equations equivalent to (8.1.9–8.1.12), and then satisfy yourself that the remaining reasoning is correct. In particular, check that the relationship to the correlation function, as given in (8.1.48) is still valid.

8.3 Linear Non-Markovian Stochastic Differential Equations

There is a quite common form of stochastic differential equation, which can be written in the form

$$\frac{du(t)}{dt} = \big(A_0 + A_1(t)\big)\,u(t). \tag{8.3.1}$$

In this equation

i) The variable of interest is $u(t)$, a column vector.

ii) A_0 is a constant matrix.

iii) $A_1(t)$ is a stochastic matrix, which need not commute with A_0, and for which $[A_1(t), A_1(t')] \neq 0$ when $t \neq t'$.

iv) It is assumed that $\langle A_1(t)\rangle = 0$.

v) In general, the object of most interest is the average $\langle u(t)\rangle$.

The lack of commutativity of $A_1(t)$ with itself at different times is the main source of the complexity of this kind of problem.

8.3.1 The Complex Oscillator with a Noisy Frequency

However, the one-variable problem, which is not affected by the issue of non-commutativity, is worthy of study in order to illustrate the nature of the approximations one makes in this kind of problem. The archetypal model is the complex harmonic oscillator with a random frequency

$$\dot{x} = \mathrm{i}\Omega\alpha(t)x, \tag{8.3.2}$$

where $\alpha(t)$ is a real random function such that $\langle \alpha(t)^2\rangle = 1$ and $\langle \alpha(t)\rangle = 0$. The explicit solution is

$$x(t) = \mathrm{e}^{\mathrm{i}\Omega\int_0^t dt'\,\alpha(t')}x_0. \tag{8.3.3}$$

This solution exists for any kind of random function $\alpha(t)$. However, it is not really a solution to the problem unless we can calculate the mean values of interest.

a) Gaussian Noise: We suppose that $\alpha(t)$ is Gaussian, then we can use the result of Ex. 7.3 (which shows how to evaluate $\langle \exp(x)\rangle$ when x is Gaussian) to say that the autocorrelation function can be written

$$\langle x(t_1)x^*(t_0)\rangle = |x_0|^2\left\langle \exp\left(\mathrm{i}\Omega\int_{t_0}^{t_1} dt'\,\alpha(t')\right)\right\rangle, \tag{8.3.4}$$

$$= |x_0|^2\exp\left(-\tfrac{1}{2}\Omega^2\int_{t_0}^{t_1}\int_{t_0}^{t_1}\langle \alpha(t')\alpha(t'')\rangle dt'\,dt''\right). \tag{8.3.5}$$

We then get a formula which can be evaluated using only the autocorrelation function of $\alpha(t)$. Supposing further that the process $\alpha(t)$ is a stationary Ornstein–Uhlenbeck process as in Sect. 7.1.1, the autocorrelation function can be written as

$$\langle \alpha(t')\alpha(t'')\rangle = \mathrm{e}^{-\gamma|t'-t''|}, \tag{8.3.6}$$

for some value of γ. The autocorrelation function of $x(t)$ is then able to be written in terms of $\tau \equiv t_1 - t_0$ as

$$\langle x(\tau)x^*(0)\rangle = |x_0|^2\exp\left[-\frac{\Omega^2}{\gamma}\left(|\tau| - \frac{1-\mathrm{e}^{-\gamma|\tau|}}{\gamma}\right)\right]. \tag{8.3.7}$$

b) Random Telegraph Noise: In this case, the noise is not Gaussian, but the autocorrelation function can still be calculated exactly. Let us suppose that $\alpha(t)$ can be described as in Sect. 6.1.3, that it takes on the values ± 1, and that the upward and downward rate constants have the same value $\frac{1}{2}\gamma$. Hence we find, as for the Gaussian case above, $\langle \alpha(t)^2 \rangle = 1$ and $\langle \alpha(t) \rangle = 0$, and the stationary autocorrelation function of $\alpha(t)$ is given by (6.1.47) as

$$\langle \alpha(t')\alpha(t'') \rangle = \mathrm{e}^{-\gamma |t'-t''|}, \tag{8.3.8}$$

exactly the same as for the Gaussian process.

As noted in Sect. 8.2.2, we can describe the system governed by the equation of motion (8.3.2) in terms of an evolution operator constructed like (8.1.9), and the operator L_1 can be any Markov process. Thus, in this case L_1 will be the evolution operator for the random telegraph process.

Let us also write $x = x_0 \mathrm{e}^{\mathrm{i}\phi}$, and describe the system in terms of ϕ, which automatically enforces the requirement that $|x(t)|$ be constant; then the process will be described by the conditional probability $P(\phi, \alpha | \phi_0, \alpha_0)$ with the evolution equation

$$\begin{aligned} \partial_t P(\phi, 1, t|\phi_0, \alpha_0, t_0) &= -\Omega \partial_\phi P(\phi, 1, t|\phi_0, \alpha_0, t_0) \\ &\quad -\tfrac{1}{2}\gamma \big(P(\phi, 1, t|\phi_0, \alpha_0, t_0) - P(\phi, -1, t|\phi_0, \alpha_0, t_0) \big), \end{aligned} \tag{8.3.9}$$

$$\begin{aligned} \partial_t P(\phi, -1, t|\phi_0, \alpha_0, t_0) &= \Omega \partial_\phi P(\phi, -1, t|\phi_0, \alpha_0, t_0) \\ &\quad +\tfrac{1}{2}\gamma \big(P(\phi, 1, t|\phi_0, \alpha_0, t_0) - P(\phi, -1, t|\phi_0, \alpha_0, t_0) \big). \end{aligned} \tag{8.3.10}$$

This combines deterministic motion of ϕ (as described in Sect. 5.1.2) with the random telegraph motion of $\alpha(t)$. The stationary autocorrelation function (8.3.4) is given by

$$\langle x(t_1) x^*(t_0) \rangle = \sum_{\alpha,\alpha_0} \int d\phi_0 \, \langle x(\alpha, t)|x_0, \alpha_0, t_0 \rangle \, x_0^* \, P_s(\phi_0, \alpha_0), \tag{8.3.11}$$

where

$$\langle x(\alpha, t)|x_0, \alpha_0, t_0 \rangle = \int d\phi \, \mathrm{e}^{\mathrm{i}\phi} P(\phi, \alpha, t|\phi_0, \alpha_0, t_0). \tag{8.3.12}$$

The stationary distribution is given by

$$P_s(\phi, \alpha) = \frac{1}{2\pi}. \tag{8.3.13}$$

The equations of motion for $\bar{x}(\alpha, t) \equiv \langle x(\alpha, t)|x_0, \alpha_0, t_0 \rangle$ follow from (8.3.7, 8.3.10)

$$\frac{d}{dt}\begin{pmatrix} \bar{x}(1,t) \\ \bar{x}(-1,t) \end{pmatrix} = \begin{pmatrix} \mathrm{i}\Omega - \frac{1}{2}\gamma & \frac{1}{2}\gamma \\ \frac{1}{2}\gamma & -\mathrm{i}\Omega - \frac{1}{2}\gamma \end{pmatrix} \begin{pmatrix} \bar{x}(1,t) \\ \bar{x}(-1,t) \end{pmatrix}, \tag{8.3.14}$$

with the initial conditions

$$\bar{x}(\alpha, t_0) = x_0 \, \delta_{\alpha,\alpha_0}. \tag{8.3.15}$$

We denote the eigenvalues of the matrix in (8.3.14) by

$$\mu_\pm \equiv -\tfrac{1}{2}\gamma \pm \sqrt{\tfrac{1}{4}\gamma^2 - \Omega^2}. \tag{8.3.16}$$

The average over the stationary probability of $\langle x(\alpha,t)|x_0,\alpha_0,t_0\rangle$ requires firstly the average over the two values of α_0, and this obeys the same equations (8.3.14), but with the initial condition

$$\sum_{\alpha_0}\langle x(\alpha,t_0)|x_0,\alpha_0,t_0\rangle = x_0. \tag{8.3.17}$$

We then sum over the two values of α to find that the autocorrelation function of $x(t)$ is able to be written in terms of $\tau \equiv t_1 - t_0$ as

$$\langle x(\tau)x^*(0)\rangle = |x_0|^2\frac{\mu_- e^{\mu_+\tau} - \mu_+ e^{\mu_-\tau}}{\mu_- - \mu_+}, \tag{8.3.18}$$

$$= |x_0|^2 e^{-\frac{1}{2}\gamma\tau}\left(\cosh\left(\sqrt{\tfrac{1}{4}\gamma^2-\Omega^2}\,\tau\right)+\frac{\frac{1}{2}\gamma}{\sqrt{\frac{1}{4}\gamma^2-\Omega^2}}\sinh\left(\sqrt{\tfrac{1}{4}\gamma^2-\Omega^2}\,\tau\right)\right), \tag{8.3.19}$$

$$= |x_0|^2 e^{-\frac{1}{2}\gamma\tau}\left(\cos\left(\sqrt{\Omega^2-\tfrac{1}{4}\gamma^2}\,\tau\right)+\frac{\frac{1}{2}\gamma}{\sqrt{\Omega^2-\frac{1}{4}\gamma^2}}\sin\left(\sqrt{\Omega^2-\tfrac{1}{4}\gamma^2}\,\tau\right)\right). \tag{8.3.20}$$

c) Comparison of the Results: The two results (8.3.7) and (8.3.18–8.3.20) have many similarities, as follows:

i) As $\tau \to 0$, in both cases $\langle x(\tau)x^*(0)\rangle \to |x_0|^2(1-\frac{1}{2}\Omega^2\tau^2)$, which can be seen to be a general result for *any* $\alpha(t)$ by taking $|t_1 - t_0|$ to be very small, and setting $\langle\alpha(t')\alpha(t'')\rangle \approx \langle\alpha^2\rangle = 1$ in (8.3.5).

ii) The limit $\gamma \to \infty$ with $\Omega^2/\gamma = \kappa$, a constant corresponds to the white noise limit, and in both cases here the autocorrelation function has the limit

$$\langle x(\tau)x(0)\rangle \to |x_0|^2 e^{-\kappa\tau}. \tag{8.3.21}$$

iii) The limit $\gamma \to \infty$ with Ω fixed corresponds to the *motional narrowing* limit, as discussed in Sect. 9.2.2, in which the autocorrelation function develops on a progressively *slower* time scale as the fluctuations become faster, and therefore it becomes more difficult for the system to follow:

$$\langle x(\tau)x(0)\rangle \sim |x_0|^2 e^{-\frac{\Omega^2\tau}{\gamma}}. \tag{8.3.22}$$

8.3.2 Approximation Methods for Multivariable Systems

In the case of multivariable systems described by an equation of the form (as discussed above)

$$\frac{du(t)}{dt} = \big(A_0 + A_1(t)\big)\,u(t), \tag{8.3.23}$$

the issue of the non-commutativity of the operators A_0 and A_1 becomes important. The available approximation methods all bear a strong resemblance to the theory presented earlier in this chapter on the white noise limit.

One way of proceeding uses the following result:

i) Integrate the equation (8.3.23) to get

$$u(t) = \mathrm{e}^{A_0 t}\left\{u(0) + \int_0^t dt'\,\mathrm{e}^{-A_0 t'} A_1(t')u(t')\right\}. \tag{8.3.24}$$

ii) Substitute this result into the last term, then take averages, getting

$$\begin{aligned}\langle u(t)\rangle = \mathrm{e}^{A_0 t}\Big\{u(0) &+ \int_0^t dt'\,\mathrm{e}^{-A_0 t'}\langle A_1(t')\rangle \exp(A_0 t')u(0)\\ &+ \int_0^t dt'\,\mathrm{e}^{-A_0 t'}\Big\langle A_1(t')\mathrm{e}^{A_0 t'}\int_0^{t'} dt''\,\mathrm{e}^{-A_0 t''}A_1(t'')u(t'')\Big\rangle\Big\}.\end{aligned} \tag{8.3.25}$$

iii) Use the fact that $\langle A_1(t)\rangle = 0$, and differentiate, finally getting an equation that does not involve the initial condition

$$\frac{d\langle u(t)\rangle}{dt} = A_0\langle u(t)\rangle + \mathrm{e}^{-A_0 t}\Big\langle A_1(t)\mathrm{e}^{A_0 t}\int_0^t dt'\,\mathrm{e}^{-A_0 t'}A_1(t')u(t')\Big\rangle. \tag{8.3.26}$$

This equation is an exact result, but the average in the second term cannot be evaluated, so it does not represent an equation which can be solved.

a) Decorrelation: The value of (8.3.26) is as a starting point for a *decorrelation approximation*. We would like to break up the average in the last term in (8.3.26) to get something like

$$\begin{aligned}&\Big\langle A_1(t)\mathrm{e}^{A_0 t}\int_0^t dt'\,\mathrm{e}^{-A_0 t'}A_1(t')u(t')\Big\rangle\\ &\qquad\longrightarrow \int_0^t dt'\Big\langle A_1(t)\mathrm{e}^{A_0(t-t')}A_1(t')\Big\rangle\langle u(t')\rangle.\end{aligned} \tag{8.3.27}$$

How to justify this procedure is the first issue; the second is to decide on the best way to implement such a procedure. It does, however, definitely yield an equation of motion for $\langle u(t)\rangle$, in the form of an integro-differential equation.

b) Parameters: The relevant parameters are the following:

i) The typical size of A_0, which we will call Ω, and which determines the time scale of the deterministic part of the time evolution.

ii) The typical size of $A_1(t)$, which we will call α, and which determines the time scale of the stochastic part of the time evolution.

iii) The correlation time τ_c associated with the decay of correlation function of the noise; that is, $\langle A_1(t)A_1(t')\rangle \to 0$ when $|t-t'| \gg \tau_c$.

c) Van Kampen's Procedure: This procedure is simple to describe, and is justified in detail by *van Kampen* [46] in his book on stochastic processes, where it is presented as a systematic expansion in $\alpha\tau_c$. It is based on the condition that, during the time interval $(t-\tau_c, t)$, the value of $u(t')$ changes mainly because of the term A_0, and that because it changes so slowly, it behaves independently of the noise in this time interval; thus the decorrelation assumption sets

$$u(t') \to \mathrm{e}^{-A_0(t-t')} u(t), \tag{8.3.28}$$

and then argues that a decorrelation should be valid. The argument is not straightforward, and we refer the reader to Van Kampen's book [46] for the details, and for the full expansion in powers of $\alpha\tau_c$. The resulting approximation is

$$\begin{aligned}&\left\langle A_1(t)\mathrm{e}^{A_0 t}\int_0^t dt'\,\mathrm{e}^{-A_0 t'} A_1(t')u(t')\right\rangle \\ &\qquad\longrightarrow \int_0^t dt' \left\langle A_1(t)\mathrm{e}^{A_0(t-t')} A_1(t')\mathrm{e}^{-A_0(t-t')}\right\rangle \langle u(t)\rangle. \end{aligned}\tag{8.3.29}$$

Finally, since we want an approximation valid on a time scale much longer than τ_c, we assume that the correlation function on the right-hand side of this equation is negligible for $t-t' \gg \tau_c$, and therefore make no significant error by setting the lower limit of the integral to $-\infty$.

The resulting approximate equation of motion is

$$\frac{d\langle u(t)\rangle}{dt} = A_0\langle u(t)\rangle + \int_{-\infty}^t dt' \left\langle A_1(t)\mathrm{e}^{A_0(t-t')} A_1(t')\mathrm{e}^{-A_0(t-t')}\right\rangle \langle u(t)\rangle. \tag{8.3.30}$$

8.3.3 Example—The Two-Dimensional Oscillator Driven by Random Telegraph Noise

In order to assess the method of van Kampen, let us consider a simple example for which an exact solution is possible using the method we used for random telegraph noise in Sect. 8.3.1b. Here we consider the two-variable complex oscillator

$$\dot{\boldsymbol{x}} = \left(\mathrm{i}\boldsymbol{\Omega} + \mathrm{i}\alpha(t)\,\boldsymbol{\Delta}\right)\boldsymbol{x}, \tag{8.3.31}$$

$$\boldsymbol{x} = \begin{pmatrix} x \\ y \end{pmatrix}, \qquad \boldsymbol{\Omega} = \begin{pmatrix} \Omega & 0 \\ 0 & -\Omega \end{pmatrix}, \qquad \boldsymbol{\Delta} = \begin{pmatrix} 0 & \delta \\ \delta & 0 \end{pmatrix}. \tag{8.3.32}$$

In the case that $\alpha(t)$ is a random telegraph process as above in Sect. 8.3.1b, and as in that example, we can derive equations of motion for

$$\langle \boldsymbol{x}(\alpha, t)|\boldsymbol{x}_0, \alpha_0, t_0\rangle \equiv \left(\bar{x}_1(t), \bar{y}_1(t), \bar{x}_2(t), \bar{y}_2(t)\right), \tag{8.3.33}$$

in the form

$$\frac{d}{dt}\begin{pmatrix} \bar{x}_1(t) \\ \bar{y}_1(t) \\ \bar{x}_2(t) \\ \bar{y}_2(t) \end{pmatrix} = \begin{pmatrix} -\frac{1}{2}\gamma + \mathrm{i}\Omega & \mathrm{i}\delta & \frac{1}{2}\gamma & 0 \\ \mathrm{i}\delta & -\frac{1}{2}\gamma - \mathrm{i}\Omega & 0 & \frac{1}{2}\gamma \\ \frac{1}{2}\gamma & 0 & -\frac{1}{2}\gamma + \mathrm{i}\Omega & -\mathrm{i}\delta \\ 0 & \frac{1}{2}\gamma & -\mathrm{i}\delta & -\frac{1}{2}\gamma - \mathrm{i}\Omega \end{pmatrix} \begin{pmatrix} \bar{x}_1(t) \\ \bar{y}_1(t) \\ \bar{x}_2(t) \\ \bar{y}_2(t) \end{pmatrix}. \tag{8.3.34}$$

The equations are much simpler in terms of variables

$$x_{\pm} \equiv \tfrac{1}{2}(x_1 \pm x_2), \qquad y_{\pm} \equiv \tfrac{1}{2}(y_1 \pm y_2), \tag{8.3.35}$$

which give

$$\frac{d}{dt}\begin{pmatrix} x_+ \\ y_- \\ y_+ \\ x_- \end{pmatrix} = \begin{pmatrix} \mathrm{i}\Omega & \mathrm{i}\delta & 0 & 0 \\ \mathrm{i}\delta & -\mathrm{i}\Omega-\gamma & 0 & 0 \\ 0 & 0 & -\mathrm{i}\Omega & \mathrm{i}\delta \\ 0 & 0 & \mathrm{i}\delta & \mathrm{i}\Omega-\gamma \end{pmatrix}\begin{pmatrix} x_+ \\ y_- \\ y_+ \\ x_- \end{pmatrix}, \tag{8.3.36}$$

that is, two independent sets of of coupled equations. The eigenvalues are

$$\lambda_{u,\pm} = -\tfrac{1}{2}\gamma \pm \sqrt{\left(\tfrac{1}{2}\gamma - \mathrm{i}\Omega\right)^2 - \delta^2}\,, \tag{8.3.37}$$

$$\lambda_{l,\pm} = -\tfrac{1}{2}\gamma \pm \sqrt{\left(\tfrac{1}{2}\gamma + \mathrm{i}\Omega\right)^2 - \delta^2}\,, \tag{8.3.38}$$

corresponding respectively to the upper and lower pair of equations.

a) The Limiting Behaviour: The limit envisaged in van Kampen's method requires both γ and Ω to become large while δ remains finite. This can be seen by looking at (8.3.30); the time scale over which the autocorrelation function of $\alpha(t)$ decays to zero is $1/\gamma$, and the time scale of $\exp(A_0 t)$ is $1/\Omega$. These must be of a similar order of magnitude for both effects to be important.

Taking a limit in which both γ and Ω become large and of the same order of magnitude

$$\lambda_{u,-} \to -\gamma - \mathrm{i}\Omega + \frac{\delta^2}{\gamma + 2\mathrm{i}\Omega}, \tag{8.3.39}$$

$$\lambda_{l,-} \to -\gamma + \mathrm{i}\Omega + \frac{\delta^2}{\gamma - 2\mathrm{i}\Omega}, \tag{8.3.40}$$

$$\lambda_{u,+} \to \mathrm{i}\Omega - \frac{\delta^2}{\gamma + 2\mathrm{i}\Omega}, \tag{8.3.41}$$

$$\lambda_{l,+} \to -\mathrm{i}\Omega - \frac{\delta^2}{\gamma - 2\mathrm{i}\Omega}. \tag{8.3.42}$$

b) Using van Kampen's Method: To use this method, according to (8.3.30) we need to evaluate

$$\int_{-\infty}^{t} dt' \left\langle \alpha(t)\boldsymbol{\Delta}\mathrm{e}^{\mathrm{i}\boldsymbol{\Omega}(t-t')}\alpha(t')\boldsymbol{\Delta}\mathrm{e}^{-\mathrm{i}\boldsymbol{\Omega}(t-t')}\right\rangle,$$

$$= \int_{-\infty}^{t} dt' \langle \alpha(t)\alpha(t')\rangle \begin{pmatrix} -\delta^2 \mathrm{e}^{-2\mathrm{i}\Omega(t-t')} & 0 \\ 0 & -\delta^2 \mathrm{e}^{2\mathrm{i}\Omega(t-t')} \end{pmatrix}, \tag{8.3.43}$$

$$= \int_{-\infty}^{t} dt' \,\mathrm{e}^{-\gamma|t-t'|} \begin{pmatrix} -\delta^2 \mathrm{e}^{-2\mathrm{i}\Omega(t-t')} & 0 \\ 0 & -\delta^2 \mathrm{e}^{2\mathrm{i}\Omega(t-t')} \end{pmatrix}, \tag{8.3.44}$$

$$= \begin{pmatrix} -\frac{\delta^2}{\gamma+2\mathrm{i}\Omega} & 0 \\ 0 & -\frac{\delta^2}{\gamma-2\mathrm{i}\Omega} \end{pmatrix}. \tag{8.3.45}$$

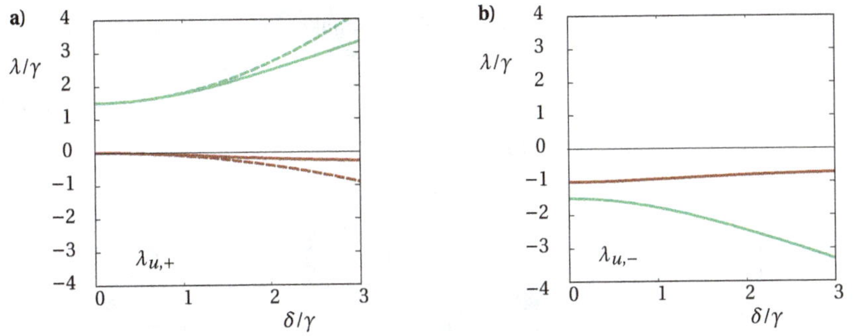

Fig. 8.2. Comparison of eigenvalues as calculated exactly (solid lines) and using the method of van Kampen (dashed lines), for $\Omega/\gamma = 1.5$ and a range of values of δ/γ. **a)** The eigenvalue $\lambda_{u,+}$, with the real part in brown and the imaginary part in green; **b)** The eigenvalue $\lambda_{u,-}$ corresponds in the limit of small δ to the process of equilibration of $\alpha(t)$, and is not given by van Kampen's procedure.

This means that the equation of motion becomes

$$\frac{d\langle x\rangle}{dt} = \left(\mathrm{i}\Omega - \frac{\delta^2}{\gamma+2\mathrm{i}\Omega}\right)\langle x\rangle, \tag{8.3.46}$$

$$\frac{d\langle y\rangle}{dt} = \left(-\mathrm{i}\Omega - \frac{\delta^2}{\gamma-2\mathrm{i}\Omega}\right)\langle y\rangle. \tag{8.3.47}$$

These two equations correspond to the behaviour given by the limits of eigenvalues in (8.3.41, 8.3.40).

c) Comparison of Exact and Approximate Results: In Fig. 8.2 we compare the exact results with those given by van Kampen's procedure. We note that:

i) The accuracy of these approximations and of those given by van Kampen's procedure is determined by the same ratio, $|\delta/(\gamma+2\mathrm{i}\Omega)|$.

ii) The exact results give two eigenvalues which do not appear in van Kampen's procedure. In the limit of small $|\delta/(\gamma+2\mathrm{i}\Omega)|$, these correspond to the approach of $\alpha(t)$ to a stationary process, in which the values $\alpha = \pm 1$ occur with equal probability.

iii) With increasing δ the difference between the two behaviours becomes less distinct.

Exercise 8.3 Driving by Asymmetric Random Telegraph Noise: Show that a random telegraph process with $\langle\alpha(t)\rangle = 0$ and $\langle\alpha(t)\alpha(t')\rangle = \mathrm{e}^{-\gamma|t-t'|}$ arises from the model of Sect. 6.1.3 for any value of a, provided the other parameters are determined by the conditions

$$ab = -1, \qquad \lambda = \frac{a\gamma}{a-b}, \qquad \mu = -\frac{b\gamma}{a-b}. \tag{8.3.48}$$

We can use this kind of noise in the two-dimensional oscillator. Show that if we define x_+ and y_+ as in (8.3.35) and the other variables by

$$x_- = \tfrac{1}{2}(ax_1 + bx_2), \qquad y_- = \tfrac{1}{2}(ay_1 + by_2), \tag{8.3.49}$$

then the equations of motion become

$$\frac{d}{dt}\begin{pmatrix} x_+ \\ y_- \\ y_+ \\ x_- \end{pmatrix} = \begin{pmatrix} \mathrm{i}\Omega & \mathrm{i}\delta & 0 & 0 \\ \mathrm{i}\delta & -\mathrm{i}\Omega-\gamma & 0 & -\mathrm{i}(a+b)\delta \\ 0 & 0 & -\mathrm{i}\Omega & \mathrm{i}\delta \\ 0 & -\mathrm{i}(a+b)\delta & \mathrm{i}\delta & \mathrm{i}\Omega-\gamma \end{pmatrix}\begin{pmatrix} x_+ \\ y_- \\ y_+ \\ x_- \end{pmatrix}. \tag{8.3.50}$$

Compute the eigenvalues of this matrix, and compare them with those given by van Kampen's method. Show that van Kampen's method is valid for small enough δ, but that for larger δ the eigenvalues depend quite strongly on $a+b$. This difference cannot arise using van Kampen's method, which depends only on the integral of the correlation function, and this is the same for any value of $a+b$ chosen to satisfy (8.3.48).

8.3.4 Driving with Other Kinds of Noise

The exercise above shows that it can quite often be necessary to go beyond the decorrelation approximation. The method used for describing the models above is generalizable whenever the noise is describable in terms of a Markov process. The random telegraph process is the simplest example of this method, where the state space is finite dimensional, and the equations of motion can be reduced to a finite (and possibly quite small) set of linear coupled equations.

In the case that the noise takes on a continuous range of values, such as in an Ornstein–Uhlenbeck process, it is usually possible to make an expansion in eigenfunctions of the evolution operator, such as are described in Sect. 5.2. The eigenfunctions form a countable set, and the resulting set of coupled differential equations forms a hierarchy which can usually be handled by truncating after a relatively small number of terms.

a) Evolution Operators: The general specification of the problem is in much the same way as we did for the white noise limit; as in Sect. 8.1.2 we consider a system written in terms of variables $\boldsymbol{x}$ and $\boldsymbol{\alpha}$, so that the conditional probability obeys the evolution equation

$$\frac{\partial P(\boldsymbol{x},\boldsymbol{\alpha},t|\boldsymbol{x}_0,\boldsymbol{\alpha}_0,t_0)}{\partial t} = (L_1+L_2+L_3)\,P(\boldsymbol{x},\boldsymbol{\alpha},t|\boldsymbol{x}_0,\boldsymbol{\alpha}_0,t_0). \tag{8.3.51}$$

In this equation:

i) L_1 describes the evolution of α, which is stochastic and represents the noise. It will have eigenfunctions $p_n(\boldsymbol{\alpha})$ such that

$$L_1 p_n(\boldsymbol{\alpha}) = -\gamma_n p_n(\boldsymbol{\alpha}), \tag{8.3.52}$$

$$\alpha_r p_n(\boldsymbol{\alpha}) = \sum_m c^r_{n,m} p_m(\boldsymbol{\alpha}). \tag{8.3.53}$$

By definition, the eigenfunction $p_0(\boldsymbol{\alpha}) \equiv p_s(\boldsymbol{\alpha})$ represents the stationary solution, that is, $\gamma_0 = 0$.

ii) L_2 describes how $\boldsymbol{x}(t)$ is driven by $\boldsymbol{\alpha}(t)$, and takes the form

$$L_2 = -\sum_{i,j,r} B^r_{i,j} \frac{\partial}{\partial x_i} x_j \alpha_r \,. \tag{8.3.54}$$

iii) L_3 describes linear motion in $\boldsymbol{x}(t)$ alone

$$L_3 = -\sum_{i,j} A_{i,j} \frac{\partial}{\partial x_i} x_j \,. \tag{8.3.55}$$

iv) The equation of motion for $\boldsymbol{x}(t)$ corresponding to these evolution operators is

$$\frac{d\boldsymbol{x}(t)}{dt} = \left(\mathsf{A} + \sum_r \mathsf{B}^r \alpha_r(t)\right) \boldsymbol{x}(t), \tag{8.3.56}$$

and $\alpha_r(t)$ are stochastic variables whose conditional probability satisfies

$$\frac{\partial p(\alpha, t|\alpha_0, t_0)}{\partial t} = L_1 p(\alpha, t|\alpha_0, t_0). \tag{8.3.57}$$

b) Equation of Motion: The quantity from which we can calculate all quantities of interest is the mean

$$\bar{x}(\boldsymbol{\alpha}, t|\boldsymbol{x}_0, t_0) \equiv \int d\boldsymbol{x} \sum_{\boldsymbol{\alpha}_0} \boldsymbol{x} P(\boldsymbol{x}, \boldsymbol{\alpha}, t|\boldsymbol{x}_0, \boldsymbol{\alpha}_0, t_0)\, p_s(\boldsymbol{\alpha}_0). \tag{8.3.58}$$

Here we average over the initial stationary distribution $p_s(\boldsymbol{\alpha}_0)$ to ensure that $\alpha(t)$ is a stationary process. From the initial condition for the conditional probability, the initial condition for $\bar{x}(\boldsymbol{\alpha}, t|\boldsymbol{x}_0, t_0)$ is

$$\bar{x}(\boldsymbol{\alpha}, t_0|\boldsymbol{x}_0, \boldsymbol{\alpha}_0, t_0) = \boldsymbol{x}_0\, p_s(\boldsymbol{\alpha}). \tag{8.3.59}$$

The corresponding equation of motion is obtained from (8.3.51) and is

$$\frac{d}{dt}\bar{x}(\boldsymbol{\alpha}, t|\boldsymbol{x}_0, t_0) = \left(\mathsf{A} + \sum_r \alpha_r \mathsf{B}^r + L_1\right) \bar{x}(\boldsymbol{\alpha}, t|\boldsymbol{x}_0, t_0). \tag{8.3.60}$$

c) Expansion in Eigenfunctions: We now expand in eigenfunctions

$$\bar{\boldsymbol{x}}(\boldsymbol{\alpha}, t|\boldsymbol{x}_0, t_0) = \sum_n \boldsymbol{X}^n(t)\, p_n(\boldsymbol{\alpha}), \tag{8.3.61}$$

and the equation of motion (8.3.60) yields the equations

$$\frac{d\boldsymbol{X}^n(t)}{dt} = \left(\mathsf{A} - \gamma_n\right) \boldsymbol{X}^n(t) + \sum_{m,r} c^r_{m,n} \mathsf{B}^r \boldsymbol{X}^m(t). \tag{8.3.62}$$

d) Quantity of Interest: This is the average over $\boldsymbol{\alpha}$ of $\bar{\boldsymbol{x}}(\boldsymbol{\alpha}, t|\boldsymbol{x}_0, t_0)$, that is

$$\langle \boldsymbol{x}(t) \rangle \equiv \int d\boldsymbol{\alpha}\, \bar{\boldsymbol{x}}(\boldsymbol{\alpha}, t|\boldsymbol{x}_0, t_0) = \boldsymbol{X}^0(\boldsymbol{x}_0, t), \tag{8.3.63}$$

since $\int d\boldsymbol{\alpha}\, p_n(\boldsymbol{\alpha}) = \delta_{n,0}$ is true for any set of eigenfunctions of an evolution operator.

Exercise 8.4 Driving with Ornstein–Uhlenbeck Noise: The Ornstein–Uhlenbeck process (as introduced in Sect. 7.1.1 and Sect. 5.1.1a), when used as a source of noise provides a simple example of the eigenfunction method. The appropriate eigenfunctions are given in Sect. 5.2.3. The autocorrelation function is of the form (8.3.6) provided that

$$k \to \gamma, \qquad D \to 2\gamma. \tag{8.3.64}$$

i) Show that the Hermite polynomial identity

$$x\mathrm{H}_n(x) = \tfrac{1}{2}\mathrm{H}_{n+1}(x) + n\mathrm{H}_{n-1}(x), \tag{8.3.65}$$

corresponds to the eigenfunction identity

$$xQ_n(x) = \left(\frac{n+1}{2}\right)^{1/2} Q_{n+1}(x) + \left(\frac{n}{2}\right)^{1/2} Q_{n-1}(x), \tag{8.3.66}$$

so that there is a single set of coefficients c_{nm}, and these are given by

$$c_{n,m} = \left(\frac{n+1}{2}\right)^{1/2} \delta_{m,n+1} + \left(\frac{n}{2}\right)^{1/2} \delta_{m,n-1}. \tag{8.3.67}$$

ii) For the model (8.3.31), show that the equations (8.3.62) become in this case

$$\begin{pmatrix} \dot{X}^1 \\ \dot{X}^2 \\ \dot{X}^3 \\ \dot{X}^4 \\ \vdots \\ \dot{Y}^1 \\ \dot{Y}^2 \\ \dot{Y}^3 \\ \dot{Y}^4 \\ \vdots \end{pmatrix} = \begin{pmatrix} \mathrm{i}\Omega & 0 & 0 & 0 & \cdots & 0 & \frac{1}{2}\mathrm{i}\delta & 0 & 0 & \cdots \\ 0 & \mathrm{i}\Omega-\gamma & 0 & 0 & \cdots & \frac{1}{2}\mathrm{i}\delta & 0 & \frac{1}{2}\mathrm{i}\sqrt{2}\delta & 0 & \cdots \\ 0 & 0 & \mathrm{i}\Omega-2\gamma & 0 & \cdots & 0 & \frac{1}{2}\mathrm{i}\sqrt{2}\delta & 0 & \frac{1}{2}\mathrm{i}\sqrt{3}\delta & \cdots \\ 0 & 0 & 0 & \mathrm{i}\Omega-3\gamma & \cdots & 0 & 0 & \frac{1}{2}\mathrm{i}\sqrt{3}\delta & 0 & \cdots \\ \vdots & \vdots & \vdots & \vdots & \ddots & \vdots & \vdots & \vdots & \vdots & \ddots \\ 0 & \frac{1}{2}\mathrm{i}\delta & 0 & 0 & \cdots & -\mathrm{i}\Omega & 0 & 0 & 0 & \cdots \\ \frac{1}{2}\mathrm{i}\delta & 0 & \frac{1}{2}\mathrm{i}\sqrt{2}\delta & 0 & \cdots & 0 & -\mathrm{i}\Omega-\gamma & 0 & 0 & \cdots \\ 0 & \frac{1}{2}\mathrm{i}\sqrt{2}\delta & 0 & \frac{1}{2}\mathrm{i}\sqrt{3}\delta & \cdots & 0 & 0 & \mathrm{i}\Omega-2\gamma & 0 & \cdots \\ 0 & 0 & \frac{1}{2}\mathrm{i}\sqrt{3}\delta & 0 & \cdots & 0 & 0 & 0 & \mathrm{i}\Omega-3\gamma & \cdots \\ \vdots & \vdots & \vdots & \vdots & \ddots & \vdots & \vdots & \vdots & \vdots & \ddots \end{pmatrix} \begin{pmatrix} X^1 \\ X^2 \\ X^3 \\ X^4 \\ \vdots \\ Y^1 \\ Y^2 \\ Y^3 \\ Y^4 \\ \vdots \end{pmatrix} \tag{8.3.68}$$

iii) Examine these equations numerically for a range of truncations of the hierarchy, and show that such a procedure is viable and valid for a wide range of parameter values.

9. Adiabatic Elimination of Fast Variables

The previous chapter considered the case of fast fluctuations of a quite specific form, namely that given by (8.2.6). In such a system, the noise variable $\xi(t)$ obeys an equation of motion which is quite independent of $u(t)$, the variable of interest. The resulting noise term in the equation of motion for u is then simply $b(u,t)\xi(t)$, and in the limit of very fast fluctuations, that is , $\xi(t) = \gamma\alpha(\gamma^2 t)$, with $\gamma \to \infty$, this leads to a white noise stochastic differential equation.

In this chapter we want to consider the more general case, where the noise variable and the variable of interest influence each other in a way which cannot be put in any simple form, such as $b(u,t)\xi(t)$. Such systems are very common in physics—the best-known case arises in the kinetic theory of gases, in which the motions of individual atoms are influenced by the motions of their nearest neighbours, and are very rapid. However, the same system at a macroscopic level behaves as a fluid, which can be described by a few local variables, such as pressure, temperature, density and fluid velocity, to which one must also add fluctuation noise terms. How do we derive the hydrodynamic equations from the microscopic equations, and how does the resulting noise arise? This is a difficult problem, whose solution was first given by *Chapman* and *Enskog*, working independently. The method depends on the identification of the *slow* variables (the hydrodynamic quantities), followed by the identification of a correct set of independent *fast* variables. The hydrodynamic equations arise as the limiting behaviour when the fast variables move infinitely more rapidly than the slow variables.

Here we will consider stochastic systems of a much simpler form, but for which the essence of the problem is the same. The mathematical procedure is essentially the same as that used in the previous chapter. The main difference lies in the correct identification of the time scales, and the appropriate variables corresponding to these time scales. The method used can be generalized to quantum-mechanical systems, and will be used extensively in the rest of this book, as well as in *Book II* and *Book III*.

9.1 Slow and Fast Variables

The selection of appropriate variables in a physical system is not purely a matter of mathematics, but rather involves an understanding of the physics of the system under study. The isolation of the correct variables is even more subtle in the case of a stochastic system. In this section we will study a simple example, based on the laser, and treat both non-stochastic and stochastic versions of this model.

9.1.1 A Simplified Laser Model

Consider firstly a system governed by deterministic equations of motion

$$\dot{\alpha} = \varepsilon\alpha - a\alpha n, \tag{9.1.1}$$

$$\dot{n} = -\kappa n + b\alpha^2, \tag{9.1.2}$$

in which all of κ, a, b are taken to be positive.

This set of equations has a structure similar to that of a large number of physical problems. One interpretation is that of a laser, which is shown in Fig. 9.1. The essential components are:

i) A single electromagnetic field mode α, for simplicity represented here as a real number, rather than as a complex number as usual.

ii) A number of two-level atoms in which the $n = n(e) - n(g)$ represents the inversion, i.e., the proportion of atoms in the upper state, more than would be there in thermodynamic equilibrium.

a) Interpretation of the Equations: There are four processes involved which have the following interpretations:

i) The term $\varepsilon\alpha$ represents the difference between the processes of *gain*, whose origin is shown in a more complex laser model to be also from the inversion of the atoms, and *loss* (out of the cavity through the mirrors). Thus, net gain corresponds to $\varepsilon > 0$, while loss corresponds to $\varepsilon < 0$.

ii) The equation (9.1.1) says that the field also increases (or decreases) in proportion to αn, depending on the sign of n.

iii) The second equation, (9.1.2), says that the number of atoms in the upper state decays ($-\kappa n$), and can also increase ($b\alpha^2$) in proportion to the intensity of the light field.

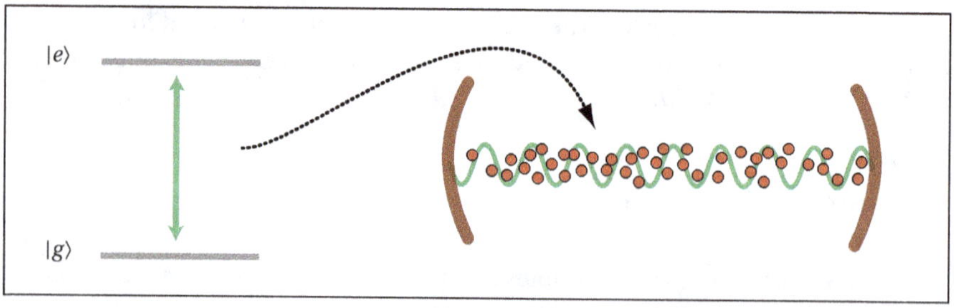

Fig. 9.1. Two-level atoms interacting with the electromagnetic field inside a cavity.

b) Stationary Solution: There are three stationary solutions to the equation of motion

$$\alpha_s = 0, \tag{9.1.3}$$

$$\alpha_s = \pm\sqrt{\frac{\kappa\varepsilon}{ab}}. \tag{9.1.4}$$

c) Deterministic Elimination of the Fast Variable: Normally the atomic relaxation constant κ is much bigger than ε, the cavity decay constant. We can then solve (9.1.2)

$$n(t) = n(0)e^{-\kappa t} + \int_0^t dt'\, e^{-\kappa(t-t')} b\alpha(t')^2. \tag{9.1.5}$$

In the limit of very large κ, we can drop the first term, and make the approximation

$$n(t) \approx \int_0^t dt'\, e^{-\kappa(t-t')} b\alpha(t')^2, \tag{9.1.6}$$

$$\approx b\alpha(t)^2 \int_0^t dt'\, e^{-\kappa(t-t')} = \frac{b\alpha(t)^2}{\kappa}. \tag{9.1.7}$$

We now substitute into the first equation to get the approximate equation of motion

$$\dot{\alpha} = \varepsilon\alpha - \frac{ab}{\kappa}\alpha^3. \tag{9.1.8}$$

What we have found is that with the fast relaxation arising from the large value of κ, the number $n(t)$ of atoms follows the value $b\alpha^2/\kappa$. This is a quasi-stationary solution, because the rate of variation of $\alpha(t)$ is so much slower than that of $n(t)$. The result is equivalent to setting $\dot{n} \to 0$ in (9.1.2), and this gives the simplest way of representing the procedure algorithmically. The variable n is conventionally said to be *adiabatically eliminated* by this procedure, yielding the equation of motion (9.1.8) which involves only α.

d) "Naive" Adiabatic Elimination: The procedure we have developed can be summarized by the following procedure:

i) Since κ is assumed to be very large, neglect the derivative term in the n equation (9.1.2).

ii) Solve this equation to get $n = b\alpha^2/\kappa$.

iii) Substitute for n into the other equation, yielding (9.1.8).

This procedure is essentially a formalization of that outlined above, and is conventionally called "naive" adiabatic elimination.

Exercise 9.1 Stability: Show that if $\varepsilon > 0$, the solution $\alpha = 0$ is stable against perturbations, while for $\alpha > 0$, the solutions $\alpha = \pm\sqrt{\kappa\varepsilon/ab}$ are stable, as illustrated in Fig. 9.2. (Set α equal to the stationary value plus δ, where δ is very small, and keep only first order terms in δ. Determine whether the equation says δ will grow or die away. Clearly, if δ grows, the stationary solution is unstable.)

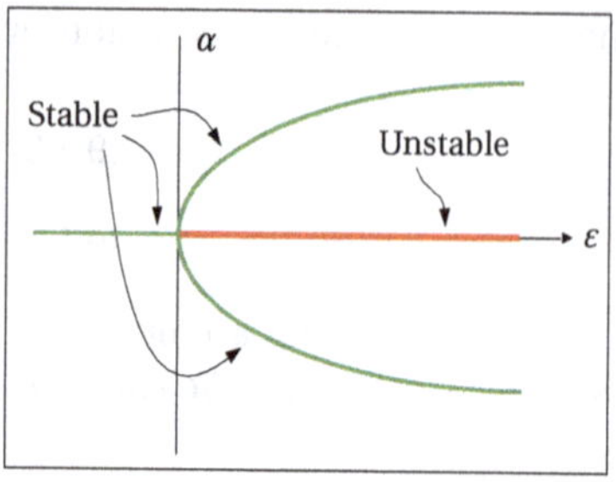

Fig. 9.2. Diagram illustrating the stable (green) and unstable (red) stationary solutions of the differential equation $\dot{\alpha} = \varepsilon\alpha - (ab/\kappa)\alpha^3$.

9.1.2 Stochastic Elimination of the Fast Variable

In a laser, the sizes of the parameters are not determined solely by the physics of the problem. The engineering aspect is also important, since the mirrors, gases and excitation processes used are carefully chosen in such a way as to produce the light in the form desired. The laser is an engineered device, not a natural phenomenon.

In the process of engineering the device, noise sources enter, related to the quantized nature of the light and the particle nature of the atoms, and the sizes of the noise sources are also related to other physical parameters. In this section we will introduce noise sources in the simplest way mathematically, and show how they affect the adiabatic elimination process. In the final analysis, it will be seen that the result depends on quite subtle considerations, and is not unique.

a) Inclusion of Noise: We will assume that the noise terms arise from the physics of the individual processes, and are independent of each other, thus we will consider stochastic equations of the form

$$d\alpha = (\varepsilon\alpha - a\alpha n)\,dt + \mathcal{M}\,dW_1(t), \tag{9.1.9}$$

$$dn = (-\kappa n + b\alpha^2)\,dt + \mathcal{N}\,dW_2(t). \tag{9.1.10}$$

(Here $\mathcal{M}$ and $\mathcal{N}$ are in principle arbitrary coefficients.)

The corresponding Fokker–Planck equation then is

$$\frac{\partial p}{\partial t} = \left[\frac{\partial}{\partial \alpha}(-\varepsilon\alpha + a\alpha n) + \tfrac{1}{2}\mathcal{M}^2\frac{\partial^2}{\partial \alpha^2} + \frac{\partial}{\partial n}(\kappa n - b\alpha^2) + \tfrac{1}{2}\mathcal{N}^2\frac{\partial^2}{\partial n^2}\right]P\,. \tag{9.1.11}$$

b) The Nature of the Limit Involved: The limit we want to consider is that of $\kappa \to \infty$, and for this to make sense in the deterministic case, the limiting equation of motion (9.1.8) shows that in this limit we will only get an interesting result if we also require the coefficients a and b to have a limiting behaviour such that

$$\lim_{\kappa\to\infty} \frac{ab}{\kappa} = A \neq 0. \tag{9.1.12}$$

If this does not happen, it can be either because A is infinite or zero; in the first case we have no limiting behaviour, and in the second all influence of the presence of the atoms disappears—a valid, but uninteresting result.

c) Changing Variables: To eliminate the fast variable n, we define new variables y, β through

$$y = \alpha, \qquad \beta = n - \frac{b}{\kappa}\alpha^2, \tag{9.1.13}$$

so that β represents the difference between n and its quasistationary value. We substitute these into the Fokker–Planck equation, changing variables by using the chain rule identities

$$\begin{aligned} \frac{\partial}{\partial \alpha} &= \frac{\partial y}{\partial \alpha}\frac{\partial}{\partial y} + \frac{\partial \beta}{\partial \alpha}\frac{\partial}{\partial \beta} = \frac{\partial}{\partial y} - \frac{2by}{\kappa}\frac{\partial}{\partial \beta}, \\ \frac{\partial}{\partial n} &= \frac{\partial y}{\partial n}\frac{\partial}{\partial y} + \frac{\partial \beta}{\partial n}\frac{\partial}{\partial \beta} = \frac{\partial}{\partial \beta}. \end{aligned} \tag{9.1.14}$$

We put these into the Fokker–Planck equation (9.1.11), to get (setting $y \to \alpha$, now that we have done the variable change)

$$\frac{\partial p}{\partial t} = (L_1^0 + L_2^0 + L_3^0)\,p, \tag{9.1.15}$$

with

$$L_1^0 = \frac{\partial}{\partial \beta}\kappa\beta + \frac{\mathcal{N}^2}{2}\frac{\partial^2}{\partial \beta^2}, \tag{9.1.16}$$

$$\begin{aligned} L_2^0 &= \frac{\partial}{\partial \alpha}a\beta\alpha - \frac{2b\alpha}{\kappa}\frac{\partial}{\partial \beta}\left(-\varepsilon\alpha + \frac{ab}{\kappa}\alpha^3 + a\alpha\beta\right), \\ &\quad -\mathcal{M}^2\left(\frac{b\alpha}{\kappa}\frac{\partial^2}{\partial\alpha\partial\beta} + \frac{\partial^2}{\partial\alpha\partial\beta}\frac{b\alpha}{\kappa}\right) + \frac{2b\alpha^2\mathcal{M}^2}{\kappa^2}\frac{\partial^2}{\partial\beta^2}, \end{aligned} \tag{9.1.17}$$

$$L_3^0 = \frac{\partial}{\partial \alpha}\left(-\varepsilon\alpha + \frac{ab}{\kappa}\alpha^3\right) + \frac{\mathcal{M}^2}{2}\frac{\partial^2}{\partial \alpha^2}. \tag{9.1.18}$$

To take the limit as $\kappa \to \infty$, we have so far only the requirement that ab/κ must approach a finite value in this limit. Physical considerations will suggest that $\mathcal{M}$, which represents the intrinsic noise arising from the gain and damping, should be independent of κ or have a finite limit as $\kappa \to \infty$, and we will for simplicity take $\mathcal{M}$ to be independent of κ. We are then left to consider the limiting behaviour of b (and thus of a because ab/κ approaches a finite value) and the noise coefficient $\mathcal{N}$. The number of possibilities is still large, so we will restrict ourselves to the two simplest and most physically interesting ones.

d) The Finite Noise Choice: Suppose $\mathcal{N}$ is proportional to $\sqrt{\kappa}$, so that in the limit $\kappa \to \infty$, the variance of β stays constant, and we can say that the typical size of the fluctuations $|\beta| \sim 1$. In order for the fluctuations in β to have any effect as they become fast, the coefficient b must be come large, so let us choose

$$\mathcal{N} = \tilde{\mathcal{N}}\sqrt{\kappa}, \tag{9.1.19}$$

$$b = \tilde{b}\sqrt{\kappa}, \tag{9.1.20}$$

$$a = \tilde{a}\sqrt{\kappa}. \tag{9.1.21}$$

The limiting forms of the differential operators are then

$$L_1^0 \to \kappa L_1 \quad = \kappa \left\{ \frac{\partial}{\partial \beta}\beta + \frac{\tilde{\mathcal{N}}^2}{2}\frac{\partial^2}{\partial \beta^2} \right\}, \tag{9.1.22}$$

$$L_2^0 \to \sqrt{\kappa} L_2 = \sqrt{\kappa}\left\{ \tilde{a}\frac{\partial}{\partial \alpha}\alpha\beta \right\}, \tag{9.1.23}$$

$$L_3^0 \to L_3 \quad = \frac{\partial}{\partial \alpha}\left(-\varepsilon\alpha + \tilde{a}\tilde{b}\alpha^3\right) + \frac{\mathcal{M}^2}{2}\frac{\partial^2}{\partial \alpha^2}. \tag{9.1.24}$$

The Fokker–Planck equation now takes the limiting form for large κ

$$\frac{\partial p}{\partial t} = \left\{ \kappa L_1 + \sqrt{\kappa} L_2 + L_3 \right\} p, \tag{9.1.25}$$

which is of the same form as (8.1.9) with $\gamma \to \sqrt{\kappa}$.

We can therefore deduce that the large κ limit gives the equation for $v \equiv \mathcal{P}p$

$$\frac{\partial v}{\partial t} = \left\{ L_3 - \mathcal{P} L_2 L_1^{-1} L_2 \right\} v, \tag{9.1.26}$$

and that this will lead to

$$\frac{\partial p(\alpha,t)}{\partial t} = \left\{ \frac{\partial}{\partial \alpha}\left(-\varepsilon\alpha + \tilde{a}\tilde{b}\alpha^3\right) + \frac{\mathcal{M}^2}{2}\frac{\partial^2}{\partial \alpha^2} + \frac{\tilde{\mathcal{N}}^2\tilde{a}^2}{2}\frac{\partial}{\partial \alpha}\alpha\frac{\partial}{\partial \alpha}\alpha \right\} p(\alpha,t). \tag{9.1.27}$$

In evaluating this expression, we have used the stationary time correlation function corresponding to L_1

$$g_\beta(\tau) = \frac{\tilde{\mathcal{N}}^2}{2}\mathrm{e}^{-|\tau|}. \tag{9.1.28}$$

The Fokker–Planck equation (9.1.27) corresponds to the *Stratonovich* stochastic differential equation we would obtain by writing the two stochastic differential equations as Langevin equations

$$\dot{\alpha} = \varepsilon\alpha - a\alpha n + \mathcal{M}\xi_1(t), \tag{9.1.29}$$

$$\dot{n} = -\kappa n + b\alpha^2 + \mathcal{N}\xi_2(t), \tag{9.1.30}$$

and applying the naive procedure to these equations, that is by setting $\dot{n} = 0$. This would yield

$$\dot{\alpha} = \varepsilon\alpha - \frac{ab\alpha^3}{\kappa} + \mathcal{M}\xi_1(t) + \frac{a\mathcal{N}}{\kappa}\alpha\xi_2(t). \tag{9.1.31}$$

However, there is a different limit in which this does not happen, which we will now investigate.

e) Infinitesimal Noise Choice: We can also consider the possibility that the noise source in L_1 is a constant as $\kappa \to \infty$, so that the typical size of β goes to zero as $\kappa \to \infty$, but at the same time the coefficient of the noise in the interaction term increases in proportion. In order to get a genuine limit, we can make the substitution

$$\eta = \beta/\sqrt{\kappa}, \tag{9.1.32}$$

and $a \sim$ constant, $b = \kappa\bar{b}$. Substituting these in (9.1.16–9.1.18)

$$L_1^0 \to \kappa L_1 \quad = \kappa\left\{\frac{\partial}{\partial\eta}\eta + \frac{\mathcal{N}^2}{2}\frac{\partial^2}{\partial\eta^2}\right\}, \tag{9.1.33}$$

$$L_2^0 \to \sqrt{\kappa}L_2 = \sqrt{\kappa}\left\{a\frac{\partial}{\partial\alpha}\alpha\eta - \bar{b}\mathcal{M}^2\left(2\alpha\frac{\partial^2}{\partial\alpha\partial\eta} + \frac{\partial}{\partial\eta}\right)\right\}, \tag{9.1.34}$$

$$L_3^0 \to L_3 \quad = \frac{\partial}{\partial\alpha}\left(-\varepsilon\alpha + a\bar{b}\alpha^3\right) + \frac{\mathcal{M}^2}{2}\frac{\partial^2}{\partial\alpha^2}. \tag{9.1.35}$$

The expression for L_2 can be simplified in two different ways by noting that in evaluating the limiting expression $\mathcal{P}L_2L_1^{-1}L_2\mathcal{P}$:

i) In the left-hand term $\mathcal{P}L_2$, the projection involves integrating over all η, so all terms involving derivatives with respect to η vanish.

ii) In the right-hand term, $L_2\mathcal{P}$, the operator L_2 acts on the stationary solution of L_1 and this is given by

$$p_s(\eta) \propto \exp(-\eta^2/\mathcal{N}^2), \tag{9.1.36}$$

so that we can make the replacement

$$\frac{\partial}{\partial\eta} \to -2\eta/\mathcal{N}^2. \tag{9.1.37}$$

iii) Together these mean that we can write

$$\mathcal{P}L_2L_1^{-1}L_2\mathcal{P} \to \mathcal{P}L_2^{(l)}L_1^{-1}L_2^{(r)}\mathcal{P}, \tag{9.1.38}$$

$$L_2^{(l)} \quad = \left\{a\frac{\partial}{\partial\alpha}\alpha\right\}\eta, \tag{9.1.39}$$

$$L_2^{(r)} \quad = \left\{a\frac{\partial}{\partial\alpha}\alpha + \frac{\bar{b}\mathcal{M}^2}{\mathcal{N}^2}\left(4\alpha\frac{\partial}{\partial\alpha} + 2\right)\right\}\eta. \tag{9.1.40}$$

We can therefore deduce that the large κ limit gives the equation for $v \equiv \mathcal{P}p$

$$\frac{\partial v}{\partial t} = \left\{L_3 - \mathcal{P}L_2^{(l)}L_1^{-1}L_2^{(r)}\right\}v, \tag{9.1.41}$$

and that this will lead to

$$\frac{\partial p(\alpha,t)}{\partial t} = \left[\frac{\partial}{\partial\alpha}\left(-\varepsilon\alpha+\tilde{a}\tilde{b}\alpha^3\right)+\frac{\mathcal{M}^2}{2}\frac{\partial^2}{\partial\alpha^2} + \frac{1}{2}\left\{a\frac{\partial}{\partial\alpha}\alpha\right\}\left\{a\mathcal{N}^2\frac{\partial}{\partial\alpha}\alpha+\bar{b}\mathcal{M}^2\left(4\alpha\frac{\partial}{\partial\alpha}+2\right)\right\}\right]p(\alpha,t). \tag{9.1.42}$$

In evaluating this expression, we have used the stationary time correlation function corresponding to L_1

$$g_\eta(\tau) = \frac{\mathcal{N}^2}{2}\mathrm{e}^{-|\tau|}. \tag{9.1.43}$$

In this case the Fokker–Planck equation obtained is not simply the Stratonovich interpretation of the naive elimination procedure, the term proportional to $\mathcal{M}^2$ in the second line is an extra term which is difficult to explain in that way.

9.2 Other Applications of the Adiabatic Elimination Method

9.2.1 A Stochastic Model of Trapped Atoms

Let us consider a system based on an optical potential, in which atoms can be trapped. A single atom can remain in the trap for quite a long time, but if two become trapped, the interaction between them is so strong that their energy is close to the top of the trap, and both atoms are rapidly expelled. The system will be modelled by a jump process, with probabilities $p(n)$, with $n = 0,1,2$. As depicted in Fig. 9.3, the master equation governing the transitions can be written as

$$\dot{p}(2) = -wp(2)+\left(t^-p(2)-t^+p(1)\right), \tag{9.2.1}$$

$$\dot{p}(1) = \left(t^-p(2)-t^+p(1)\right)+\left(r^+p(0)-r^-p(1)\right), \tag{9.2.2}$$

$$\dot{p}(0) = wp(2)+\left(r^-p(1)-r^+p(0)\right). \tag{9.2.3}$$

Writing

$$\boldsymbol{p} = \begin{pmatrix} p(2) \\ p(1) \\ p(0) \end{pmatrix}, \tag{9.2.4}$$

we can write this master equation as

$$\dot{\boldsymbol{p}} = (wL_1+L_2+L_3)\boldsymbol{p}, \tag{9.2.5}$$

$$L_1 = \begin{pmatrix} -1 & 0 & 0 \\ 0 & 0 & 0 \\ 1 & 0 & 0 \end{pmatrix}, \quad L_2 = \begin{pmatrix} -t^- & t^+ & 0 \\ t^- & -t^+ & 0 \\ 0 & 0 & 0 \end{pmatrix}, \quad L_3 = \begin{pmatrix} 0 & 0 & 0 \\ 0 & -r^- & r^+ \\ 0 & r^- & -r^+ \end{pmatrix}. \tag{9.2.6}$$

The limit of interest is when the two-body loss process becomes very fast, and this becomes a problem of exactly the same kind as the adiabatic elimination problem.

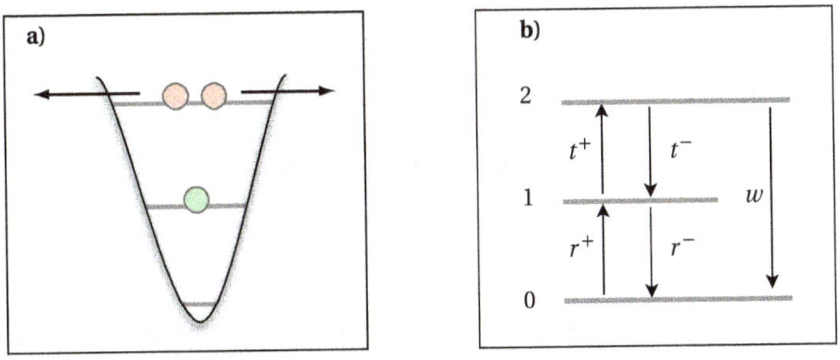

Fig. 9.3. a) Schematic diagram of trapped atoms, in which the repulsive forces between two atoms in the trap cause the escape of both of them from the trap; **b)** Processes for the atoms in a trap as in (9.2.1–9.2.2). Here $t^{\pm}$, and $r^{\pm}$ represent two-way processes in which one atom can be added to or lost from the number of atoms in the trap. The process w represents the two-body loss process.

a) Stationary State of L_1: The solution of the equation $\dot{\boldsymbol{p}} = wL_1\boldsymbol{p}$ is

$$p(2,t) = e^{-wt}p(2,0), \tag{9.2.7}$$
$$p(1,t) = p(1,0), \tag{9.2.8}$$
$$p(0,t) = p(0,0) + (1 - e^{-wt})p(2,0). \tag{9.2.9}$$

Letting $t \to \infty$, we find

$$\boldsymbol{p}_s = \begin{pmatrix} 0 \\ p(1,0) \\ p(0,0) + p(2,0) \end{pmatrix} = \begin{pmatrix} 0 & 0 & 0 \\ 0 & 1 & 0 \\ 1 & 0 & 1 \end{pmatrix} \boldsymbol{p}(0) \equiv \mathcal{P}\boldsymbol{p}(0). \tag{9.2.10}$$

b) Projectors: We see therefore that

$$\mathcal{P} = \begin{pmatrix} 0 & 0 & 0 \\ 0 & 1 & 0 \\ 1 & 0 & 1 \end{pmatrix}, \qquad \mathcal{Q} = \begin{pmatrix} 1 & 0 & 0 \\ 0 & 0 & 0 \\ -1 & 0 & 0 \end{pmatrix}. \tag{9.2.11}$$

c) Elimination Procedure: With the projector $\mathcal{P}$ as defined by this process, we find that neither $\mathcal{P}L_2\mathcal{P}$ nor $\mathcal{P}L_3\mathcal{P}$ is zero. This is only a minor technical problem, since the same procedure as used in Sect. 8.1.2e gives analogous results; firstly let us introduce a more compact notation

$$L_u \equiv L_2 + L_3. \tag{9.2.12}$$

We then use the notation as in Sect. 8.1.2e,

$$\boldsymbol{v}(t) = \mathcal{P}\boldsymbol{p}(t), \tag{9.2.13}$$
$$\tilde{\boldsymbol{v}}(s) = \int_0^\infty e^{-st}\boldsymbol{v}(t)\,dt. \tag{9.2.14}$$

One then finds

$$s\tilde{\boldsymbol{v}}(s)-\boldsymbol{v}(0)$$
$$=(\mathcal{P}L_u\mathcal{P})\tilde{\boldsymbol{v}}(s)+(\mathcal{P}L_u\mathcal{Q})\big(s-\mathcal{Q}L_u-wL_1\big)^{-1}(\mathcal{Q}L_u\mathcal{P})\tilde{\boldsymbol{v}}(s), \tag{9.2.15}$$
$$\sim(\mathcal{P}L_u\mathcal{P})\tilde{\boldsymbol{v}}(s)-\frac{1}{w}(\mathcal{P}L_u\mathcal{Q})L_1^{-1}(\mathcal{Q}L_u\mathcal{P})\tilde{\boldsymbol{v}}(s). \tag{9.2.16}$$

d) Evaluation of $L_1^{-1}\mathcal{Q}$: The operator L_1^{-1} is defined only in the subspace into which $\mathcal{Q}$ projects, and this space is one dimensional. By straightforward calculation $L_1\mathcal{Q}=\mathcal{Q}$, hence $L_1^{-1}\mathcal{Q}=\mathcal{Q}$.

e) Evaluation of the Stochastic Equation: It is simply a matter of matrix multiplication to evaluate the operator on the right-hand side of (9.2.16), and this gives the stochastic equation

$$\dot{\boldsymbol{v}}=\begin{pmatrix}0 & 0 & 0\\ r^+ & -(t^++r^-) & r^+\\ -r^+ & t^++r^- & -r^+\end{pmatrix}\boldsymbol{v}+\frac{1}{w}\begin{pmatrix}0 & 0 & 0\\ 0 & t^+(t^--r^+) & 0\\ 0 & -t^+(t^--r^+) & 0\end{pmatrix}\boldsymbol{v}. \tag{9.2.17}$$

By definition (9.2.13) $v(2,t)=0$, and the remaining two equations give the master equation for the lower two occupations

$$\dot{p}(1,t)=-\left[r^-+t^+-\frac{t^+(t^--r^+)}{w}\right]p(1,t)+r^+p(0,t), \tag{9.2.18}$$

$$\dot{p}(0,t)=\left[r^-+t^+-\frac{t^+(t^--r^+)}{w}\right]p(1,t)-r^+p(0,t). \tag{9.2.19}$$

f) Conclusion: This is now a simple random telegraph process, as described in Sect. 6.1.3.

i) To zero order, that is, in the limit $w\to\infty$, the transition from $1\to 2$ becomes equivalent to a transition from $1\to 0$ at the same rate t^+, a result which is intuitively obvious.

ii) The next order is a correction to the rate, which may have either sign. It represents the small number of transitions $1\to 2\to 1$, but is also influenced by the transition $0\to 1$ in a way which is hard to explain intuitively.

iii) Later in this book, in Sect. 20.2, we will show how to treat a similar, but fully quantum-mechanical, system. The formalism is very similar, but the result is rather different.

9.2.2 Motional Narrowing

The concept of *motional narrowing* arose in the theory of nuclear magnetic resonance, in which the spin of a proton is influenced by a static magnetic field, and a fluctuating field which arises from all the other magnetic moments of the constituents of the fluid of which it forms a part. The motion involved is that of the resonant proton itself through the fluid, as well as the motion of all of the other molecules in the fluid.

The equations of motion for the spin $\boldsymbol{s}$ in a magnetic field $\boldsymbol{B}(t)$ can written as

$$\dot{\boldsymbol{s}} = g\boldsymbol{s} \times \boldsymbol{B}(t). \tag{9.2.20}$$

We write $\boldsymbol{s} \equiv (x, y, z)$, and consider a magnetic field composed of a constant strong field along the z-axis, and a smaller fluctuating field along the x-axis thus

$$\boldsymbol{B}(t) = \boldsymbol{B}_s + \boldsymbol{B}_f(t), \tag{9.2.21}$$

$$\boldsymbol{B}_s = (0, 0, \Omega/g), \tag{9.2.22}$$

$$\boldsymbol{B}_f(t) = (B_f(t), 0, 0), \tag{9.2.23}$$

so that the equations of motion become

$$\dot{x} = -\Omega y, \tag{9.2.24}$$

$$\dot{y} = \Omega x - gzB_f(t), \tag{9.2.25}$$

$$\dot{z} = gyB_f(t). \tag{9.2.26}$$

The quantity $B_f(t)$ represents the fluctuating magnetic field. These equations are a form of the *Bloch equations*, which occur extensively in a quantum-optical context. The parameter Ω is the precession frequency of the spin in the magnetic field, and g is the magnetic moment.

The speed of the the fluctuations of the magnetic field can be introduced by writing the field in a form similar to that introduced in the beginning of the chapter

$$B_f(t) = B_0 \alpha(\gamma^2 t), \tag{9.2.27}$$

where we now consider the limit of large γ at *fixed amplitude* B_0 of $\alpha(t)$. Thus, the factor of $1/\gamma$ introduced in (8.1.4) is not present here. The equations of motion in the case that γ is large (but not infinite) are clearly the Stratonovich stochastic differential equations

$$(\mathbf{S})\, dx = -\Omega y\, dt, \tag{9.2.28}$$

$$(\mathbf{S})\, dy = \Omega x\, dt - \frac{gzB_0}{\gamma^2}\, dW(t), \tag{9.2.29}$$

$$(\mathbf{S})\, dz = \frac{gyB_0}{\gamma^2}\, dW(t). \tag{9.2.30}$$

At fixed B_0 the effect is to produce stochastic differential equations in which the noise becomes very small as the speed of the fluctuations increases. Even if the magnitude B_0 of the fluctuations is quite large, their effect becomes smaller as they become more rapid. This means that the spectrum of fluctuations of the spin will become narrower—hence the name "motional narrowing".

The effect is a consequence of what we have seen all through this chapter; that white noise requires both rapid fluctuations and large amplitude. If the large amplitude condition is not fulfilled, the rapid fluctuations simply average to nothing.

III Fields, Quanta and Atoms

10. Ideal Bose and Fermi Systems

The properties of systems of Bosons and Fermions even when they do not interact are surprisingly wide ranging. In this chapter we will give a brief outline of the standard description in terms of creation and destruction operators, in a language sufficiently general to allow the discussion of photons, phonons and atoms, and the kinds of quantum states that they can have.

10.1 The Quantum Gas

We want to consider systems of many particles, which may obey either Bose or Fermi statistics, and where the concept of a "quantum gas" is understood to include all assemblies of particles in which a description in terms of the location or the momentum of particular quanta forms a relevant concept. Thus, there will be an underlying quantum field of which the Bosons or Fermions are quanta. This may be a fundamental field, such as that of light, or a derived field, like that of sound. It may be a matter wave field, in which the quanta are indestructible atoms, and therefore for which the conservation of particle number becomes important. All of these fields display the same basic properties of coherence, partial coherence or incoherence, and of interference, all in the absence of interactions.

10.1.1 Bosons

Quantized Bosons are described by introducing annihilation operators a_i with Bosonic commutation relations

$$[a_i, a_j^\dagger] = \delta_{i,j}, \quad [a_i, a_j] = [a_i^\dagger, a_j^\dagger] = 0. \tag{10.1.1}$$

The complete set of quantum states can be written

$$|n_1, n_2, \dots, n_i, \dots\rangle, \tag{10.1.2}$$

where n_i is the number of particles corresponding to the i-th mode. The identity of the particles is thus enforced by the fact that it is only the *numbers* of them with each momentum which are required to specify the state completely. The action of the creation and destruction operators on these states is

$$a_i|\dots n_i \dots\rangle = \sqrt{n_i}|\dots n_i - 1 \dots\rangle, \tag{10.1.3}$$

$$a_i^\dagger|\dots n_i \dots\rangle = \sqrt{n_i + 1}|\dots n_i + 1 \dots\rangle. \tag{10.1.4}$$

The states are normalized so that

$$\langle n_1', n_2', \dots, n_i', \dots | n_1, n_2, \dots, n_i, \dots \rangle = \delta_{n_1,n_1'} \delta_{n_2,n_2'} \cdots \delta_{n_i,n_i'} \cdots \tag{10.1.5}$$

From the definitions (10.1.3, 10.1.4), it follows that these quantum states are eigenvectors of the *number operators*

$$N_i \equiv a_i^\dagger a_i, \tag{10.1.6}$$

so that

$$N_i |\dots n_i \dots\rangle = n_i |\dots n_i \dots\rangle. \tag{10.1.7}$$

Thus the N_i form a complete set of commuting observables for the many-particle system. The formalism thus gives a description entirely in terms of particle numbers.

10.1.2 Fermions

In the case that the particles obey Fermi–Dirac statistics we introduce operators b_i, $b_i^\dagger$, which are Fermion destruction and creation operators, and these have *anticommutation relations*

$$[b_i, b_j^\dagger]_+ = \delta_{i,j}, \quad [b_i, b_j]_+ = [b_i^\dagger, b_j^\dagger]_+ = 0, \tag{10.1.8}$$

where the *anticommutator* is defined by

$$[A, B]_+ \equiv AB + BA. \tag{10.1.9}$$

For Fermions we can also use states of the form (10.1.2), but the only values of the n_i are 1 or 0, which follows from the action of the Fermion operators on these states

$$b_i |\dots n_i \dots\rangle = \sqrt{n_i} |\dots n_i - 1 \dots\rangle, \tag{10.1.10}$$

$$b_i^\dagger |\dots n_i \dots\rangle = \sqrt{1 - n_i} |\dots n_i + 1 \dots\rangle. \tag{10.1.11}$$

From these it follows that

$$b_i^2 = \tfrac{1}{2}[b_i, b_i]_+ = 0, \tag{10.1.12}$$

$$\begin{aligned} N_i^2 &= b_i^\dagger b_i b_i^\dagger b_i = b_i^\dagger b_i - b_i^\dagger b_i^\dagger b_i b_i \\ &= b_i^\dagger b_i = N_i, \end{aligned} \tag{10.1.13}$$

and thus the only eigenvalues of N_i are 0 and 1, leading to Pauli's exclusion principle.

10.1.3 The Hamiltonian and Total Number Operators

a) Hamiltonian: The physics we wish to consider enters by the choice of Hamiltonian, and for non-interacting particles, we should choose Hamiltonians of the form

$$H_{\text{Bose}} = \sum_i \hbar\omega_i a_i^\dagger a_i, \tag{10.1.14}$$

$$H_{\text{Fermi}} = \sum_i \hbar\omega_i b_i^\dagger b_i. \tag{10.1.15}$$

b) Total Number: The operator for the total number of particles takes the two corresponding forms

$$N_{\text{Bose}} = \sum_i a_i^\dagger a_i, \tag{10.1.16}$$

$$N_{\text{Fermi}} = \sum_i b_i^\dagger b_i. \tag{10.1.17}$$

10.2 Thermal States

Statistical mechanics makes use of three principal ensembles, the microcanonical, the canonical and the grand canonical. These almost always give results which are experimentally distinguishable, but the most convenient for theoretical treatment are the canonical and the grand canonical ensembles, and it is on these that we shall focus our attention.

10.2.1 Canonical Ensemble

In the canonical ensemble, the density operator is taken to be

$$\rho_{\text{can}} = \frac{1}{Z(T)} \exp\left(-\frac{H}{k_B T}\right), \tag{10.2.1}$$

where the partition function

$$Z(T) = \text{Tr}\left\{e^{-H/k_B T}\right\}, \tag{10.2.2}$$

ensures the density operator has unit trace.

a) Bosons: In the case that the Hamiltonian corresponds to that of an assembly of Bosons, that is, to (10.1.14), the density operator factorizes into the direct product

$$\rho_{\text{can}} = \prod_i \frac{1}{Z_i(T)} \exp\left(-\frac{\hbar\omega_i}{k_B T} a_i^\dagger a_i\right). \tag{10.2.3}$$

This form has two very important useful properties:

i) The factorizable form means the individual oscillator modes are statistically independent.

ii) The quadratic nature of the Hamiltonian means that the density operator has the quantum analogue of a *Gaussian* probability distribution.

b) Fermions: Even though the individual creation and annihilation operators anticommute, the number operators commute, so that the canonical density operator takes exactly the same form, but with $a_i \to b_i$, and with the consequent factorizability of the density operator. The density operator has the form of a Fermi analogue of a Gaussian probability distribution.

10.2.2 The Grand Canonical Ensemble

Bosons fall naturally into two categories, depending on whether the relevant interactions do or do not conserve the total number of quanta. For example, neither photons nor phonons are conserved, while Bosonic atoms such as ^{4}He are conserved (provided we are only interested in energies at which antiparticles are unlikely).

When dealing with an assembly of Fermions or conserved Bosons the total number operator also plays a role in the density operator, and this is known as the grand canonical ensemble, in which the density operator takes the form

$$\rho_{\text{gc}} = \frac{1}{Z(\mu, T)} \exp\left(-\frac{H - \mu N}{k_B T}\right), \tag{10.2.4}$$

where we now have the grand canonical partition function

$$Z(\mu, T) = \text{Tr}\left\{e^{-(H-\mu N)/k_B T}\right\}. \tag{10.2.5}$$

a) Matrix Elements and Normalization: We can now write the density operator explicitly in terms of creation and destruction operators.

i) *Bosons*: For Bose operators we have

$$\rho_{\text{gc}}^{\text{Bose}} = \prod_i \frac{1}{Z_i(\mu, T)} \exp\left(\frac{\mu - \hbar\omega_i}{k_B T} a_i^\dagger a_i\right). \tag{10.2.6}$$

This has exactly the same desirable factorizability and Gaussian properties as the canonical density operator. For compactness, let us introduce the notation

$$\lambda_i = \exp\left(\frac{\mu - \hbar\omega_i}{k_B T}\right). \tag{10.2.7}$$

In terms of this notation, the matrix elements of the density operator are

$$\langle \ldots m_i \ldots | \rho_{\text{gc}}^{\text{Bose}} | \ldots n_i \ldots \rangle = \mathcal{N} \prod_i \delta_{n_i m_i} \lambda_i^{n_i}, \quad n_i, m_i = 0, 1, 2, 3, \ldots, \tag{10.2.8}$$

and normalizing the density operator gives

$$\mathcal{N} = \frac{1}{Z(\mu, T)} = \prod_i (1 - \lambda_i). \tag{10.2.9}$$

ii) *Fermions*: In this case for Fermi operators we have

$$\rho_{\text{gc}}^{\text{Fermi}} = \prod_i \frac{1}{Z_i(\mu, T)} \exp\left(\frac{\mu - \hbar\omega_i}{k_B T} b_i^\dagger b_i\right). \tag{10.2.10}$$

In the case of Fermions, the density operator matrix elements become

$$\langle \ldots m_i \ldots | \rho_{\text{gc}}^{\text{Fermi}} | \ldots n_i \ldots \rangle = \mathcal{N} \prod_i \delta_{n_i m_i} \lambda_i^{n_i}, \quad n_i, m_i = 0, 1, \tag{10.2.11}$$

with λ_i still given by (10.2.7).

Normalizing the density operator gives

$$\mathcal{N} = \frac{1}{Z(\mu,T)} = \prod_i \frac{1}{1+\lambda_i}\,. \tag{10.2.12}$$

The density operator of the form (10.2.10) forms the prototype of the *Fermi–Gaussian state*.

b) Vacuum State: For the vacuum state, all of the occupations $n_i = 0$, which occurs when all of $\lambda_i \to 0$. The vacuum state is thus a limiting form of a quantum Gaussian state of both the Fermi and Bose kind.

10.3 Fluctuations in the Ideal Bose Gas

Since the grand canonical density operator factorizes into individual density operators for each mode, it is possible to consider the system in terms of individual modes. Therefore in this section, we will consider only a single mode of frequency ω, with

$$\lambda = \exp\left(\frac{\mu - \hbar\omega}{k_B T}\right), \tag{10.3.1}$$

and as above, the density operator matrix elements are

$$\langle n|\rho|m\rangle = \delta_{mn}(1-\lambda)\lambda^n. \tag{10.3.2}$$

10.3.1 Moments of the Number Operator

From the density operator (10.3.2), these moments are

$$\langle N^q \rangle = (1-\lambda)\left(\lambda\frac{d}{d\lambda}\right)^q \frac{1}{1-\lambda}\,. \tag{10.3.3}$$

a) Mean Number: Using this formula, the mean value of $N = a^\dagger a$ is

$$\langle N \rangle \equiv \bar{n} = \frac{\lambda}{1-\lambda} = \frac{1}{e^{(\hbar\omega-\mu)/k_B T} - 1}\,. \tag{10.3.4}$$

Corresponding to this we can express $\bar{n}$ in terms of λ as

$$\lambda = \frac{\bar{n}}{1+\bar{n}}\,. \tag{10.3.5}$$

b) Variance: The second moment is given by using (10.3.3) as

$$\langle N^2 \rangle = \frac{2\lambda^2}{(1-\lambda)^2} + \frac{\lambda}{1-\lambda} = 2\bar{n}^2 + \bar{n}\,. \tag{10.3.6}$$

The variance is therefore

$$\mathrm{var}[N] = \bar{n} + \bar{n}^2. \tag{10.3.7}$$

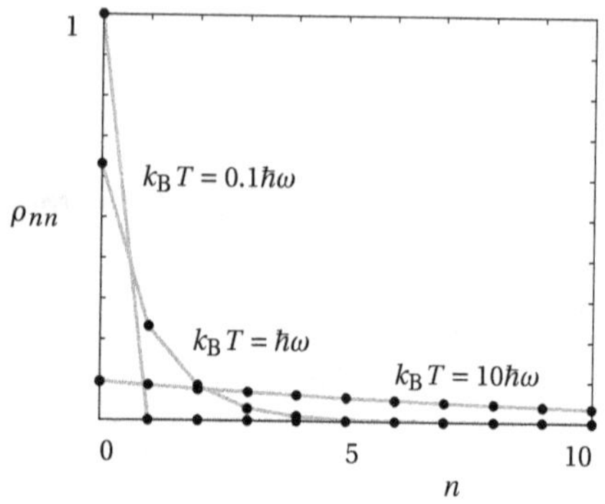

Fig. 10.1. Occupation probabilities ρ_{nn} for various temperatures, showing the broad, power-law distributions at higher temperatures.

This variance corresponds to a broad distribution, as expected from the power-law formula for the occupation probabilities given by the diagonal elements of (10.3.2), as shown in Fig. 10.1.

c) Normally Ordered Moments: These are defined as

$$\left\langle (a^\dagger)^p a^q \right\rangle = \mathrm{Tr}\left\{\rho_{gc}(a^\dagger)^p a^q\right\}, \tag{10.3.8}$$

$$= \delta_{p,q}(1-\lambda)\sum_n \lambda^n n(n-1)\ldots(n-q+1), \tag{10.3.9}$$

$$= \delta_{p,q}(1-\lambda)\lambda^q \frac{d^q}{d\lambda^q}\frac{1}{1-\lambda} \tag{10.3.10}$$

$$= \delta_{p,q}\, q!\left(\frac{\lambda}{1-\lambda}\right)^q \equiv q!\,\bar{n}^q\,\delta_{p,q}. \tag{10.3.11}$$

10.3.2 Many Modes

Suppose we have a many-mode system with operators a_i and corresponding frequencies ω_i. Thus we have

$$\langle a_i^\dagger a_j\rangle = \bar{n}_i\delta_{ij}. \tag{10.3.12}$$

The more interesting moment is

$$\langle a_i^\dagger a_j^\dagger a_k a_l\rangle = \delta_{ik}\delta_{jl}(1-\delta_{ij})\langle a_i^\dagger a_j^\dagger a_i a_j\rangle + \delta_{il}\delta_{jk}(1-\delta_{ij})\langle a_i^\dagger a_j^\dagger a_j a_i\rangle$$
$$+\delta_{ij}\delta_{ik}\delta_{il}\langle a_i^\dagger a_i^\dagger a_i a_i\rangle. \tag{10.3.13}$$

The reasoning that leads to this equation is as follows: Since the modes are independent, a non-zero result can only occur if every creation operator is matched by a destruction operator, which can happen in two ways if $i \neq j$, but only in one way when $i = j$. The two moments on the first line are then factorizable; $\langle a_i^\dagger a_j^\dagger a_i a_j\rangle = \bar{n}_i\bar{n}_j$, while we use the result (10.3.11) to get $\langle a_i^\dagger a_i^\dagger a_i a_i\rangle = 2\bar{n}_i^2$ for the final term. Putting these all together we get

$$\langle a_i^\dagger a_j^\dagger a_k a_l \rangle = \{\delta_{ik}\delta_{jl} + \delta_{il}\delta_{jk}\}\, \bar{n}_i \bar{n}_j. \tag{10.3.14}$$

10.4 Bosonic Quantum Gaussian Systems

A quantum Gaussian density operator is one which can be written in the form

$$\rho = \mathcal{N}\exp\left\{-\sum_{ij}\left(A_{ij}a_i^\dagger a_j + B_{ij}a_i^\dagger a_j^\dagger + B_{ij}^* a_i a_j\right) + \sum_i\left(c_i a_i^\dagger + c_i^* a_i\right)\right\}. \tag{10.4.1}$$

This provides a natural generalization of a classical Gaussian probability distribution. It is natural to diagonalize the density operator as follows:

i) The linear term can be transformed away by a c-number shift α_i

$$a_i \to a_i + \alpha_i. \tag{10.4.2}$$

ii) The quadratic form in the exponent, which must be positive definite of course, can be diagonalized by a linear transformation of the form

$$a_i \to \sum_{ij}\left(C_{ij}a_i + S_{ij}a_j^\dagger\right), \tag{10.4.3}$$

which is a generalized form of the *Bogoliubov* transformation, which is covered fully in *Book III.*

iii) These linear transformations can be implemented as unitary transformations, so that in the end, the density operator takes a diagonal form like

$$\rho_{\text{diag}} = \mathcal{N}\exp\left\{-\sum_i \kappa_i a_i^\dagger a_i\right\}, \tag{10.4.4}$$

where the eigenvalues $\kappa_i > 0$ characterize the density operator.

Exercise 10.1 Implementation of a Unitary Transformation: In the case that the linear transformation (10.4.3) is infinitesimally close to the identity, determine the unitary transformation which generates it.

10.4.1 Hartree–Fock Factorization

Suppose we have two operators defined as linear combinations of the a_i;

$$A \equiv \sum_i A_i a_i, \qquad B \equiv \sum_i B_i a_i, \tag{10.4.5}$$

where the A_i, B_i are c-number coefficients. If the density operator has the diagonal Gaussian form (10.4.4), from (10.3.14) it follows quite straightforwardly that

$$\langle A^\dagger B^\dagger B A \rangle = \langle A^\dagger A \rangle \langle B^\dagger B \rangle + \langle A^\dagger B \rangle \langle B^\dagger A \rangle. \tag{10.4.6}$$

This is known as the *Hartree–Fock* factorization.

The two terms are known as the *Hartree term*, $\langle A^\dagger A\rangle\langle B^\dagger B\rangle$, in which the operator is paired with its Hermitian conjugate, and the *Fock term*, $\langle A^\dagger B\rangle\langle B^\dagger A\rangle$, where the pairing is with the Hermitian conjugate of the other operator. In this context, two operators under consideration are usually field operators $\psi(\boldsymbol{x}_1)$, $\psi(\boldsymbol{x}_2)$, of the same field at two different positions.

10.4.2 Generalized Hartree–Fock Factorization for Quantum Gaussian Density Operators

Hartree–Fock factorization can be put in a form which is valid for any quantum Gaussian density operator and for operators expressed as arbitrary linear combinations of creation and destruction operators.

a) Diagonal Quantum Gaussian Density Operator: Using the same methods as in the previous subsection, it is straightforward to show for a diagonal quantum Gaussian density operator that

$$\langle a_i^\dagger a_j a_k^\dagger a_l\rangle = \delta_{ij}\delta_{kl}\bar{n}_i\bar{n}_k + \delta_{il}\delta_{jk}\bar{n}_i(\bar{n}_j+1), \tag{10.4.7}$$

$$= \langle a_i^\dagger a_j\rangle\langle a_k^\dagger a_l\rangle + \langle a_i^\dagger a_l\rangle\langle a_j a_k^\dagger\rangle. \tag{10.4.8}$$

The second form exhibits the general rule: If the density operator is thermal and A, B, C and D are linear combinations of creation and destruction operators, i.e., are of the form

$$A = \sum_i \left(A_i^{(-)} a_i + A_i^{(+)} a_i^\dagger\right) \text{ etc.,} \tag{10.4.9}$$

then Hartree–Fock factorization takes the form

$$\boxed{\langle ABCD\rangle = \langle AB\rangle\langle CD\rangle + \langle AC\rangle\langle BD\rangle + \langle AD\rangle\langle BC\rangle.} \tag{10.4.10}$$

This result is the sum of all possible factorizations into pair averages in which the order of the operators within the individual pair averages is the same as that in the four-operator average.

b) General Quantum Gaussian Density Operator: We will consider a quantum Gaussian density operator of the form (10.4.1), but with no linear component, that is, $c_i = 0$. We can diagonalize this density operator by an appropriate linear transformation of the form (10.4.3). In this case, if the operators A, B, C and D are of the form (10.4.9), this linear transformation will give a new expression of them in terms of the transformed creation and destruction operators. However, these will still be linear combinations of the form (10.4.10), so that Hartree–Fock factorization in the form (10.4.10) will be true.

c) Higher-Order Factorization—Wick's Theorem: A corresponding result is true for any even number of operators $A, B, C, D, E, \ldots$. This is often called *Wick's theorem*, although it is not quite in the form first enunciated by *Wick* [47].

It is important to remember that all of these results, including Wick's theorem, are true *only* for quantum Gaussian density operators, and that for most interacting systems such density operators are normally not valid. To use these results more generally one must make approximations.

10.5 Coherent States

The study of both coherent light and of ultra-cold Bose gases is necessarily the study of highly occupied states of Bose particles. The two most important kinds of such states are *number states* and *coherent states*. Number states are eigenstates of the number operator $N = a^\dagger a$. Since in quantum-optical situations, photons are not conserved, a photon number state is very hard to create and preserve, whereas a coherent state is much more robust.

In an ultra-cold gas, the fact that the total number of atoms in a gas is absolutely conserved might lead one think that number states would be the most relevant for their study. Surprisingly, this is not so, and the coherent state $|\alpha\rangle$, which satisfies the defining eigenvalue equation

$$a|\alpha\rangle = \alpha|\alpha\rangle, \tag{10.5.1}$$

and does not correspond to a definite number of particles, still provides a valuable tool for the study of ultra-cold gases.

10.5.1 Properties of the Coherent States

a) Expression in Terms of Number States: The only solution to the defining equation (10.5.1), which also satisfies $\langle\alpha|\alpha\rangle = 1$ is (up to a phase)

$$|\alpha\rangle = \exp\left(-\tfrac{1}{2}|\alpha|^2\right)\sum_{n=0}^{\infty}\frac{\alpha^n}{\sqrt{n!}}|n\rangle. \tag{10.5.2}$$

b) Action of Creation and Destruction Operators: Using $a|n\rangle = \sqrt{n}|n-1\rangle$ and $a^\dagger|n\rangle = \sqrt{n+1}|n+1\rangle$, it is straightforward to show that

$$\left.\begin{aligned}
a|\alpha\rangle &= \alpha|\alpha\rangle,\\
a^\dagger|\alpha\rangle &= \left(\alpha^* + \tfrac{1}{2}\frac{\partial}{\partial\alpha}\right)|\alpha\rangle,\\
\langle\alpha|a^\dagger &= \langle\alpha|\alpha^*,\\
\langle\alpha|a &= \left(\alpha + \tfrac{1}{2}\frac{\partial}{\partial\alpha^*}\right)\langle\alpha|.
\end{aligned}\right\} \tag{10.5.3}$$

c) Unitary Transformation of the Vacuum: We can also show that

$$|\alpha\rangle = \exp\left(\alpha a^\dagger - \alpha^* a\right)|0\rangle. \tag{10.5.4}$$

This involves the use of the Baker–Hausdorff formula: For any two operators A and B, such that $[A,B]$ commutes with both of them, one can write

$$\exp(A+B) = \exp(A)\exp(B)\exp\left(-\tfrac{1}{2}[A,B]\right), \tag{10.5.5}$$
$$= \exp(B)\exp(A)\exp\left(\tfrac{1}{2}[A,B]\right), \tag{10.5.6}$$

and this is proved in *Quantum Noise.* Using this identity, we see that (10.5.4) gives

$$|\alpha\rangle = \exp\left(-\tfrac{1}{2}|\alpha|^2\right)\exp\left(\alpha a^\dagger\right)\exp\left(-\alpha^* a\right)|0\rangle, \tag{10.5.7}$$

and noting that $a|0\rangle = 0$, we see

$$|\alpha\rangle = \exp\left(-\tfrac{1}{2}|\alpha|^2\right)\sum\frac{\alpha^n(a^\dagger)^n}{n!}|0\rangle, \tag{10.5.8}$$

which yields (10.5.2) when we use the expression

$$|n\rangle \equiv \frac{a^{\dagger n}}{\sqrt{n!}}|0\rangle. \tag{10.5.9}$$

d) Scalar Product:

$$\langle\alpha|\beta\rangle = \exp\left(\alpha^*\beta - \tfrac{1}{2}|\alpha|^2 - \tfrac{1}{2}|\beta|^2\right), \tag{10.5.10}$$
$$|\langle\alpha|\beta\rangle|^2 = \exp\left(-|\alpha-\beta|^2\right). \tag{10.5.11}$$

Notice that no two coherent states are actually orthogonal to each other. However, if α and β are significantly different from each other, the two states are almost orthogonal.

e) Completeness Formula:

$$1 = \frac{1}{\pi}\int d^2\alpha |\alpha\rangle\langle\alpha|. \tag{10.5.12}$$

Here,

$$\alpha = \alpha_x + \mathrm{i}\alpha_y, \qquad d^2\alpha = d\alpha_x d\alpha_y, \tag{10.5.13}$$

and the integral is over the whole complex plane.

Exercise 10.2 Trace in Terms of Coherent States: Show that the resolution of the identity (10.5.12) implies that

$$\mathrm{Tr}\{A\} = \frac{1}{\pi}\int d^2\alpha\,\langle\alpha|A|\alpha\rangle. \tag{10.5.14}$$

f) Normal Products: In evaluating matrix elements, *normal products* of operators in which all destruction operators stand to the right of creation operators, are useful. Thus,

$$\langle\alpha|a^\dagger a a^\dagger|\beta\rangle = \langle\alpha|a^\dagger a^\dagger a + a^\dagger[a, a^\dagger]|\beta\rangle, \tag{10.5.15}$$

$$= \langle\alpha|a^\dagger a^\dagger a + a^\dagger|\beta\rangle, \tag{10.5.16}$$

$$= (\alpha^{*2}\beta + \alpha^*)\langle\alpha|\beta\rangle. \tag{10.5.17}$$

The symbol :: around an expression means that it is to be considered a normal product; thus,

$$:(a+a^\dagger)(a+a^\dagger):= a^{\dagger 2} + a^2 + 2a^\dagger a. \tag{10.5.18}$$

From (10.5.1) it follows that the matrix element between coherent states $\langle\alpha|$ and $|\beta\rangle$ of any normally ordered function $F(a^\dagger, a)$ of creation and destruction operators is given by $F(\alpha^*, \beta)$. Thus, for example,

$$\langle\alpha| : (a+a^\dagger)^3 : |\beta\rangle = (\beta + \alpha^*)^3. \tag{10.5.19}$$

g) Poissonian Number Distribution of Coherent States: The state $|n\rangle$ is known as an *n-quantum state* and the probability of observing n quanta in a coherent state $|\alpha\rangle$ is

$$P_\alpha(n) = |\langle n|\alpha\rangle|^2 = \left|\exp(-\tfrac{1}{2}\alpha^2)\frac{\alpha^n}{\sqrt{n!}}\right|^2 = \frac{e^{-|\alpha|^2}|\alpha|^{2n}}{n!}, \tag{10.5.20}$$

which is a Poisson distribution with mean $|\alpha|^2$. Since the number n corresponds to the eigenvalue of the number operator N, we have

$$\langle N\rangle = \langle\alpha|N|\alpha\rangle = \sum_n nP(n) = |\alpha|^2, \tag{10.5.21}$$

$$\langle N^2\rangle = \langle\alpha|a^\dagger a a^\dagger a|\alpha\rangle = \langle\alpha|a^\dagger a^\dagger a a + a^\dagger a|\alpha\rangle = |\alpha|^4 + |\alpha|^2. \tag{10.5.22}$$

Hence,

$$\mathrm{var}\,[N] = |\alpha|^2 = \langle N\rangle, \tag{10.5.23}$$

as required for a Poisson.

Exercise 10.3 Action of $\exp(\lambda a^\dagger a)$ on a Coherent State: If $\lambda = \gamma + \mathrm{i}\nu$, show that the definition (10.5.2) of the coherent state implies that

$$\exp(\lambda a^\dagger a)|\alpha\rangle = \exp\left(\tfrac{1}{2}|\alpha|^2(e^{2\gamma} - 1)\right)\left|\alpha e^\lambda\right\rangle. \tag{10.5.24}$$

10.5.2 The Harmonic Oscillator

The harmonic oscillator Hamiltonian can be written

$$H_{\rm HO} = \hbar\omega\left(a^\dagger a + \tfrac{1}{2}\right), \tag{10.5.25}$$

and the result of Ex. 10.3 shows that the coherent state wavefunction

$$|\psi_c, t\rangle \equiv \exp(-\mathrm{i}H_{\rm HO}t/\hbar)|\alpha_0\rangle, \tag{10.5.26}$$

$$= \mathrm{e}^{-\mathrm{i}\omega t/2}\left|\mathrm{e}^{-\mathrm{i}\omega t}\alpha_0\right\rangle, \tag{10.5.27}$$

is a solution of the equations of motion.

Let us consider the situation when we drive this with an external field. This can be done by the Hamiltonian

$$H_{\rm Driven} = \hbar\left\{\omega\left(a^\dagger a + \tfrac{1}{2}\right) + g^*(t)a + g(t)a^\dagger\right\}. \tag{10.5.28}$$

The Heisenberg equation of motion for $a(t)$ is

$$\frac{da(t)}{dt} = \frac{\mathrm{i}}{\hbar}[H_{\rm Driven}, a(t)] \tag{10.5.29}$$

$$= -\mathrm{i}\omega a(t) - \mathrm{i}g(t), \tag{10.5.30}$$

which has the solution

$$a(t) = \mathrm{e}^{-\mathrm{i}\omega t}a(0) - \mathrm{i}\int_0^t dt'\,\mathrm{e}^{-\mathrm{i}\omega(t-t')}g(t'), \tag{10.5.31}$$

$$\equiv \mathrm{e}^{-\mathrm{i}\omega t}a(0) + \alpha(t). \tag{10.5.32}$$

a) Initial Vacuum State: If the initial state of the system corresponds to $|\mathrm{vac}, 0\rangle$, the vacuum of $a(0)$, then initially

$$a(0)|\mathrm{vac}, 0\rangle = 0. \tag{10.5.33}$$

Thus, at the time t we have

$$a(t)|\mathrm{vac}, 0\rangle = \alpha(t)|\mathrm{vac}, 0\rangle. \tag{10.5.34}$$

This means that this state is equivalent to the coherent state of argument $\alpha(t)$ with respect to the the Heisenberg operator at time t, that is

$$|\mathrm{vac}, 0\rangle \equiv |\alpha(t), t\rangle. \tag{10.5.35}$$

Exercise 10.4 Schrödinger Picture: Show that it follows from the Heisenberg picture result (10.5.34), that in the Schrödinger picture the states are described in terms of coherent states of the time-independent operators $a, a^\dagger$, and the solution of the Schrödinger equation

$$\mathrm{i}\hbar\frac{d|\psi, t\rangle}{dt} = H_{\rm Driven}|\psi, t\rangle, \tag{10.5.36}$$

is the coherent state $|\alpha(t)\rangle$.

b) Initial Coherent State: If the Schrödinger picture initial state is the coherent state $|\alpha_0\rangle$, then it is straightforward to show using (10.5.32) that the solution of the Schrödinger equation is

$$|\psi, t\rangle = \left|\alpha_0 e^{-i\omega t} + \alpha(t)\right\rangle. \tag{10.5.37}$$

10.6 Fluctuations in Systems of Fermions

If we take instead Fermi particles the results are, in a formal mathematical sense, surprisingly similar to those for Bosons. Of course the physical consequences of the differences that are in fact present, are profound.

10.6.1 The Single-Mode System

a) Moments: As in (10.3.4), we find

$$\bar{n} = \frac{\lambda}{1+\lambda} = \frac{1}{e^{\beta(\hbar\omega-\mu)}+1}. \tag{10.6.1}$$

Also, for consistency, we note that

$$\langle b^\dagger b^\dagger b b\rangle = \frac{1}{1+\lambda}\lambda^2 \frac{d^2}{d\lambda^2}(1+\lambda) = 0, \tag{10.6.2}$$

a result which is obvious from the anticommutation relations.

b) Variance: Using the anticommutation relations it follows that

$$0 = \langle b^\dagger b^\dagger b b\rangle = -\langle b^\dagger b b^\dagger b\rangle + \langle b^\dagger b\rangle, \tag{10.6.3}$$

so that $\langle N^2\rangle = \bar{n}$ and thus

$$\mathrm{var}[N] = \bar{n} - \bar{n}^2, \tag{10.6.4}$$

which differs from (10.3.7), the corresponding result for Bosons, only by the sign of the term in $\bar{n}^2$.

10.6.2 Fermi–Gaussian Systems

We define a Fermi–Gaussian density operator in a similar way to that for Bosons, except that there are no Fermi mean fields. A Fermi–Gaussian density operator therefore has the form

$$\rho = \mathcal{N}\exp\left\{-\sum_{ij}\left(A_{ij} b_i^\dagger b_j + B_{ij} b_i^\dagger b_j^\dagger + B_{ij}^* b_i b_j\right)\right\}. \tag{10.6.5}$$

a) Diagonal Fermi–Gaussian Density Operator: This takes the form

$$\rho_{\mathrm{diag}} = \mathcal{N}\exp\left\{-\sum_i \kappa_i b_i^\dagger b_i\right\}, \tag{10.6.6}$$

where the eigenvalues $\kappa_i > 0$ characterize the density operator.

b) Fermi–Gaussian Factorization: When we evaluate $\langle b_i^\dagger b_j^\dagger b_k b_l\rangle$, we note that when $i = j$ this vanishes, so that we can write

$$\langle b_i^\dagger b_j^\dagger b_k b_l\rangle = \left(1-\delta_{ij}\right)\left\{\delta_{ik}\delta_{jl}\langle b_i^\dagger b_j^\dagger b_i b_j\rangle + \delta_{il}\delta_{jk}\langle b_i^\dagger b_j^\dagger b_j b_i\rangle\right\} \tag{10.6.7}$$

$$= \left(1-\delta_{ij}\right)\left\{-\delta_{ik}\delta_{jl}\langle b_i^\dagger b_i b_j^\dagger b_j\rangle + \delta_{il}\delta_{jk}\langle b_i^\dagger b_i b_j^\dagger b_j\rangle\right\}. \tag{10.6.8}$$

Noting $b_i^\dagger b_i = N_i$, which commutes with N_j, we see that the final result is

$$\langle b_i^\dagger b_j^\dagger b_k b_l\rangle = \{\delta_{il}\delta_{jk} - \delta_{ik}\delta_{jl}\}\,\bar{n}_i\bar{n}_j. \tag{10.6.9}$$

We have been able to drop the factor $\left(1-\delta_{ij}\right)$, since its coefficient vanishes when $i = j$. Again, this result differs from the Bose result (10.3.14) only by a sign in one term.

10.6.3 Hartree–Fock Factorization in Systems of Fermions

The essential results for the development of Hartree–Fock factorization can be derived from the results for a diagonal Fermi–Gaussian operator in much the same way as the corresponding rules are derived for Bosons described by a diagonal quantum Gaussian density operator. The differences arise from the use of anti-commutation relations rather than commutation relations.

The result is that for Fermions the factorization rule is very similar to that for Bosons: If A, B, C and D are linear combinations of Fermi creation and destruction operators, then Hartree–Fock factorization takes the form

$$\langle ABCD\rangle = \langle AB\rangle\langle CD\rangle - \langle AC\rangle\langle BD\rangle + \langle AD\rangle\langle BC\rangle. \tag{10.6.10}$$

The sign of each term on the right-hand side is determined by the signature of the permutation required to take the order $ABCD$ of the operators to that of the order of the operators written in the particular term. The generalization to higher-order moments is straightforward.

10.7 Two-Level Systems and Pauli Matrices

Another building block of quantum systems is provided by the two-level system, which considers the most elementary kind of system which allows transition between energy levels. There is an obvious analogy to a single Fermionic system, which also has two levels, corresponding to the absence or presence of a Fermion. Systems of several two-level systems often occur, as for example a model of an assembly of atoms, but these differ from assemblies of Fermions, since the anti-commutation relations between different Fermions are not appropriate for different atoms.

A two-level system is described by a wavefunction which can be written as a two-component matrix

$$u = \begin{pmatrix} u^{(+)} \\ u^{(-)} \end{pmatrix}, \tag{10.7.1}$$

and all operators can be expressed in terms of the Pauli matrices

$$\sigma_x = \begin{pmatrix} 0 & 1 \\ 1 & 0 \end{pmatrix}, \qquad \sigma_y = \begin{pmatrix} 0 & -\mathrm{i} \\ \mathrm{i} & 0 \end{pmatrix}, \qquad \sigma_z = \begin{pmatrix} 1 & 0 \\ 0 & -1 \end{pmatrix}, \tag{10.7.2}$$

and the identity matrix. The Hamiltonian for such a two-level system then can always be written

$$H = \tfrac{1}{2}\hbar\Omega \boldsymbol{\sigma}\cdot\boldsymbol{n}, \tag{10.7.3}$$

where $\boldsymbol{n}$ is a unit vector—if this vector is along the z axis, then

$$H = \tfrac{1}{2}\hbar\Omega\sigma_z = \begin{pmatrix} \frac{1}{2}\hbar\Omega & 0 \\ 0 & -\frac{1}{2}\hbar\Omega \end{pmatrix}. \tag{10.7.4}$$

10.7.1 Pauli Matrix Properties

i) $$\sigma_i\sigma_j = \delta_{ij} + \mathrm{i}\sum_k \epsilon_{ijk}\sigma_k. \tag{10.7.5}$$

ii) $$(\boldsymbol{\sigma}\cdot\boldsymbol{A})(\boldsymbol{\sigma}\cdot\boldsymbol{B}) = \boldsymbol{A}\cdot\boldsymbol{B} + \mathrm{i}\boldsymbol{\sigma}\cdot\boldsymbol{A}\times\boldsymbol{B}, \tag{10.7.6}$$
where $\boldsymbol{A}$ and $\boldsymbol{B}$ are vectors (either c-number or operator) which commute with the Pauli matrices.

iii) $$[\sigma_i, \sigma_j] = 2\mathrm{i}\sum_k \epsilon_{ijk}\sigma_k. \tag{10.7.7}$$

iv) $$[\sigma_i, \sigma_j]_+ = 2\delta_{ij}. \tag{10.7.8}$$

v) If G is any two-dimensional matrix
$$G = \tfrac{1}{2}\left\{ \mathbf{1}\,\mathrm{Tr}\,\{G\} + \sum_i \sigma_i \mathrm{Tr}\,\{\sigma_i G\} \right\}. \tag{10.7.9}$$

vi) $$\exp\{\mathrm{i}\theta\boldsymbol{\sigma}\cdot\boldsymbol{n}\} = \cos\theta + \mathrm{i}\boldsymbol{\sigma}\cdot\boldsymbol{n}\sin\theta. \tag{10.7.10}$$

vii) If
$$\sigma^+ \equiv \tfrac{1}{2}(\sigma_x + \mathrm{i}\sigma_y) = \begin{pmatrix} 0 & 1 \\ 0 & 0 \end{pmatrix}, \tag{10.7.11}$$

$$\sigma^- \equiv \tfrac{1}{2}(\sigma_x - \mathrm{i}\sigma_y) = \begin{pmatrix} 0 & 0 \\ 1 & 0 \end{pmatrix}, \tag{10.7.12}$$

then

$$(\sigma^+)^2 = (\sigma^-)^2 = 0, \quad \sigma_z^2 = 1, \tag{10.7.13}$$

$$\sigma^+\sigma^- = \tfrac{1}{2}(1+\sigma_z), \tag{10.7.14}$$

$$\sigma^-\sigma^+ = \tfrac{1}{2}(1-\sigma_z), \tag{10.7.15}$$

$$\sigma_z\sigma^+ = -\sigma^+\sigma_z = \sigma^+, \tag{10.7.16}$$

$$\sigma_z\sigma^- = -\sigma^-\sigma_z = -\sigma^-. \tag{10.7.17}$$

viii) For any matrix G

$$G = \tfrac{1}{2}\mathbf{1}\,\mathrm{Tr}\,\{G\} + \tfrac{1}{2}\sigma_z \mathrm{Tr}\,\{\sigma_z G\} + \sigma^+ \mathrm{Tr}\,\{\sigma^- G\} + \sigma^- \mathrm{Tr}\,\{\sigma^+ G\}. \tag{10.7.18}$$

ix)

$$\exp(a + \boldsymbol{b}\cdot\boldsymbol{\sigma}) = \mathrm{e}^a \left\{ \cosh|\boldsymbol{b}| + \frac{\boldsymbol{\sigma}\cdot\boldsymbol{b}}{|\boldsymbol{b}|} \sinh|\boldsymbol{b}| \right\}, \tag{10.7.19}$$

where a and $\boldsymbol{b}$ are c-numbers.

11. Quantum Fields

The creation and destruction operators introduced in the previous chapter almost always occur in the context of quantum fields, even though these fields may have quite variable interpretations. For example, the electromagnetic field can be quantized by use of the canonical quantization picture, and this leads naturally to the interpretation of the quantized amplitude of a field mode in terms of a destruction or creation operator. On the other hand, the description of conserved particles (such as atoms) as Bosons or Fermions leads to the construction of a corresponding quantized matter wave field operator, which obeys the Schrödinger equation. The result is that matter-wave duality becomes universally enmeshed with the concept of field quantization.

11.1 Kinds of Quantum Field

Any field, either classical and quantum, can be expressed as a linear combination of linearly independent *mode functions*. These mode functions can in principle be quite arbitrary, but the most useful modes are the *plane wave* modes. In terms of these plane wave modes, the universal concepts shared by all quantized fields are:

i) The association of each mode, labelled by an index i, with a wavevector $\boldsymbol{k}_i$.

ii) The mode function for each $\boldsymbol{k}_i$.

iii) The expression of the field operator in terms of the creation and destruction operators, and the mode functions.

iv) The choice of Hamiltonian.

In the following we shall formulate the quantization procedure for some of the most common quantum fields. For all of these, we will use box normalization, in which the fields are supposed to be confined within a box of volume V, which is considered to be very large.

11.1.1 Matter Wave Fields

We will consider a spinless particle of mass m, whose modes can be completely characterized by the momentum vector $\hbar\boldsymbol{k}$. Thus for the matter wave field, the mode functions are

$$u_{\boldsymbol{k}}(\boldsymbol{x}) = \frac{1}{\sqrt{V}} \mathrm{e}^{\mathrm{i}\boldsymbol{k}\cdot\boldsymbol{x}}. \tag{11.1.1}$$

a) Bosons: These are described in terms of a field operator in the Schrödinger picture, which we will write in the form

$$\psi(\boldsymbol{x}) = \sum_{\boldsymbol{k}} a_{\boldsymbol{k}} u_{\boldsymbol{k}}(\boldsymbol{x}). \tag{11.1.2}$$

b) Commutation Relations for Boson Fields: As a result of the creation and destruction operator commutation relations (as in (10.1.1)), we can derive the field operator commutation relations

$$[\psi(\boldsymbol{x}), \psi^\dagger(\boldsymbol{x}')] = \delta(\boldsymbol{x}-\boldsymbol{x}'), \tag{11.1.3}$$

$$[\psi(\boldsymbol{x}), \psi(\boldsymbol{x}')] = [\psi^\dagger(\boldsymbol{x}), \psi^\dagger(\boldsymbol{x}')] = 0. \tag{11.1.4}$$

These can be interpreted as meaning that the field operator $\psi^\dagger(\boldsymbol{x})$ creates a Bose particle at the point $\boldsymbol{x}$.

c) Hamiltonian: We will use the notation for the mode frequency

$$\hbar\omega_{\boldsymbol{k}} \equiv \frac{\hbar^2 \boldsymbol{k}^2}{2m}, \tag{11.1.5}$$

and the Hamiltonian becomes

$$H_{\text{matter}} = \sum_{\boldsymbol{k}} \hbar\omega_{\boldsymbol{k}} a^\dagger_{\boldsymbol{k}} a_{\boldsymbol{k}}, \tag{11.1.6}$$

$$= \int d^3\boldsymbol{x}\, \psi^\dagger(\boldsymbol{x}) \left(-\frac{\hbar^2\nabla^2}{2m}\right) \psi(\boldsymbol{x}). \tag{11.1.7}$$

d) Field Operators in the Heisenberg Picture: In the Heisenberg picture the field operator takes the form

$$\psi(\boldsymbol{x}, t) = \sum_{\boldsymbol{k}} a_{\boldsymbol{k}} u_{\boldsymbol{k}}(\boldsymbol{x}) \mathrm{e}^{-\mathrm{i}\omega_{\boldsymbol{k}} t}. \tag{11.1.8}$$

In this equation, the $a_{\boldsymbol{k}}$ are the Schrödinger picture operators. The time dependence is entirely represented in the arguments of the exponentials.

e) Equation of Motion: Using the commutation relations (11.1.3, 11.1.4), one can derive the equation of motion for the quantum field, which is identical in form to the free particle Schrödinger equation:

$$\mathrm{i}\hbar \frac{\partial\psi(\boldsymbol{x}, t)}{\partial t} = -\frac{\hbar^2\nabla^2\psi(\boldsymbol{x}, t)}{2m}. \tag{11.1.9}$$

f) Fermions: The expression of Fermi fields in terms of mode functions is exactly the same as for Bose fields, apart from the operator substitution $a_{\boldsymbol{k}} \to b_{\boldsymbol{k}}$ where $b_{\boldsymbol{k}}$ are Fermi operators. These satisfy, as in (10.1.8), *anticommutation* relations, which lead to the field anticommutation relations

$$[\psi(\boldsymbol{x}), \psi^\dagger(\boldsymbol{x}')]_+ = \delta(\boldsymbol{x}-\boldsymbol{x}'), \tag{11.1.10}$$

$$[\psi(\boldsymbol{x}), \psi(\boldsymbol{x}')]_+ = [\psi^\dagger(\boldsymbol{x}), \psi^\dagger(\boldsymbol{x}')]_+ = 0. \tag{11.1.11}$$

The field operator $\psi^\dagger(\boldsymbol{x})$ can be taken as the creation operator of a Fermi particle at $\boldsymbol{x}$. Even though the commutation relations have been replaced by anticommutation relations, the Hamiltonian and equations of motion take exactly the same form as for the Boson field, namely (11.1.7) and (11.1.9).

11.1.2 Sound Waves

Sound waves occur in a range of media—solid, liquid or gaseous—so here we will simply give a generic and simplified description of their formulation and quantization. We introduce a variable $f(\boldsymbol{x},t)$ which represents the deviation of the medium from its equilibrium state. For example, for sound waves in an elastic solid, this is a measure of the strain, while for a gas it might be the deviation of the pressure or the density from equilibrium. We take the resulting potential and kinetic energy in the forms

$$E_{\rm pot} \equiv \frac{1}{2\kappa}\int d^3\boldsymbol{x}\,(\nabla f)^2, \tag{11.1.12}$$

$$E_{\rm kin} \equiv \frac{\lambda}{2}\int d^3\boldsymbol{x}\left(\frac{\partial f}{\partial t}\right)^2. \tag{11.1.13}$$

Here we have introduced κ, which is a measure of elasticity, and λ, which is a measure of inertia.

The wave equation resulting from these can be deduced using Lagrange's equations, and is

$$\frac{\partial^2 f}{\partial t^2} = c_{\rm s}^2\nabla^2 f, \tag{11.1.14}$$

with the speed of sound $c_{\rm s}$ being given by

$$c_{\rm s} = \frac{1}{\sqrt{\kappa\lambda}}. \tag{11.1.15}$$

a) Mode Functions: We now introduce mode functions and frequencies

$$u_{\boldsymbol{k}}(\boldsymbol{x}) \equiv \frac{1}{\sqrt{V}}\,\mathrm{e}^{i\boldsymbol{k}\cdot\boldsymbol{x}}, \tag{11.1.16}$$

$$\omega_{\boldsymbol{k}} = |\boldsymbol{k}|c_{\rm s}, \tag{11.1.17}$$

and write the operator expansion

$$f(\boldsymbol{x},t) = f^{(+)}(\boldsymbol{x},t) + f^{(-)}(\boldsymbol{x},t), \tag{11.1.18}$$

where $f^{(+)}(\boldsymbol{x},t)$ is called the *positive frequency* part of the sound field, and the corresponding *negative frequency* part is $f^{(-)}(\boldsymbol{x},t)$. These are defined by

$$f^{(+)}(\boldsymbol{x},t) \equiv \sum_{\boldsymbol{k}}\sqrt{\frac{\hbar}{2\lambda\omega_{\boldsymbol{k}}}}\,a_{\boldsymbol{k}}u_{\boldsymbol{k}}(\boldsymbol{x})\mathrm{e}^{-i\omega_{\boldsymbol{k}}t}, \tag{11.1.19}$$

$$f^{(-)}(\boldsymbol{x},t) \equiv \sum_{\boldsymbol{k}}\sqrt{\frac{\hbar}{2\lambda\omega_{\boldsymbol{k}}}}\,a_{\boldsymbol{k}}^\dagger u^*(\boldsymbol{x})_{\boldsymbol{k}}\mathrm{e}^{i\omega_{\boldsymbol{k}}t}. \tag{11.1.20}$$

By construction, the frequency-wavenumber relation (11.1.17) assures that $f(\boldsymbol{x},t)$ and $f^{(\pm)}(\boldsymbol{x},t)$ are all solutions of the wave equation (11.1.14). The wave equation is to be interpreted as a Heisenberg equation of motion for the field operator $f(\boldsymbol{x},t)$.

b) Commutation Relations: We interpret $a_{\boldsymbol{k}}^{\dagger}$ and $a_{\boldsymbol{k}}$ as creation and destruction operators, and introduce the commutation relations

$$\left[a_{\boldsymbol{k}}, a_{\boldsymbol{k}'}^{\dagger}\right] = \delta_{\boldsymbol{k},\boldsymbol{k}'}. \tag{11.1.21}$$

From these we can deduce the field commutation relations

$$\left[\frac{\partial f(\boldsymbol{x},t)}{\partial t}, f(\boldsymbol{x}',t)\right] = \frac{i\hbar}{\lambda}\delta(\boldsymbol{x}-\boldsymbol{x}'). \tag{11.1.22}$$

These commutation relations can also be obtained using an appropriate Lagrangian to derive the canonical co-ordinates and momenta, which are then quantized by imposing canonical commutation relations.

c) Hamiltonian: The Hamiltonian is obtained from the kinetic and potential energies (11.1.12, 11.1.13), and takes the form

$$H_{\text{sound}} = \tfrac{1}{2}\int d^3\boldsymbol{x}\left\{\frac{(\nabla f)^2}{\kappa} + \lambda\left(\frac{\partial f}{\partial t}\right)^2\right\}, \tag{11.1.23}$$

$$= \sum_{\boldsymbol{k}} \hbar\omega_{\boldsymbol{k}}\left(a_{\boldsymbol{k}}^{\dagger}a_{\boldsymbol{k}} + \tfrac{1}{2}\right). \tag{11.1.24}$$

d) Equation of Motion: Using the Hamiltonian and the commutation relations, the Heisenberg equations of motion can be derived for both $f(\boldsymbol{x},t)$ and its time derivative $\partial f(\boldsymbol{x},t)/\partial t$, which behaves as the canonical momentum conjugate to $f(\boldsymbol{x},t)$, because of the commutation relation (11.1.22) between them. These yield (11.1.14), the sound wave equation.

e) Comparison to Matter Wave Field Operators: The sound field $f(\boldsymbol{x},t)$ contains the creation and destruction operators in equal proportions, while the matter wave fields $\psi(\boldsymbol{x},t)$, $\psi^{\dagger}(\boldsymbol{x},t)$ consists only of either destruction or creation operators. The explicit inclusion of the factor $1/\sqrt{\omega_{\boldsymbol{k}}}$ means that there is no simple commutation relation between $f^{(+)}$ and $f^{(-)}$. However, this factor is also the reason that $f(\boldsymbol{x},t)$, and $\partial f(\boldsymbol{x},t)/\partial t$ obey the canonical commutation relation (11.1.22), since it cancels with the factor of $\omega_{\boldsymbol{k}}$ which arises from the time derivative.

11.1.3 The Electromagnetic Field

The electromagnetic field equations in free space (that is, with no charge or current sources, and no dielectric or permeable materials) are the quantized versions of Maxwell's equations. Their quantization is similar to that for sound waves, but is considerably more technically complex because of the vector nature of the fields involved.

a) Heisenberg Equations of Motion: In this case it is simpler to work in the Heisenberg picture, and for the electromagnetic field operators the Heisenberg equations of motion are Maxwell's equations

$$\nabla\cdot\boldsymbol{D} = 0, \tag{11.1.25}$$

$$\nabla\cdot\boldsymbol{B} = 0, \tag{11.1.26}$$

$$\nabla\times\boldsymbol{E} = -\frac{\partial\boldsymbol{B}}{\partial t}, \tag{11.1.27}$$

$$\nabla\times\boldsymbol{H} = \frac{\partial\boldsymbol{D}}{\partial t}, \tag{11.1.28}$$

where $\boldsymbol{B}=\mu_0\boldsymbol{H}, \boldsymbol{D}=\varepsilon_0\boldsymbol{E}$, where μ_0, ε_0 are the magnetic permeability and electric permittivity of free space and $\mu_0\varepsilon_0=c^{-2}$.

b) Potentials: The electromagnetic field is best represented in terms of the vector and scalar potentials $\boldsymbol{A}$ and ϕ, in terms of which

$$\boldsymbol{B} = \nabla\times\boldsymbol{A}, \tag{11.1.29}$$

$$\boldsymbol{E} = -\nabla\phi-\frac{\partial\boldsymbol{A}}{\partial t}. \tag{11.1.30}$$

However there is no unique set of $\boldsymbol{A}$ and ϕ which specify a given $\boldsymbol{B}$ and $\boldsymbol{E}$ since a *gauge transformation*

$$\boldsymbol{A}' = \boldsymbol{A}+\nabla\chi, \tag{11.1.31}$$

$$\phi' = \phi-\frac{\partial\chi}{\partial t}, \tag{11.1.32}$$

does not change the measurable fields $\boldsymbol{B}$ and $\boldsymbol{E}$.

c) Coulomb Gauge: For the purposes of optics the *Coulomb gauge* is convenient; this is defined by the choice of a *time-independent* ϕ, and a transverse vector potential, thus

$$\phi(\boldsymbol{x},t) \to \phi(\boldsymbol{x}), \tag{11.1.33}$$

$$\nabla\cdot\boldsymbol{A}(\boldsymbol{x},t) = 0. \tag{11.1.34}$$

Thus, we can write

$$\boldsymbol{B}(\boldsymbol{x},t) = \nabla\times\boldsymbol{A}(\boldsymbol{x},t), \tag{11.1.35}$$

$$\boldsymbol{E}(\boldsymbol{x},t) = -\nabla\phi(\boldsymbol{x})-\frac{\partial\boldsymbol{A}(\boldsymbol{x},t)}{\partial t}, \tag{11.1.36}$$

with the transversality or Coulomb gauge condition (11.1.34).

This particular choice of gauge is not the only one—relativistically invariant choices are also possible, and are much more appropriate for more advanced work.

d) Wave Equation: The expansion in modes is essentially the same kind of expansion used for the one component sound field of Sect. 11.1.2, with some complications introduced by the vector nature of the fields, and the transversality condition. Substituting (11.1.35) in (11.1.27) we find that $\boldsymbol{A}(\boldsymbol{x},t)$ satisfies the wave equation

$$\nabla^2 \boldsymbol{A}(\boldsymbol{x},t) = \frac{1}{c^2}\frac{\partial^2 \boldsymbol{A}(\boldsymbol{x},t)}{\partial t^2}. \tag{11.1.37}$$

Maxwell's equations, and the wave equation (11.1.37) for the the vector potential, will be the Heisenberg equations of motion which should arise from the correct choice of Hamiltonian operator. The electrostatic potential $\phi(\boldsymbol{x})$ on the other hand remains as an unquantized c-number.

e) Expansion in Mode Functions: As for sound waves, we separate the vector potential into positive and negative frequency terms

$$\boldsymbol{A}(\boldsymbol{x},t) = \boldsymbol{A}^{(+)}(\boldsymbol{x},t) + \boldsymbol{A}^{(-)}(\boldsymbol{x},t). \tag{11.1.38}$$

Here $\boldsymbol{A}^{(+)}(\boldsymbol{x},t)$ contains only Fourier components with positive frequency, i.e., only terms which vary as $\mathrm{e}^{-\mathrm{i}\omega t}$ for $\omega > 0$, and $\boldsymbol{A}^{(-)}(\boldsymbol{x},t)$ contains amplitudes which vary as $\mathrm{e}^{\mathrm{i}\omega t}$. We take $\boldsymbol{A}$ to be Hermitian so $\boldsymbol{A}^{(-)}(\boldsymbol{x},t) = \{\boldsymbol{A}^{(+)}(\boldsymbol{x},t)\}^{\dagger}$.

f) Mode Functions: The positive frequency part of vector potential is expanded in terms of the discrete set of orthogonal mode functions and destruction operators as

$$\boldsymbol{A}^{(+)}(\boldsymbol{x},t) = \sum_k a_k \boldsymbol{u}_k(\boldsymbol{x})\mathrm{e}^{-\mathrm{i}\omega_k t}. \tag{11.1.39}$$

The set of vector mode functions $\boldsymbol{u}_k(\boldsymbol{x})$ which correspond to frequency ω_k will satisfy the wave equation

$$\left(\nabla^2 + \frac{\omega_k^2}{c^2}\right)\boldsymbol{u}_k(\boldsymbol{x}) = 0. \tag{11.1.40}$$

The mode functions are also required to satisfy the transversality condition which arises from (11.1.34)

$$\nabla\cdot\boldsymbol{u}_k(\boldsymbol{x}) = 0. \tag{11.1.41}$$

They also form an orthonormal set

$$\int_V \boldsymbol{u}_k^*(\boldsymbol{x})\cdot\boldsymbol{u}_{k'}(\boldsymbol{x})\,d^3\boldsymbol{x} = \delta_{kk'}, \tag{11.1.42}$$

which is complete within the chosen volume. Plane wave mode functions may be written as

$$\boldsymbol{u}_k(\boldsymbol{x}) = \frac{1}{\sqrt{V}}\hat{\boldsymbol{e}}^{(\lambda)}\exp(\mathrm{i}\boldsymbol{k}\cdot\boldsymbol{x}), \tag{11.1.43}$$

where $\hat{\boldsymbol{e}}^{(\lambda)}$ are the unit polarization vectors satisfying

$$\boldsymbol{k}\cdot\hat{\boldsymbol{e}}^{(\lambda)} = 0, \qquad \hat{\boldsymbol{e}}^{(\lambda)*}\cdot\hat{\boldsymbol{e}}^{(\lambda')} = \delta_{\lambda\lambda'}. \tag{11.1.44}$$

(For plane polarization $\hat{\boldsymbol{e}}^{(\lambda)}$ can be chosen real, but this is not possible for circular polarization.) The mode index k describes several discrete variables, the polarization index ($\lambda = 1,2$), and the three Cartesian components of the propagation vector $\boldsymbol{k}$. Thus, we have the equivalence

$$k \longleftrightarrow (\boldsymbol{k}, \lambda),$$

and as a consequence of the wave equation (11.1.40)

$$\omega_k = c|\boldsymbol{k}|. \tag{11.1.45}$$

The polarization vector $\hat{\boldsymbol{e}}^{(\lambda)}$ is required to be perpendicular to $\boldsymbol{k}$ by the transversality condition (11.1.41).

g) Expansion of Field Operators: The vector potential operator may now be written in the form

$$\boldsymbol{A}(\boldsymbol{x}, t) = \sum_k \sqrt{\frac{\hbar}{2\omega_k \varepsilon_0}} \left(a_k \boldsymbol{u}_k(\boldsymbol{x}) e^{-i\omega_k t} + a_k^\dagger \boldsymbol{u}_k^*(\boldsymbol{x}) e^{i\omega_k t} \right). \tag{11.1.46}$$

The corresponding form for the quantized electric field (i.e., excluding the part $-\nabla\phi(\boldsymbol{x})$ arising from the static scalar potential) is

$$\boldsymbol{E}_{\text{rad}}(\boldsymbol{x}, t) = -\frac{\partial \boldsymbol{A}(x, t)}{\partial t}, \tag{11.1.47}$$

$$= i \sum_k \sqrt{\frac{\hbar\omega_k}{2\varepsilon_0}} \left(a_k \boldsymbol{u}_k(\boldsymbol{x}) e^{-i\omega_k t} - a_k^\dagger \boldsymbol{u}_k^*(\boldsymbol{x}) e^{i\omega_k t} \right). \tag{11.1.48}$$

Thus $\boldsymbol{E}_{\text{rad}}(\boldsymbol{x}, t)$ acts as a variable canonically conjugate to the vector potential $\boldsymbol{A}(\boldsymbol{x}, t)$. The commutation relation does not take a simple delta function form, however, because of the transversality condition.

h) Hamiltonian: The Hamiltonian for the electromagnetic field is given by

$$H_{\text{EM}} = \int \left(\frac{\varepsilon_0 \boldsymbol{E}^2}{2} + \frac{\boldsymbol{B}^2}{2\mu_0} \right) d^3\boldsymbol{x}. \tag{11.1.49}$$

By substituting for $\boldsymbol{E}$ and $\boldsymbol{B}$ using the expression (11.1.48) for $\boldsymbol{E}$ and a similar one for $\boldsymbol{B}$, and by making use of the conditions (11.1.40, 11.1.41), this Hamiltonian can be reduced to the form

$$H_{\text{EM}} = \sum_{\boldsymbol{k},\lambda} c|\boldsymbol{k}| \left(a_{\boldsymbol{k},\lambda}^\dagger a_{\boldsymbol{k},\lambda} + \tfrac{1}{2} \right). \tag{11.1.50}$$

11.1.4 Monochromatic Electromagnetic Waves

No wave is truly monochromatic, so what we mean here is a wave in which the frequencies of interest are around a value Ω, so that we can write an expression for the fields in the form

$$\boldsymbol{A}(\boldsymbol{x}, t) = \left(\frac{\hbar}{2\Omega\varepsilon_0} \right)^{1/2} \left(e^{-i\Omega t} \boldsymbol{\Psi}(\boldsymbol{x}, t) + e^{i\Omega t} \boldsymbol{\Psi}^\dagger(\boldsymbol{x}, t) \right), \tag{11.1.51}$$

$$\boldsymbol{E}(\boldsymbol{x}, t) = i \left(\frac{\hbar\Omega}{2\varepsilon_0} \right)^{1/2} \left(e^{-i\Omega t} \boldsymbol{\Psi}(\boldsymbol{x}, t) - e^{i\Omega t} \boldsymbol{\Psi}^\dagger(\boldsymbol{x}, t) \right), \tag{11.1.52}$$

in which

$$\Psi(x,t) \equiv \sum_k a_k u_k(x) e^{-i\tilde{\omega}_k t}, \tag{11.1.53}$$

$$\tilde{\omega}_k \equiv \omega_k - \Omega. \tag{11.1.54}$$

The intensity of the electromagnetic field is given by the energy density, as expressed in (11.1.49). In the case of a monochromatic field this will contain terms proportional to $\exp(\pm 2i\Omega t)$, which oscillate so rapidly compared to the other terms that they are not measurable by any ordinary detector, which must average over periods far longer than that of a single optical cycle. The end result is that the operator for the intensity of the detectable electromagnetic field is the rate of photon counting which is determined by the mean value of the intensity operator

$$I(x,t) = \left(\frac{\hbar\Omega}{2\varepsilon_0}\right)\Psi^\dagger(x,t)\cdot\Psi(x,t) + \left(\frac{\hbar c^2}{2\Omega}\right)\left(\nabla\times\Psi^\dagger(x,t)\right)\cdot\left(\nabla\times\Psi(x,t)\right). \tag{11.1.55}$$

The vector nature of the field makes for the somewhat complicated formula, but qualitatively, this is very similar to the number density operator for a matter field, $\psi^\dagger(x,t)\psi(x,t)$.

11.1.5 States of Quantized Fields

The dynamical states of a quantized field may be described by assigning an appropriate quantum state to each of the modes—these modes may be assigned and described independently. All of the apparatus developed for Boson or Fermion operators in Chap. 10 can then be applied to the specification of the relevant quantum states.

In this section we will consider the electromagnetic field only—the results for other fields are very similar.

a) Vacuum State: This is the state $|0\rangle$ with no photons in it, i.e., such that

$$a_k|0\rangle = 0 \text{ for all } k. \tag{11.1.56}$$

In this state, although

$$\langle E(x,t)\rangle = \langle 0|E(x,t)|0\rangle = 0, \tag{11.1.57}$$

the correlation functions are not zero, in fact

$$\langle 0|E(x,t)E(x',t')|0\rangle = \sum_{k,k'}\left(\frac{\hbar^2\omega_{k'}\omega_k}{4\varepsilon_0^2}\right)^{\frac{1}{2}} u_k(x)u^*_{k'}(x')e^{-i(\omega_k t-\omega_{k'}t')}\delta_{k,k'}, \tag{11.1.58}$$

$$= \sum_k \left(\frac{\hbar\omega_k}{2\varepsilon_0 L^3}\right) e_k^{\lambda *} e_k^{\lambda} e^{-i\omega_k(t-t')+ik(x-x')}. \tag{11.1.59}$$

This represents *vacuum fluctuations*, in which even in the vacuum there is a non-vanishing fluctuating electromagnetic field, whose average is zero.

b) Coherent State: If every mode of the electromagnetic field is in a *coherent state* $|\boldsymbol{\alpha}\rangle$, such that

$$a_k|\boldsymbol{\alpha}\rangle = \alpha_k|\boldsymbol{\alpha}\rangle, \tag{11.1.60}$$

then this is a coherent state of the *positive frequency part* electromagnetic field operator;

$$\boldsymbol{E}^{(+)}(\boldsymbol{x},t)|\boldsymbol{\alpha}\rangle = \mathcal{E}(\boldsymbol{x},t)|\boldsymbol{\alpha}\rangle, \tag{11.1.61}$$

where $\mathcal{E}(\boldsymbol{x},t)$ is a function given by

$$\mathcal{E}(\boldsymbol{x},t) = \mathrm{i}\sum_k \left(\frac{\hbar\omega_k}{2\varepsilon_0}\right)^{1/2} \alpha_k \mathbf{u}_k(\boldsymbol{x})\mathrm{e}^{-\mathrm{i}\omega_k t}. \tag{11.1.62}$$

The mean value of the electric field in this state is clearly $\mathcal{E}(\boldsymbol{x},t) + \mathcal{E}^*(\boldsymbol{x},t)$.

c) Number State: A number state is one with a definite number of quanta of one mode, thus

$$|k_1, n_1, k_2, n_2, \ldots\rangle = \frac{(a_1^\dagger)^{n_1}(a_2^\dagger)^{n_2}\ldots}{\sqrt{n_1!n_2!\ldots}}|0\rangle. \tag{11.1.63}$$

Such a state is quite hard to create. The mean field in such a state is zero, but the mean square field is non-zero. Number states are, however, the most useful *basis* states for calculations.

Exercise 11.1 Mean Square Electric Field in a Number State: What is the mean square electric field in a number state with n quanta in the mode $k = (\boldsymbol{k}, \lambda)$?

11.2 Coherence and Correlation Functions

The principal difference between quantum mechanics and classical mechanics arises from the wave nature of matter, leading to the importation into mechanics of the characteristically optical phenomena of interference, diffraction, and the concept of coherence as a way of characterizing these phenomena quantitatively. The elementary object of classical wave theory is a wave with a well-defined phase and amplitude, and interference is regarded as arising when two such waves are superposed. Since the phase is considered to be well-defined, interference minima will appear at positions determined by the relative phase between the two waves.

Such ideal wave sources are found essentially only in radio waves, where the phase of the wave can be determined by that of the oscillator at the source. In a laser, the coherent field which emerges has a very stable phase, but it is not determined by any kind of classical oscillator. Thermal sources of light, which are the most common sources, must be treated statistically.

In the remainder of this chapter we will consider the case of scalar fields, such as matter fields and sound fields, and will consider the kinds of correlations and interference that can occur between modes, and thus between the field at different points in space. Time correlations make more sense when there is an underlying dynamical theory which describes how the fields interact with each other or other objects, and these will be dealt with in Chap. 13, and also in *Book II*.

11.2.1 Interference of Classical Waves

Classically, we can imagine fields which are only statistically known, in the sense that the phase and the amplitude are random variables with a certain probability distribution. We can then consider combining together such random fields at points $\boldsymbol{x}_1$ and $\boldsymbol{x}_2$ by some appropriate interference experiment so that the field at some point $\boldsymbol{r}$ is given by

$$\psi(\boldsymbol{r}) = \psi(\boldsymbol{x}_1) + \psi(\boldsymbol{x}_2). \tag{11.2.1}$$

The mean particle intensity at $\boldsymbol{r}$, where the beams are combined, is

$$I(\boldsymbol{r}) = \langle |\psi(\boldsymbol{r})|^2 \rangle. \tag{11.2.2}$$

Thus total intensity at $\boldsymbol{r}$ is

$$I(\boldsymbol{r}) = \langle |\psi(\boldsymbol{x}_1)|^2 \rangle + \langle |\psi(\boldsymbol{x}_2)|^2 \rangle + 2\,\mathrm{Re}\left\{ \langle \psi^*(\boldsymbol{x}_2)\psi(\boldsymbol{x}_1) \rangle \right\}. \tag{11.2.3}$$

a) First-Order Correlation Function: The quantity

$$G_1(\boldsymbol{x}_1, \boldsymbol{x}_2) \equiv \langle \psi^*(\boldsymbol{x}_1)\psi(\boldsymbol{x}_2) \rangle, \tag{11.2.4}$$

is a measure of correlations at different spatial points. It is usual to normalize by dividing by the intensity, so we define the *first-order correlation function* as

$$g_1(\boldsymbol{x}_1, \boldsymbol{x}_2) \equiv \frac{\langle \psi^*(\boldsymbol{x}_1)\psi(\boldsymbol{x}_2) \rangle}{\sqrt{I(\boldsymbol{x}_1) I(\boldsymbol{x}_2)}}, \tag{11.2.5}$$

so that (11.2.3) becomes

$$I(\boldsymbol{r}) = I(\boldsymbol{x}_1) + I(\boldsymbol{x}_2) + 2\sqrt{I(\boldsymbol{x}_1) I(\boldsymbol{x}_2)}\,\mathrm{Re}\left\{ g_1(\boldsymbol{x}_1, \boldsymbol{x}_2) \right\}. \tag{11.2.6}$$

Exercise 11.2 Effect of a Random Phase: The most important effects come from the phases of the fields. Suppose that

$$\psi(\boldsymbol{x}_i) = a(\boldsymbol{x}_i) e^{i\Phi(\boldsymbol{x}_i) + i\delta(\boldsymbol{x}_i)}, \tag{11.2.7}$$

where the $a(\boldsymbol{x}_i)$ are non-random and the $\delta(\boldsymbol{x}_i)$ are mutually Gaussian with mean zero. Show that

$$g_1(\boldsymbol{x}_1, \boldsymbol{x}_2) = e^{-i\Phi(\boldsymbol{x}_1) + i\Phi(\boldsymbol{x}_2)} \exp\left(-\tfrac{1}{2} \left\langle \left(\delta(\boldsymbol{x}_1) - \delta(\boldsymbol{x}_2) \right)^2 \right\rangle \right). \tag{11.2.8}$$

Show that even if the amplitudes $a(\boldsymbol{x}_i)$ are moderately random, the first-order correlation function is not greatly altered.

b) Second-Order Correlation Function: Even when the first-order correlation function is zero, one can obtain interference effects by correlating *intensities*, as was first done by *Hanbury Brown* and *Twiss* [48]. Instead of combining wave sources from two different points and measuring the intensity, one measures the intensity at two different points and correlates these intensities. This defines the *intensity correlation function* or *second-order correlation function*

$$G_2(\boldsymbol{x}_1,\boldsymbol{x}_2) \equiv \langle|\psi(\boldsymbol{x}_1)|^2\psi(\boldsymbol{x}_2)|^2\rangle. \tag{11.2.9}$$

A corresponding normalized correlation function is also defined by

$$g_2(\boldsymbol{x}_1,\boldsymbol{x}_2) \equiv \frac{G_2(\boldsymbol{x}_1,\boldsymbol{x}_2)}{I(\boldsymbol{x}_1)I(\boldsymbol{x}_2)}. \tag{11.2.10}$$

Exercise 11.3 Gaussian Fields: If the fields are jointly Gaussian with no mean field then

$$G_2(\boldsymbol{x}_1,\boldsymbol{x}_2) = I(\boldsymbol{x}_1)I(\boldsymbol{x}_2) + \left|\langle\psi^*(\boldsymbol{x}_1)\psi(\boldsymbol{x}_2)\rangle\right|^2 + \left|\langle\psi(\boldsymbol{x}_1)\psi(\boldsymbol{x}_2)\rangle\right|^2. \tag{11.2.11}$$

In thermal situations, particularly in optics, the correlation function $\langle\psi(\boldsymbol{x}_1)\psi(\boldsymbol{x}_2)\rangle$ vanishes, and the last term is then omitted. In this case we get

$$g_2(\boldsymbol{x}_1,\boldsymbol{x}_2) = 1 + \left|g_1(\boldsymbol{x}_1,\boldsymbol{x}_2)\right|^2. \tag{11.2.12}$$

c) Interference Effects Using Intensity Correlations: Consider a field made by adding two plane waves

$$\psi(\boldsymbol{x}) = r_1 e^{i\phi_1} e^{i\boldsymbol{k}_1\cdot\boldsymbol{x}} + r_2 e^{i\phi_2} e^{i\boldsymbol{k}_2\cdot\boldsymbol{x}}. \tag{11.2.13}$$

The intensity at $\boldsymbol{x}$ is

$$I(x) = r_1^2 + r_2^2 + 2r_1r_2\cos\left(\phi_1 - \phi_2 + (\boldsymbol{k}_1 - \boldsymbol{k}_2)\cdot\boldsymbol{x}\right). \tag{11.2.14}$$

We can now take the intensity correlation function

$$\begin{aligned} G_2(\boldsymbol{x}_1,\boldsymbol{x}_2) = {} & |a_1|^4 + |a_2|^4 + 2|a_1|^2|a_2|^2 \\ & +2\left(|a_1|^2 + |a_2|^2\right)\mathrm{Re}\left\{a_2^* a_1\left(e^{i(\boldsymbol{k}_1-\boldsymbol{k}_2)\cdot\boldsymbol{x}_1} + e^{i(\boldsymbol{k}_1-\boldsymbol{k}_2)\cdot\boldsymbol{x}_2}\right)\right\} \\ & +4\,\mathrm{Re}\left\{a_1 a_2^* e^{i(\boldsymbol{k}_1-\boldsymbol{k}_2)\cdot\boldsymbol{x}_1}\right\}\mathrm{Re}\left\{a_1 a_2^* e^{i(\boldsymbol{k}_1-\boldsymbol{k}_2)\cdot\boldsymbol{x}_2}\right\}, \end{aligned} \tag{11.2.15}$$

where

$$a_1 = r_1 e^{i\phi_1}, \qquad a_2 = r_2 e^{i\phi_2}. \tag{11.2.16}$$

Now let us consider the case that the amplitudes a_1 and a_2 have random phases. For convenience we write

$$\phi_1 \to \theta, \qquad \phi_2 \to \theta - \phi, \tag{11.2.17}$$

and average the correlation functions over these phases:

i) The intensity averaged over these angles simply becomes $r_1^2 + r_2^2$, and thus shows no interference fringes.

ii) For the intensity correlation function, the ensemble average of the second line of (11.2.15) vanishes, and the ensemble average of the last line is

$$\frac{1}{4\pi^2}\int_0^{2\pi} d\theta \int_0^{2\pi} d\phi\, 4r_1^2 r_2^2 \cos\big(\phi + (\boldsymbol{k}_1 - \boldsymbol{k}_2)\cdot\boldsymbol{x}_1\big)\cos\big(-\phi + (\boldsymbol{k}_1 - \boldsymbol{k}_2)\cdot\boldsymbol{x}_2\big),$$
$$= 2r_1^2 r_2^2 \cos\big((\boldsymbol{k}_1 - \boldsymbol{k}_2)\cdot(\boldsymbol{x}_1 - \boldsymbol{x}_2)\big). \tag{11.2.18}$$

Thus, the ensemble average of all of (11.2.15) is

$$G_2(\boldsymbol{x}_1, \boldsymbol{x}_2)|_{\text{Ensemble}} = \big(r_1^2 + r_2^2\big)^2 + 2r_1^2 r_2^2 \cos\big((\boldsymbol{k}_1 - \boldsymbol{k}_2)\cdot(\boldsymbol{x}_1 - \boldsymbol{x}_2)\big). \tag{11.2.19}$$

iii) This means that we can take two fields with randomized phases, and there will be no interference pattern visible in the intensity, because $G_1(\boldsymbol{x}_1, \boldsymbol{x}_2)$ vanishes, but that nevertheless there is a clearly visible interference pattern in the intensity correlation function. This pattern has the same spatial frequency which would be observed in the intensity if the phases were not completely randomized.

Exercise 11.4 Fringe Visibility: The fringe visibility of the pattern given by (11.2.19), defined by the ratio of maximum to minimum of $G_2(\boldsymbol{x}_1, \boldsymbol{x}_2)|_{\text{Ensemble}}$, is

$$v = \frac{G_2^{\max} - G_2^{\min}}{G_2^{\max} + G_2^{\min}} = \frac{2r_1^2 r_2^2}{\big(r_1^2 + r_2^2\big)^2}. \tag{11.2.20}$$

This has a maximum value of 50%, which happens when $r_1 = r_2$.

d) Physical Interpretation: In practice we imagine the random phasing of the amplitudes to arise as a result of time-averaging over a long period, during which the amplitudes have a definite phase at any given time. Thus, if one did an accurate time-resolved measurement of the intensity, one would see the interference pattern (11.2.14) with time-dependent phases—the pattern would be seen to jitter back and forth, but always remain of the same shape. Thus, the average of this pattern would give the structureless result $r_1^2 + r_2^2$. However the periodic structure of $I(\boldsymbol{x})$ still reveals itself in the intensity correlation function (11.2.19).

11.2.2 Quantum Interference

When considering interference of fields in quantum theory, we need only one *field operator* $\psi(\boldsymbol{x})$, whose modes are populated in such a way as to correspond to interfering modes of the same field. Let us consider therefore a quantum field in which there are only two occupied modes, corresponding to two wavenumbers $\boldsymbol{k}_1$ and $\boldsymbol{k}_2$;

$$\psi(\boldsymbol{x}) = a_1 e^{i\boldsymbol{k}_1\cdot\boldsymbol{x}} + a_2 e^{i\boldsymbol{k}_2\cdot\boldsymbol{x}} + \text{unoccupied modes}. \tag{11.2.21}$$

The correlation functions can be defined in various ways because of choices of operator ordering. The quantum correlation functions conventionally used are *normally ordered*, so that

$$G_1(\boldsymbol{x}_1,\boldsymbol{x}_2) \equiv \langle \psi^\dagger(\boldsymbol{x}_1)\psi(\boldsymbol{x}_2)\rangle, \tag{11.2.22}$$

$$G_2(\boldsymbol{x}_1,\boldsymbol{x}_2) \equiv \langle \psi^\dagger(\boldsymbol{x}_1)\psi^\dagger(\boldsymbol{x}_2)\psi(\boldsymbol{x}_2)\psi(\boldsymbol{x}_1)\rangle. \tag{11.2.23}$$

Because of the normal ordering, these are zero in the vacuum state.

The *intensity correlation* is defined as the correlation function of the operator intensity

$$I(\boldsymbol{x}) \equiv \psi^\dagger(\boldsymbol{x})\psi(\boldsymbol{x}), \tag{11.2.24}$$

and it follows that

$$\langle I(\boldsymbol{x}_1)I(\boldsymbol{x}_2)\rangle = G_2(\boldsymbol{x}_1,\boldsymbol{x}_2) + \delta(\boldsymbol{x}_1-\boldsymbol{x}_2)\langle I(\boldsymbol{x}_1)\rangle. \tag{11.2.25}$$

a) Coherent States: The coherent states which are so useful in quantum optics exist for any kind of Boson operators, and are described fully in Chap. 15. Thus, the coherent state can be expressed as

$$|\alpha\rangle \equiv e^{-|\alpha|^2/2}\sum_n^\infty \frac{\alpha^n}{\sqrt{n!}}|n\rangle, \tag{11.2.26}$$

and this leads to the defining equation

$$a|\alpha\rangle = \alpha|\alpha\rangle. \tag{11.2.27}$$

Such a state is a superposition of states of different particle number, and when these particles are atoms, which are massive and conserved, this cannot represent anything other than a mathematical construction. In contrast, photons can be created and destroyed easily, are not massive, and thus the actual creation of such a state is feasible, and indeed this is a good description of the electromagnetic field of a laser.

b) Interference of Coherent States: If the quantum states of modes 1 and 2 are *coherent states*, written $|\alpha_1,\alpha_2\rangle$, then the density shows interference:

$$\langle\psi^\dagger(\boldsymbol{x})\psi(\boldsymbol{x})\rangle = \langle\alpha_1,\alpha_2|\left(a_1^\dagger e^{-i\boldsymbol{k}_1\cdot\boldsymbol{x}} + a_2^\dagger e^{-i\boldsymbol{k}_2\cdot\boldsymbol{x}}\right)\left(a_1 e^{i\boldsymbol{k}_1\cdot\boldsymbol{x}} + a_2 e^{i\boldsymbol{k}_2\cdot\boldsymbol{x}}\right)|\alpha_1,\alpha_2\rangle, \tag{11.2.28}$$

$$= |\alpha_1|^2 + |\alpha_2|^2 + 2\,\mathrm{Re}\left\{\alpha_2^*\alpha_1 e^{i(\boldsymbol{k}_1-\boldsymbol{k}_2)\cdot\boldsymbol{x}}\right\}. \tag{11.2.29}$$

If we set

$$\alpha_1 \to r_1 e^{i\phi_1}, \qquad \alpha_2 \to r_2 e^{i\phi_2}, \tag{11.2.30}$$

this is identical to the classical result for interference of fields with well-defined amplitude and phase.

Similarly, when we consider the density correlation function for the same coherent states, we get

$$\begin{aligned} G_2(\boldsymbol{x}_1,\boldsymbol{x}_2)|_{\text{Coherent}} = {} & |\alpha_1|^4 + |\alpha_2|^4 + 2|\alpha_1|^2|\alpha_2|^2 \\ & +2\left(|\alpha_1|^2+|\alpha_2|^2\right)\mathrm{Re}\left\{\alpha_2^*\alpha_1\left(e^{i(\boldsymbol{k}_1-\boldsymbol{k}_2)\cdot\boldsymbol{x}_1} + e^{i(\boldsymbol{k}_1-\boldsymbol{k}_2)\cdot\boldsymbol{x}_2}\right)\right\} \\ & +4\,\mathrm{Re}\left\{\alpha_1\alpha_2^* e^{i(\boldsymbol{k}_1-\boldsymbol{k}_2)\cdot\boldsymbol{x}_1}\right\}\mathrm{Re}\left\{\alpha_1\alpha_2^* e^{i(\boldsymbol{k}_1-\boldsymbol{k}_2)\cdot\boldsymbol{x}_2}\right\}, \end{aligned} \tag{11.2.31}$$

and making the same replacements (11.2.30), this is also identical to the corresponding classical result (11.2.15).

c) Number States: Let us consider the bivariate number states $|n_1, n_2\rangle$, and use the usual results

$$a_1|n_1, n_2\rangle = \sqrt{n_1}\,|n_1, n_2\rangle, \text{ etc.} \tag{11.2.32}$$

The intensity at the point $\boldsymbol{x}$ becomes

$$\langle n_1, n_2|\psi^\dagger(\boldsymbol{x})\psi(\boldsymbol{x})|n_1, n_2\rangle = n_1 + n_2, \tag{11.2.33}$$

while the correlation function becomes

$$G_2(\boldsymbol{x}_1, \boldsymbol{x}_2)|_{\text{Number}} = \langle a_1^\dagger a_1^\dagger a_1 a_1\rangle + \langle a_1^\dagger a_2^\dagger a_1 a_2\rangle + \langle a_2^\dagger a_1^\dagger a_1 a_2\rangle + \langle a_2^\dagger a_2^\dagger a_2 a_2\rangle$$
$$+ \langle a_2^\dagger a_1^\dagger a_1 a_2\rangle e^{-i(\boldsymbol{k}_1-\boldsymbol{k}_2)\cdot(\boldsymbol{x}_1-\boldsymbol{x}_2)} + \langle a_1^\dagger a_2^\dagger a_1 a_2\rangle e^{i(\boldsymbol{k}_1-\boldsymbol{k}_2)\cdot(\boldsymbol{x}_1-\boldsymbol{x}_2)}, \tag{11.2.34}$$
$$= (n_1 + n_2)^2 + 2n_1 n_2 \cos\big((\boldsymbol{k}_1 - \boldsymbol{k}_2)\cdot(\boldsymbol{x}_1 - \boldsymbol{x}_2)\big) - n_1 - n_2. \tag{11.2.35}$$

If we set

$$n_1 \to r_1^2, \qquad n_2 \to r_2^2, \tag{11.2.36}$$

the result for the intensity is exactly the same as that for the classical interference of two fields with random phases. The result for G_2 is almost the same, but the quantum result is reduced by a term $n_1 + n_2$. For large occupations n_1, n_2, this becomes negligible, and we can regard the interference between number states as being similar to that from a thermalized classical source.

Exercise 11.5 Fringe Visibility: Show that for the case of number states the visibility is

$$v = \frac{2n_1 n_2}{(n_1 + n_2)^2 - n_1 - n_2}. \tag{11.2.37}$$

When n_1, n_2 are large, this is slightly bigger than the classical result (11.2.20). For the technically difficult, but physically conceivable case of $n_1 = n_2 = 1$

$$G(\boldsymbol{x}_1, \boldsymbol{x}_2) = 2 + 2\cos\big((\boldsymbol{k}_1 - \boldsymbol{k}_2)\cdot(\boldsymbol{x}_1 - \boldsymbol{x}_2)\big), \tag{11.2.38}$$

and the fringe visibility increases to 100%.

Exercise 11.6 Fermion Interference: If a_1, a_2 are *Fermi* operators, show that

$$G(\boldsymbol{x}_1, \boldsymbol{x}_2) = 2 - 2\cos\big((\boldsymbol{k}_1 - \boldsymbol{k}_2)\cdot(\boldsymbol{x}_1 - \boldsymbol{x}_2)\big), \tag{11.2.39}$$

and that fringe visibility is again 100%.

d) Correlated State of Fixed Total Number: Let us consider the projection of the bivariate coherent state $|\alpha_1, \alpha_2\rangle$ onto states of fixed total number of atoms. This is easy to write down directly from the definition (11.2.26) as

$$|M\rangle \equiv e^{-(|\alpha_1|^2+|\alpha_2|^2)/2} \sum_{n=0}^{M} \frac{\alpha_1^n \alpha_2^{M-n}}{\sqrt{n!(M-n)!}} |n, M-n\rangle. \tag{11.2.40}$$

Exercise 11.7 Expression in Terms of Coherent States: Show that if

$$\alpha_1 = r_1 e^{i\phi_1}, \qquad \alpha_2 = r_2 e^{i\phi_2}, \tag{11.2.41}$$

then we can write

$$|M\rangle = \frac{1}{2\pi}\int_0^{2\pi} d\theta\, e^{-iM\theta} |\alpha_1 e^{i\theta}, \alpha_2 e^{i\theta}\rangle, \tag{11.2.42}$$

$$\langle M|M\rangle = \frac{(r_1^2+r_2^2)^M e^{-r_1^2-r_2^2}}{M!}. \tag{11.2.43}$$

Show that if i, j, k, l take on the values 1,2,

$$\langle M|a_i^\dagger a_j|M\rangle = \frac{M\alpha_i^*\alpha_j}{|\alpha_1|^2+|\alpha_2|^2}\langle M|M\rangle. \tag{11.2.44}$$

Similarly

$$\langle M|a_i^\dagger a_j^\dagger a_k a_l|M\rangle = \frac{M(M-1)\alpha_i^*\alpha_j^*\alpha_k\alpha_l}{\left(|\alpha_1|^2+|\alpha_2|^2\right)^2}\langle M|M\rangle. \tag{11.2.45}$$

Using the results (11.2.44, 11.2.45) both G_1 and G_2 can be evaluated from the coherent state results (11.2.29, 11.2.31). If $M \gg 1 \Longrightarrow M(M-1) \approx M^2$, these are of the same form as the classical coherent results (11.2.14, 11.2.16) with the correspondences

$$a_1 = \frac{\alpha_1\sqrt{M}}{\sqrt{|\alpha_1|^2+|\alpha_2|^2}}, \qquad a_2 = \frac{\alpha_2\sqrt{M}}{\sqrt{|\alpha_1|^2+|\alpha_2|^2}}. \tag{11.2.46}$$

The Poissonian nature of (11.2.43) means that the dominant contribution to the bivariate coherent state comes from $M \approx |\alpha_1|^2 + |\alpha_2|^2$, so this result is almost indistinguishable from that of the bivariate coherent state.

11.2.3 Summary—Phase and Interference

The classical picture of coherence, correlation and phase sees a wave as having a definite phase, and incoherence as arising as an ensemble or time average over fluctuations of this phase. That is, phase and amplitude exist simultaneously, but may be obscured by fluctuations. The quantum concepts are quite different, but the results for correlation functions and interference are not as different as one might expect. The interference effects found from number states are very similar to those found from classical ensembles with random phases, and the results from interference between correlated states of fixed number are barely distinguishable from either the classical or the results from bivariate coherent states.

The bivariate coherent state representation of interfering beams of bosons is the most convenient one available, but is misleading in a situation where it is clear that no absolute phase exists—the only phase reference for a matter wave is another matter wave. This is in contrast to a classically generated electric field, whose phase is directly related to the phase of the oscillator driving it. Furthermore, the actual strength of the electric field can be measured and its peaks determined; this is truly an absolute phase measurement. For optical fields this is in principle still true, but the frequencies are so high that the direct measurement of the field is not easy. In matter waves the phase we want to talk about is essentially the same as that of a Schrödinger wavefunction, and this truly has no absolute meaning. Nevertheless, it is awkward to do quantum mechanics in such a way as to avoid absolute phases, so we accept them as a convenience which has no physical consequences.

Therefore, from time to time we will choose to use matter wave fields which do have an absolute phase, and this is in particular the case with a Bose–Einstein condensate. If we then take a coherent state $|\alpha_{\boldsymbol{k}}\rangle$ for each mode of the matter wave field, then a mean field arises

$$\Psi(\boldsymbol{x}, t) = \sum_{\boldsymbol{k}} \alpha_{\boldsymbol{k}} \mathrm{e}^{\mathrm{i}(\boldsymbol{k}\cdot\boldsymbol{x} - \mathrm{i}\omega_{\boldsymbol{k}} t)}, \tag{11.2.47}$$

and since this is merely a complex function, it has a phase. However, any actual physics takes place only in subspaces of fixed total number. These are correlated states of fixed total number, that is, multivariate versions of those in Sect. 11.2.2c, but in the case of highly occupied states, such as in a Bose–Einstein condensate, there is no measurable difference between the results predicted by these states and those given by multivariate coherent states.

12. Atoms, Light and their Interaction

In practice, the electromagnetic interaction between atoms and light is manifested in two main ways, coherently or incoherently. The incoherent aspect is seen as damping, noise and radiation, with the principal quantum-mechanical effect being seen as the decay of an excited atom. The coherent aspect arose with the development of the laser as a laboratory tool, and is seen principally in the manipulation of atomic states by the application of appropriately tuned and timed laser pulses. This can be done with great flexibility and precision, and can be used to exert mechanical forces on atoms, leading to the trapping and cooling of atoms.

These two aspects are the extremes. It is never possible to eliminate the incoherent aspects completely, so that a full theory always requires their inclusion, even when the interaction is principally coherent. This is especially true if precise manipulation of the quantum state of an atom is the aim, as in the study of quantum information.

As in the classical case, in quantum mechanics there is a close connection between damping and noise, but in addition, the underlying probabilistic nature of quantum mechanics introduces noise which is purely quantum-mechanical, and not directly connected with damping.

This chapter therefore studies the simplest atomic system—an atom approximated by only two energy levels—interacting with the simplest forms of the electromagnetic field. The two topics are:

i) *The Decay of an Excited Atom*: Here the atom interacts with an electromagnetic field initially in the vacuum state. We do this because the formalism for quantum noise and damping can become quite elaborate, and the motivation for this formalism can consequently become rather obscure. The radiation of a single photon from a single atom—itself simplified to a system with only two energy levels—provides a useful elementary physical example, introducing the concepts necessary for a full formulation of quantum stochastic processes. A more detailed development of quantum stochastic formalism is presented in the following two chapters, and this is extended and completed in *Book II*.

ii) *The Manipulation of the Quantum State of an Atom Using a Coherent Optical Field*: Here we omit the quantization of the electromagnetic field, and study the possible manipulations of the state of a two-level atom that can be executed using coherent optical pulses.

12.1 Interaction with the Quantized Radiation Field

We will use the quantized electromagnetic field as formulated in Sect. 11.1.3, and the main task of this section is to introduce an appropriate formulation of the interaction with a quantized atom.

12.1.1 Hamiltonian and Schrödinger Equation

a) Use of the Schrödinger Picture: For computational purposes the Schrödinger picture is a more appropriate starting point than the Heisenberg picture. This means that the electromagnetic field operators have a mode expansion like those in (11.1.46, 11.1.48), but without the time-dependent exponentials, namely

$$\boldsymbol{A}_S(\boldsymbol{x}) = \sum_k \left(\frac{\hbar}{2\omega_k \varepsilon_0}\right)^{\frac{1}{2}} \left(a_k \mathbf{u}_k(\boldsymbol{x}) + a_k^\dagger \mathbf{u}_k^*(\boldsymbol{x})\right), \tag{12.1.1}$$

$$\boldsymbol{E}_S(\boldsymbol{x}) = \mathrm{i}\sum_k \left(\frac{\hbar\omega_k}{2\varepsilon_0}\right)^{\frac{1}{2}} \left(a_k \mathbf{u}_k(\boldsymbol{x}) - a_k^\dagger \mathbf{u}_k^*(\boldsymbol{x})\right). \tag{12.1.2}$$

However, in the remainder of this treatment in the Schrödinger picture, we will not explicitly write the subscript S—all operators will implicitly be in the Schrödinger picture.

b) Interaction Hamiltonian: We consider an electron interacting with the quantized electromagnetic field, and bound to a nucleus by an electrostatic field. Using the Coulomb gauge as in Sect. 11.1.3, the Hamiltonian for such a system is

$$H = \frac{\left(\boldsymbol{p} - e\boldsymbol{A}(\boldsymbol{x})\right)^2}{2m} + e\phi(\boldsymbol{x}) + H_{EM}. \tag{12.1.3}$$

In this equation:

i) The problem is formulated in the Schrödinger picture, so the vector potential operator has no time dependence.

ii) The static potential $\phi(\boldsymbol{x})$ represents the Coulomb potential of the nucleus.

iii) H_{EM} is the Hamiltonian for the quantized electromagnetic field, as defined in equations (11.1.49) and (11.1.50).

We rewrite the first two parts of (12.1.3) as a Hamiltonian for the bound electron, and a Hamiltonian for the interaction with the quantized electromagnetic field. Thus, we write

$$H \equiv H_0 + V_{\mathrm{Int}} + H_{EM}, \tag{12.1.4}$$

in which

$$H_0 = \frac{\boldsymbol{p}^2}{2m} + e\phi(\boldsymbol{x}), \tag{12.1.5}$$

$$V_{\text{Int}} = -\frac{e}{2m}\Big(\boldsymbol{p}\cdot\boldsymbol{A}(\boldsymbol{x}) + \boldsymbol{A}(\boldsymbol{x})\cdot\boldsymbol{p}\Big) + \frac{e^2}{2m}\boldsymbol{A}(\boldsymbol{x})\cdot\boldsymbol{A}(\boldsymbol{x}). \tag{12.1.6}$$

Note that the choice of the Coulomb gauge condition (11.1.34) leads to

$$\boldsymbol{p}\cdot\boldsymbol{A}(\boldsymbol{x}) - \boldsymbol{A}(\boldsymbol{x})\cdot\boldsymbol{p} = -i\hbar\nabla\cdot\boldsymbol{A}(\boldsymbol{x}) = 0, \tag{12.1.7}$$

so that the order of factors is not important.

c) Simplifications and Approximations: Formulating the interaction in terms of the vector potential can be awkward, and a reformulation in terms of the electric field and a dipole moment has many advantages, particularly in making approximations. The simplest explanation of the transformation is based on the Lagrangian formulation, in which the electromagnetic interaction of a particle with an electromagnetic field is achieved by the interaction Lagrangian

$$\mathcal{L}_{\text{Int}} = e\dot{\boldsymbol{x}}\cdot\boldsymbol{A}(\boldsymbol{x},t). \tag{12.1.8}$$

It is always possible to add a total time derivative to the Lagrangian, so we add such a term so that

$$\mathcal{L}_{\text{Int}} \longrightarrow e\dot{\boldsymbol{x}}\cdot\boldsymbol{A}(\boldsymbol{x},t) - \frac{d}{dt}\Big(e\boldsymbol{x}\cdot\boldsymbol{A}(\boldsymbol{x},t)\Big), \tag{12.1.9}$$

$$= -e\boldsymbol{x}\cdot\frac{d\boldsymbol{A}(\boldsymbol{x},t)}{dt}, \tag{12.1.10}$$

$$= -e\boldsymbol{x}\cdot\left(\frac{\partial A(\boldsymbol{x},t)}{\partial t} + (\dot{\boldsymbol{x}}\cdot\nabla)\boldsymbol{A}(\boldsymbol{x},t)\right). \tag{12.1.11}$$

It is at this stage that we can make approximations based on the physics of the situation. The second term in this equation can usually be neglected since:

i) The first term is of order of magnitude $\omega\boldsymbol{A}$ for a monochromatic field.

ii) The term $\nabla\boldsymbol{A}$ is of order of magnitude $vecA/c\omega$, and if the typical speed of the atom is v, the second term is of order of magnitude v/c smaller than the first term, and is thus almost always negligible.

iii) We therefore use the approximate form for the interaction Lagrangian

$$\mathcal{L}_{\text{Int}} = e\boldsymbol{x}\cdot\boldsymbol{E}(\boldsymbol{x},t). \tag{12.1.12}$$

iv) This form depends on the derivative of field $\boldsymbol{A}(\boldsymbol{x},t)$ and this means that the field canonical momentum variable is no longer $\epsilon_0\boldsymbol{E}$, but becomes

$$\boldsymbol{D}(\boldsymbol{x},t) \equiv \epsilon_0\boldsymbol{E}(\boldsymbol{x},t) - e\boldsymbol{x}(t)\delta\big(\boldsymbol{x}-\boldsymbol{x}(t)\big). \tag{12.1.13}$$

The Hamiltonian then becomes

$$\mathcal{H} = H_0 + \int d^3\boldsymbol{x}\left\{\frac{\boldsymbol{D}(\boldsymbol{x},t)^2}{2\epsilon_0} + \frac{\boldsymbol{B}(\boldsymbol{x},t)^2}{2\mu_0}\right\} - \frac{e\boldsymbol{x}(t)\cdot\boldsymbol{D}\big(x(t),t\big)}{\epsilon_0} + \frac{e^2\boldsymbol{x}(t)^2\,\delta(0)}{\epsilon_0}. \tag{12.1.14}$$

The final term requires interpretation, and we should take the delta function as really being some representation of the size over which the interaction takes place, thus we can reasonably set

$$\delta(0) \longrightarrow \frac{1}{4\pi R^3}, \tag{12.1.15}$$

where R represents the effective size of the interacting system. The term then gives a correction to the electrostatic potential which is a factor $\approx \left(x(t)/R\right)^3$ times the Coulomb potential. The correction is an artifact of our neglect of the second term in (12.1.11), and should not be taken as fundamental. We shall neglect it from now on.

d) The Electric Dipole Approximation: It is $\boldsymbol{D}$ and $\boldsymbol{A}$ which are basic to the Hamiltonian (12.1.14), but since $\boldsymbol{D} = \epsilon_0 \boldsymbol{E}$ except at the position of the atom, we shall not make the distinction, and as our basic Hamiltonian we shall take (12.1.14) with the substitution $\boldsymbol{D} \longrightarrow \epsilon_0 \boldsymbol{E}$, which is equivalent to the neglect of the delta function term.

Furthermore, in practice the wavelength of the light is thousands of times larger than the size of an atom, so that the dependence on position of the interaction Hamiltonian is negligible. Assuming then that the atom is located at $\boldsymbol{x} = 0$, we may write

$$V_{\text{Int}}(t) \rightarrow -e\boldsymbol{x}\cdot\boldsymbol{E}(0). \tag{12.1.16}$$

This approximation is known as the *electric dipole approximation.*

e) Atom-Field Hamiltonian in the Electric Dipole Approximation: The Hamiltonian for the full system, after making approximations and simplifications, becomes

$$H = H_{\text{Atom}} + H_{\text{EM}} + H_{\text{Int}}, \tag{12.1.17}$$

with

$$H_{\text{Atom}} = \frac{\boldsymbol{p}^2}{2m} + e\phi(\boldsymbol{x}), \tag{12.1.18}$$

$$H_{\text{EM}} = \sum_k \hbar\omega_k \left(a_k^\dagger a_k + \tfrac{1}{2}\right), \tag{12.1.19}$$

$$H_{\text{Int}} = -\boldsymbol{d}\cdot\boldsymbol{E}(0). \tag{12.1.20}$$

Here, we have defined the *electric dipole moment operator*

$$\boldsymbol{d} \equiv e\boldsymbol{x}. \tag{12.1.21}$$

12.1.2 The Two-Level Atom Approximation

If the energy eigenstates of the atom are written $|i\rangle$ with energy eigenvalue E_i, then we can write

$$H_{\text{Atom}} = \sum_i E_i |i\rangle\langle i|, \tag{12.1.22}$$

$$e\boldsymbol{x} = \sum_{f,i} \boldsymbol{d}_{if} |f\rangle\langle i|. \tag{12.1.23}$$

We now make a drastic approximation—we will neglect all but *two* eigenstates, an excited state $|e, E_e\rangle$ and a ground state $|g, E_g\rangle$. This should give a basic understanding of how transitions happen from one state to another.

We can write

$$|e\rangle = \begin{pmatrix}1\\0\end{pmatrix}, \qquad |g\rangle = \begin{pmatrix}0\\1\end{pmatrix}, \tag{12.1.24}$$

and consequently in the two-state approximation

$$H_{\text{Atom}} = \begin{pmatrix}E_e & 0\\ 0 & E_g\end{pmatrix}, \tag{12.1.25}$$

$$e\boldsymbol{x} = \begin{pmatrix}0 & \boldsymbol{d}_{eg}\\ \boldsymbol{d}^*_{eg} & 0\end{pmatrix}. \tag{12.1.26}$$

We now use the expression (11.1.48) for the electric field operator, and find that (noting that we are in the Schrödinger picture, so that the time dependences in the electromagnetic field operators are omitted) the interaction Hamiltonian becomes

$$H_{\text{Int}} = i\hbar\sigma^- \sum_k \left(a_k^\dagger \kappa_k^* - a_k \bar{\kappa}_k^*\right) - i\hbar\sigma^+ \sum_k \left(a_k \kappa_k - a_k^\dagger \bar{\kappa}_k\right), \tag{12.1.27}$$

and in which

$$\kappa_k \equiv -\sqrt{\frac{\omega_k}{2\hbar\varepsilon_0}}\, \boldsymbol{d}_{eg} \cdot \mathbf{u}_k(0), \qquad \bar{\kappa}_k \equiv -\sqrt{\frac{\omega_k}{2\hbar\varepsilon_0}}\, \boldsymbol{d}_{eg} \cdot \mathbf{u}_k^*(0). \tag{12.1.28}$$

12.1.3 The Rotating Wave Approximation

A term like $\sigma^+ a_k$ removes a photon from the radiation field, and raises the atom from the ground state to the excited state, and the Hermitian conjugate term $\sigma^- a_k^\dagger$ does the reverse. Provided the energy of the photon is similar to the energy difference between the two atomic eigenstates, this process is *resonant*, and consequently is very significant. On the other hand, the term $\sigma^- a_k$ lowers the atom's energy while also removing a photon from the electromagnetic field, which cannot be resonant, and therefore is not an important process. The *rotating wave approximation* corresponds to omitting terms which cannot be resonant. After making this approximation, the interaction Hamiltonian becomes

$$H_{\text{Int}} = i\hbar\left(\sigma^- \sum_k a_k^\dagger \kappa_k^* - \sigma^+ \sum_k a_k \kappa_k\right). \tag{12.1.29}$$

We can also define the *transition frequency* by

$$\omega = \frac{E_e - E_g}{\hbar}, \tag{12.1.30}$$

then choose the zero of energy so that $E_g = 0$, and write

$$H_{\text{Atom}} \to \hbar\omega|e\rangle\langle e| = \tfrac{1}{2}\hbar\omega(1+\sigma_z). \tag{12.1.31}$$

Thus we arrive at the *Hamiltonian for a two-level atom interacting with the electromagnetic field in the rotating wave approximation*

$$H = \hbar\omega|e\rangle\langle e| + \sum_k \hbar\omega_k\, a_k^\dagger a_k + i\hbar\left(\sigma^- \sum_k a_k^\dagger \kappa_k^* - \sigma^+ \sum_k a_k \kappa_k\right). \tag{12.1.32}$$

12.1.4 "Stripped-Down" Quantum Electrodynamics

The Hamiltonian (12.1.32), and the various approximations we have introduced in Sect. 12.1.1–Sect. 12.1.3 represent a "stripped-down" version of the quantum electrodynamics originally introduced in Sect. 12.1.1. This description is adequate for most quantum-optical phenomena, and when it is not, it can be usually be extended appropriately. It will be used almost exclusively in the remainder of this book.

12.2 Decay of an Excited Atom

In this section we will show how atomic decay, and hence spectral line broadening, arise from the solution of the Schrödinger equation for the Hamiltonian, in an adaptation of a method formulated by *Weisskopf* and *Wigner* [49] in 1930. Although approximations are made, it will become clear that the irreversibility implicit in atomic decay arises as a result of the continuous energy spectrum of the electromagnetic field rather than the approximations—thus Hamiltonian evolution and irreversibility are not incompatible.

12.2.1 Wavefunction and Initial Condition

We suppose that the initial state consists of no photons in the electromagnetic field, and the atom in the excited state, as illustrated in Fig. 12.1. From this state, the rotating wave Hamiltonian can only create states of the form

$$|g,k\rangle \equiv a_k^\dagger\sigma^-|e,0\rangle = a_k^\dagger|g,0\rangle, \tag{12.2.1}$$

where:

i) $|g,0\rangle$ is the state with no photons, and the atom in the ground state.

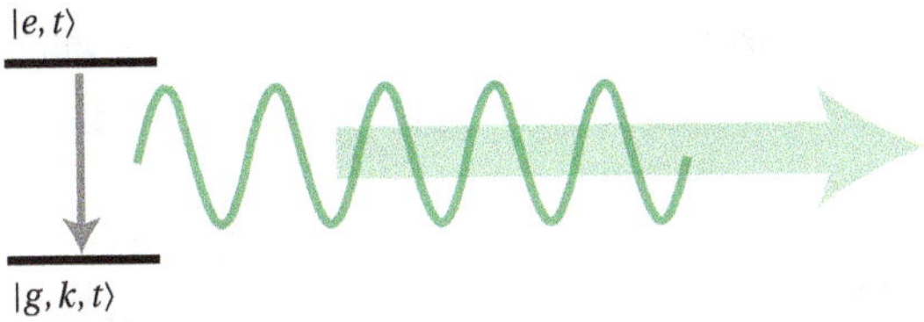

Fig. 12.1. Decay of a two-level atom.

ii) $|e,0\rangle$ is the state with no photons, and the atom in the excited state.

If we write the time-dependent state of the system in terms of these states as

$$|\Psi, t\rangle = u(e,t)|e,0\rangle + \sum_k u(g,k,t)|g,k\rangle, \tag{12.2.2}$$

then the Schrödinger equation yields equations of motion for the coefficients in the form

$$i\hbar\dot{u}(e,t) = \hbar\omega u(e,t) - i\hbar\sum_k \kappa_k u(g,k,t), \tag{12.2.3}$$

$$i\hbar\dot{u}(g,k,t) = \hbar\omega_k u(g,k,t) + i\hbar\kappa_k^* u(e,t). \tag{12.2.4}$$

The assumed initial condition of an excited atom and no photons means

$$u(e,0) = 1, \qquad u(g,k,0) = 0. \tag{12.2.5}$$

a) Solving the Equations: The second equation can be integrated to give

$$u(k,g,t) = e^{-i\omega_k t}u(k,g,0) + \int_0^t dt'\, e^{-i\omega_k(t-t')}\kappa_k^* u(e,t'). \tag{12.2.6}$$

The initial condition $u(k,g,0) = 0$ means that the first term is zero, so we substitute the resulting expression in the first equation to get

$$i\hbar\dot{u}(e,t) = \hbar\omega u(e,t) - i\hbar\int_0^t dt'\, u(e,t')\left\{\sum_k |\kappa_k|^2 e^{-i\omega_k(t-t')}\right\}. \tag{12.2.7}$$

This can be simplified if we substitute

$$v(e,t) = u(e,t)e^{i\omega t}, \tag{12.2.8}$$

and it becomes

$$\dot{v}(e,t) = -\int_0^t dt'\, v(e,t')\left\{\sum_k |\kappa_k|^2 e^{-i(\omega_k-\omega)(t-t')}\right\}. \tag{12.2.9}$$

b) The Markov Approximation: The function inside the curly brackets

$$\gamma(t-t') \equiv \sum_k |\kappa_k|^2 e^{-i(\omega_k-\omega)(t-t')}, \tag{12.2.10}$$

is central to the understanding of the decay. We can replace the summation by an integral over frequency, and a density of states

$$\sum_k |\kappa_k|^2 \to \int d\omega_k\, g(\omega_k)|\kappa(\omega_k)|^2, \tag{12.2.11}$$

where $|\kappa(\omega_k)|^2$ is the average over polarizations of $|\kappa_k|^2$ at the appropriate value of $|\boldsymbol{k}|$. Thus

$$\gamma(t-t') \to \int_0^\infty d\omega_k\, g(\omega_k)|\kappa(\omega_k)|^2 \mathrm{e}^{-\mathrm{i}(\omega_k-\omega)(t-t')}. \tag{12.2.12}$$

The significant features of this equation are:

i) The frequency range of the term $\mathrm{e}^{-\mathrm{i}(\omega_k-\omega)(t-t')}$ in the integral is $(-\infty,\omega)$.

ii) ω is itself a very high frequency.

iii) The coefficient of the exponential is a very smooth function.

These mean that $\gamma(t-t')$ is very small except when $|t-t'|$ is very small. Thus, we can neglect the time variation of $v(e,t')$ inside the integral in (12.2.9), and write

$$\dot{v}(e,t) \approx -v(e,t)\int_0^t \gamma(t-t')\,dt'. \tag{12.2.13}$$

The approximation that leads to the equation of motion in this form is known as the *Markov approximation*—it enables one to replace the integro-differential equation (12.2.9), which involves $v(e,t')$ for all $t'<t$, by the simpler first-order differential equation (12.2.13). It now remains to evaluate the coefficient on the right-hand side.

c) Significance of the Markov Approximation: The Markov approximation, as it arises here, is the simplest example of how this concept arises in a quantum-mechanical context. It is relevant because there are very different time scales involved. The optical frequencies represent very fast time scales, whereas the decay happens very slowly by comparison. The basic structures of the damping constant and frequency shift which arise here are the same as those found in the more systematic formulations of damping and noise found in the remainder of this book.

d) Evaluation of the Coefficient: Now change variables to $\tau = t-t'$, and evaluate the integral as $\int_0^t d\tau\,\gamma(\tau)$. In this form, for typical values of t the function $\gamma(\tau)$ is essentially zero when $\tau > t$ so we can set the upper limit of the integral to ∞ with very little error, and the integral can then be written

$$\int_0^\infty \gamma(\tau)\,d\tau = \lim_{\varepsilon\to0+}\int_0^\infty d\tau \int_0^\infty d\omega_k\, g(\omega_k)|\kappa(\omega_k)|^2 \mathrm{e}^{-\mathrm{i}(\omega_k-\omega)\tau-\varepsilon\tau}. \tag{12.2.14}$$

Here we have inserted a convergence factor $\mathrm{e}^{-\varepsilon\tau}$, where $\varepsilon>0$ is set equal to zero at the end of the calculation. With this factor inserted, the order of integrals can be interchanged, and we get

$$\int_0^\infty \gamma(\tau)\,d\tau = \lim_{\varepsilon\to0}\int_0^\infty d\omega_k \frac{g(\omega_k)|\kappa(\omega_k)|^2}{\mathrm{i}(\omega_k-\omega)+\varepsilon}, \tag{12.2.15}$$

$$= \lim_{\varepsilon\to0}\int_0^\infty d\omega_k\, g(\omega_k)|\kappa(\omega_k)|^2\left[\frac{\varepsilon}{(\omega_k-\omega)^2+\varepsilon^2} - \frac{\mathrm{i}(\omega_k-\omega)}{(\omega_k-\omega)^2+\varepsilon^2}\right], \tag{12.2.16}$$

$$\equiv \tfrac{1}{2}\Gamma + \mathrm{i}\delta\omega. \tag{12.2.17}$$

In the limit $\varepsilon \to 0$, we can write, for any reasonably well-behaved function $f(z)$

$$\int \frac{\varepsilon}{z^2+\varepsilon^2} f(z)\,dz \to \pi f(0), \tag{12.2.18}$$

$$\int \frac{z}{z^2+\varepsilon^2} f(z)\,dz \to \mathrm{P}\int \frac{dz\,f(z)}{z}, \tag{12.2.19}$$

where $\mathrm{P}\int$ means the principal value integral. This is often summarized in the formula

$$\int_0^\infty d\tau\,\mathrm{e}^{-\mathrm{i}z\tau} = \pi\delta(z) + \mathrm{i}\frac{\mathrm{P}}{z}. \tag{12.2.20}$$

Exercise 12.1 Principal Value Integral: The principal value integral is defined mathematically by

$$\mathrm{P}\int_{-a}^{b} \frac{dz\,f(z)}{z} \equiv \lim_{\delta\to 0+}\left\{\int_{-a}^{-\delta}\frac{dz\,f(z)}{z} + \int_{\delta}^{b}\frac{dz\,f(z)}{z}\right\}, \tag{12.2.21}$$

where a and b are positive. Show that, when the principal value integral exists, (12.2.19) gives the same result for reasonably well-behaved functions.

e) Decay Constant and Lineshift: We can now put together the results we have derived in (12.2.14)–(12.2.21), and thus write (12.2.13) in the form

$$\dot{v}(e,t) = -\left(\tfrac{1}{2}\Gamma + \mathrm{i}\delta\omega\right)v(e,t), \tag{12.2.22}$$

$$\Gamma \equiv 2\pi g(\omega)|\kappa(\omega)|^2, \tag{12.2.23}$$

$$\delta\omega \equiv -\mathrm{i}\mathrm{P}\int \frac{d\omega_k\,g(\omega_k)|\kappa(\omega_k)|^2}{\omega-\omega_k}. \tag{12.2.24}$$

Using the initial condition (12.2.5), this leads to the solution

$$u(e,t) = \mathrm{e}^{-\mathrm{i}(\omega+\delta\omega)t-\Gamma t/2}. \tag{12.2.25}$$

Thus, the interaction with the quantized electromagnetic field leads to an exponential decay of the excited state population $|u(e,t)|^2$, with a lifetime given by $1/\Gamma$. The quantity Γ is called the *decay constant* or the *linewidth*, while $\delta\omega$ is known as the *lineshift*.

The formula for the lineshift as given in (12.2.24) is divergent unless a cutoff is imposed. In our drastically approximated treatment, in which only a single transition is treated, a cutoff would be necessary to eliminate the effect of other transitions, and would thus be a relatively small frequency. A full treatment using relativistic quantum electrodynamics still gives a divergence, which can be remedied using renormalization theory—the resulting shift is known as the *Lamb shift*. The order of magnitude of the shift calculated this way is still much the same as that of the linewidth Γ, and is usually very small.

The decay constant is a much more important parameter, since it leads to qualitatively different behaviour, namely, the decay of excited states and the consequent broadening of the spectral line. Furthermore, the formula (12.2.23) derived for it does not involve the use of cutoffs or renormalization.

12.2.2 Solutions for Atomic Decay and Radiated Field

So far we have studied mainly the state of the atom. The full solutions in this degree of approximation are implicit in the method of solution chosen, and are very illuminating. It is even possible to put these solutions in a form which looks very like that found classically for the radiation from an electric dipole.

a) Full Solution for Field and Atom: Having obtained the solution for $u(e,t)$, we can go back to (12.2.6) to get

$$u(k,g,t) = \kappa_k^* \int_0^t dt'\, \mathrm{e}^{-\mathrm{i}\omega_k(t-t')}\mathrm{e}^{-(\mathrm{i}\omega+\delta\omega+\Gamma/2)t'}, \tag{12.2.26}$$

$$= \mathrm{i}\kappa_k^* \left(\frac{\mathrm{e}^{-\mathrm{i}\omega_k t} - \mathrm{e}^{-\mathrm{i}(\omega+\delta\omega)t-\Gamma t/2}}{\omega_k - \omega - \delta\omega + \mathrm{i}\Gamma/2} \right). \tag{12.2.27}$$

The full solution is now given by (12.2.2), namely

$$|\Psi,t\rangle = u(e,t)|e,0\rangle + \sum_k u(g,k,t)|g,k\rangle, \tag{12.2.28}$$

with the values of $u(e,t)$ and $u(g,k,t)$ given by (12.2.25, 12.2.27).

b) Entanglement between the Atom and the Electromagnetic Field: The quantum state $|\Psi,t\rangle$ is an *entangled state.* This is a state representing two distinct physical subsystems A and L (in this case the atom and the electromagnetic field) in which it is not possible to write the state as a simple direct product $|a\rangle \otimes |l\rangle$, no matter what bases are chosen for each of the subsystems.

In this case, we *can* write

$$|e,0\rangle = |e\rangle \otimes |0\rangle, \qquad |g,k\rangle = |g\rangle \otimes |k\rangle, \tag{12.2.29}$$

but there is no way of factoring $|\Psi,t\rangle$ itself in the same way into a product of the form $|\text{atom}\rangle \otimes |\text{light}\rangle$. The simplest consequence of this is that if a *photon* is detected, then the *atom* is in the ground state. Even if the photon is detected 100 metres away from the atom, it is *instantly* known that the atom is in the ground state.

Quantum entanglement is the foundation of quantum information and quantum computation, and is treated more extensively in *Book II.*

c) The Radiated Electromagnetic Field: In this state it is clear that

$$\langle\Psi,t|a_k|\Psi,t\rangle = 0, \tag{12.2.30}$$

$$\langle\Psi,t|\sigma^{\pm}|\Psi,t\rangle = 0, \tag{12.2.31}$$

$$\langle\Psi,t|\sigma_z|\Psi,t\rangle = |u(e,t)|^2 - \sum_k |u(g,k,t)|^2 = 2|u(e,t)|^2 - 1. \tag{12.2.32}$$

Since $\langle\Psi,t|a_k|\Psi,t\rangle = 0$, there is no mean electric field, nor any polarization, which would require non-vanishing $\langle\Psi,t|\sigma^{\pm}|\Psi,t\rangle$. There is of course a mean square field, and a mean Poynting vector, since these are bilinear in creation and destruction operators.

One would like to get a reasonably simple idea of the propagating radiated electric field, and we can do this by noting that

$$\langle\Psi,t|\sigma^{+}a_k|\Psi,t\rangle = u^*(e,t)u(g,k,t) \neq 0. \tag{12.2.33}$$

This is a consequence of the entanglement property noted above; it says that there can be a non-vanishing electromagnetic field if the atom is in the ground state. Making the replacement $\sigma^{+} \to \sigma^{-}$ in this equation gives zero, showing that it is not possible for there to be a non-vanishing electromagnetic field if the atom is in the excited state.

With these solutions we can now calculate the mean value

$$\langle \boldsymbol{E}_S^{(+)}(\boldsymbol{x},t)\sigma_S^{+}(t)\rangle = \langle\Psi,t|\boldsymbol{E}_S^{(+)}(\boldsymbol{x})\sigma_S^{+}|\Psi,t\rangle. \tag{12.2.34}$$

In this expression, the subscript S emphasizes that we are in the Schrödinger picture, and the operator is not time-dependent. To evaluate this expression we need the expression (12.1.2) and this leads us to evaluate

$$\langle a_k(t)\sigma^{+}(t)\rangle = \langle a_k^{\dagger}(t)\sigma^{-}(t)\rangle^* = \langle\Psi,t|a_k|\Psi,t\rangle = u^*(e,t)u(g,k,t), \tag{12.2.35}$$

$$= \frac{\kappa_k^* e^{-\Gamma t/2}}{\hbar}\left\{\frac{e^{-i(\omega_k-\omega-\delta\omega)t} - e^{-\Gamma t/2}}{(\omega_k-\omega-\delta\omega)+i\Gamma/2}\right\}. \tag{12.2.36}$$

The mean value we want is then

$$\langle \boldsymbol{E}^{(+)}(\boldsymbol{x},t)\sigma^{+}(t)\rangle = i\sum_k \left(\frac{\hbar\omega_k}{2\varepsilon_0}\right)^{\frac{1}{2}} \left\langle a_k(t)\sigma^{+}(t)\right\rangle \mathbf{u}_k(\boldsymbol{x}). \tag{12.2.37}$$

The expression can be evaluated, though there is a lot of algebra involved. The final result is

$$\langle \boldsymbol{E}^{(+)}(\boldsymbol{x},t)\sigma^{+}(t)\rangle = \frac{e^{(i(\omega+\delta\omega)-\Gamma/2)t}\,\nabla\times\left(\boldsymbol{d}_{eg}\times\nabla\left(\theta(ct-r)e^{(i(\omega+\delta\omega)-\Gamma/2)(r-ct)/c}\right)\right)}{2(2\pi)^3\epsilon_0 r}, \tag{12.2.38}$$

where c is the speed of light and $r = |\boldsymbol{x}|$. The first factor in braces is the time dependence appropriate to σ^{+}, while the remaining factor is a propagating damped outgoing wave.

Exercise 12.2 Calculation of the Radiated Field: Verify the result (12.2.38).

12.3 The Two-Level Atom in a Strong Classical Driving Field

Atoms can be manipulated in a wide variety of ways by the application of strong coherent optical fields, for which the quantum nature of the optical field plays a rather unimportant rôle, and thus may be considered to be classical. In this section we will firstly show how transfer from one state to another can be effected, and then discuss the basic principle of an optical trapping potential for a two-level system.

12.3.1 Interaction Hamiltonian

We will develop our treatment in close analogy with that given in Sect. 12.1.1 and Sect. 12.1.2 for the interaction with a quantized electromagnetic field. As in those sections, we start in the Schrödinger picture, but in this case there is no electromagnetic Hamiltonian, and the time dependence of the classical electromagnetic field must be included explicitly in the Schrödinger picture Hamiltonian. Thus, this takes the form

$$H = H_{\text{Atom}} + H_{\text{Int}}, \tag{12.3.1}$$

with

$$H_{\text{Atom}} = \frac{\boldsymbol{p}^2}{2m} + e\phi(\boldsymbol{x}), \tag{12.3.2}$$

$$H_{\text{Int}} = -\boldsymbol{d}\cdot\boldsymbol{E}(0,t). \tag{12.3.3}$$

Here the electric dipole moment operator is, as previously,

$$\boldsymbol{d} \equiv e\boldsymbol{x}. \tag{12.3.4}$$

a) Two-Level Approximation: The interaction with a classical field can then be made with the electric dipole approximation and the two-level system approximation, leading to the Hamiltonian

$$H = \begin{pmatrix} E_e & \boldsymbol{d}_{eg}\cdot\boldsymbol{E}(0,t) \\ \boldsymbol{d}^*_{eg}\cdot\boldsymbol{E}(0,t) & E_g \end{pmatrix}. \tag{12.3.5}$$

b) Monochromatic Electromagnetic Field: We consider first the case of a strictly monochromatic electromagnetic field, which we introduce in the form (chosen for the convenience of the calculations)

$$\boldsymbol{d}_{eg}\cdot\boldsymbol{E}(0,t) = 2\mathrm{i}\mathcal{E}\cos(\omega t). \tag{12.3.6}$$

c) Schrödinger Equation in the Interaction Picture: We introduce an interaction picture by writing the wavefunction in the form

$$|\psi,t\rangle = A_e(t)\mathrm{e}^{\mathrm{i}\omega_e t}|e\rangle + A_g(t)\mathrm{e}^{\mathrm{i}\omega_g t}|g\rangle = \begin{pmatrix} A_e(t)\mathrm{e}^{\mathrm{i}\omega_e t} \\ A_g(t)\mathrm{e}^{\mathrm{i}\omega_g t} \end{pmatrix}, \tag{12.3.7}$$

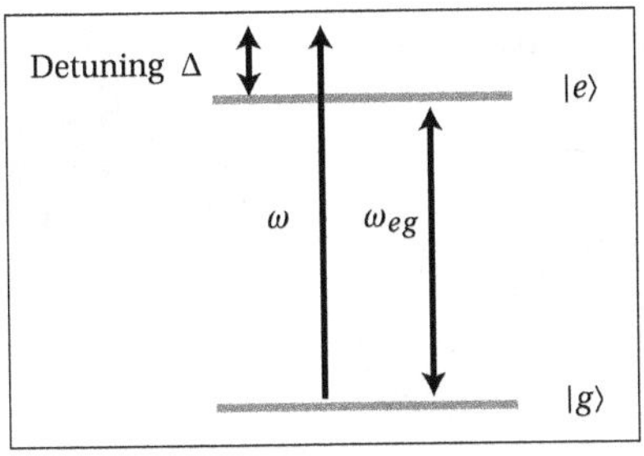

Fig. 12.2. Near-resonant excitation of a two-level system.

and the Schrödinger equation in the interaction picture becomes

$$\dot{A}_e(t) = \frac{2\mathcal{E}}{\hbar} e^{i\omega_{eg} t} \cos\omega t \, A_g(t), \tag{12.3.8}$$

$$\dot{A}_g(t) = -\frac{2\mathcal{E}^*}{\hbar} e^{-i\omega_{eg} t} \cos\omega t \, A_e(t). \tag{12.3.9}$$

Here we use the notation

$$\hbar\omega_{eg} \equiv E_e - E_g. \tag{12.3.10}$$

The driven atom and its parameters are illustrated in Fig. 12.2.

12.3.2 Solution of the Schrödinger Equation

a) Rotating Wave Approximation: We can write

$$e^{i\omega_{eg} t} \cos\omega t = \tfrac{1}{2} e^{i(\omega_{eg}+\omega)t} + \tfrac{1}{2} e^{i(\omega_{eg}-\omega)t}. \tag{12.3.11}$$

We want to consider the situation in which $\omega_{eg} \approx \omega$, so that the first term oscillates very rapidly with time compared with the second, and we can assume that it averages to zero on the time scale of the solutions of the equations of motion. In this case, the equations simplify to

$$\dot{A}_e(t) = \frac{\mathcal{E}}{\hbar} e^{i(\omega_{eg}-\omega)t} A_g(t), \tag{12.3.12}$$

$$\dot{A}_g(t) = -\frac{\mathcal{E}^*}{\hbar} e^{-i(\omega_{eg}-\omega)t} A_e(t). \tag{12.3.13}$$

b) Rabi Frequency and Detuning: We can simplify the equations by defining:

The detuning	$\Delta \equiv \omega - \omega_{eg}$,	(12.3.14)
The Rabi frequency	$\Omega_R \equiv \left\lvert \frac{2\mathcal{E}}{\hbar} \right\rvert = \left\lvert \frac{\boldsymbol{d}_{eg} \cdot \boldsymbol{E}}{\hbar} \right\rvert$,	(12.3.15)
The phase of $\mathcal{E}$ by	$\mathcal{E} \equiv \lvert\mathcal{E}\rvert e^{i\phi}$.	(12.3.16)

c) Rabi Hamiltonian: We can remove the explicit time dependence of the coefficients by writing

$$a_e(t) \equiv A_e(t)\mathrm{e}^{\mathrm{i}(\Delta t-\lambda)}, \qquad a_g(t) \equiv A_g(t)\mathrm{e}^{\mathrm{i}\lambda}, \qquad \lambda = \phi/2 - \pi/4, \tag{12.3.17}$$

which leaves the equations in the *Rabi Hamiltonian* form

$$i\hbar\frac{d}{dt}\begin{pmatrix} a_e(t) \\ a_g(t)\end{pmatrix} = H_{\text{Rabi}}(\Omega_R,\Delta)\begin{pmatrix} a_e(t) \\ a_g(t)\end{pmatrix}, \tag{12.3.18}$$

in which

$$H_{\text{Rabi}}(\Omega_R,\Delta) \equiv \hbar\begin{pmatrix} -\Delta & -\frac{1}{2}\Omega_R \\ -\frac{1}{2}\Omega_R & 0\end{pmatrix}. \tag{12.3.19}$$

12.3.3 Optical Pulses

Using pulsed laser fields, it is possible to manipulate the quantum state of a two-level system, and this forms the basis of concept of *quantum state engineering.* In this case, the equation of motion remains the same as (12.3.18), but the Rabi frequency becomes $\Omega_R(t)$, a time-dependent quantity.

The most important special case of a pulse shape is the *rectangular pulse*

$$\mathcal{E}(t) = \begin{cases} \mathcal{E}, & \text{for } 0 \leqslant t \leqslant T, \\ 0, & \text{otherwise.}\end{cases} \tag{12.3.20}$$

Here the step function must be understood as being "slow" on the optical time scale that is, the step function is an idealization of the turning on and off of an optical field, both of which take place over many optical cycles.

a) Validity of the Rotating Wave Approximation: The condition for the off resonant terms to be negligible requires that all time dependences retained and of interest must be on a slow time scale. This means that

$$\Omega_R \ll \omega \approx \omega_{eg}, \tag{12.3.21}$$

$$|\Delta| \ll \omega \approx \omega_{eg}, \tag{12.3.22}$$

and as well that $\mathcal{E}(t)$ be slowly varying on the optical time scale, which requires

$$\left|\frac{1}{\mathcal{E}(t)}\frac{d\mathcal{E}(t)}{dt}\right| \ll \omega. \tag{12.3.23}$$

This means that a "instantaneous" turn on of a square pulse as in (12.3.20) has to be interpreted as being slow on the optical time scale, but fast on the time scales given by Ω_R and Δ.

Under these conditions, the coefficients $a_e(t)$ and $a_g(t)$ are also slowly varying on the optical time scale.

b) On-Resonance Rabi Oscillations: For $\Delta = 0$ we have

$$i\frac{d}{dt}\begin{pmatrix} a_e \\ a_g \end{pmatrix} = \begin{pmatrix} 0 & -\frac{1}{2}\Omega_R(t) \\ -\frac{1}{2}\Omega_R(t) & 0 \end{pmatrix}\begin{pmatrix} a_e \\ a_g \end{pmatrix}. \tag{12.3.24}$$

These equations can be solved exactly by writing them in terms of the total angle given by the time integral of the Rabi frequency. Thus we introduce the variable

$$\tau(t) \equiv \int_{-\infty}^{t} \Omega_R(t)\, dt, \tag{12.3.25}$$

as a new variable instead of time, and we then solve the equation exactly in the form

$$\begin{pmatrix} a_e(t) \\ a_g(t) \end{pmatrix} = U_t \begin{pmatrix} a_e(-\infty) \\ a_g(-\infty) \end{pmatrix}, \tag{12.3.26}$$

$$\text{where } U_t \equiv \begin{pmatrix} \cos\frac{1}{2}\tau(t) & -i\sin\frac{1}{2}\tau(t) \\ -i\sin\frac{1}{2}\tau(t) & \cos\frac{1}{2}\tau(t) \end{pmatrix}. \tag{12.3.27}$$

The probability of being in the excited state, having initially been in the ground state, oscillates, a phenomenon called *Rabi oscillations*:

$$P_{e\leftarrow g}(t) = \sin^2\tfrac{1}{2}\tau(t) = \tfrac{1}{2}\big(1 - \cos\tau(t)\big). \tag{12.3.28}$$

c) Particular Pulses:

i) A *π-pulse* is defined by a choice of the function $\mathcal{E}(t)$ so that the total angle generated from beginning to the end of the pulse is π, that is

$$\tau^{\pi}(\infty) = \int_{-\infty}^{\infty} \Omega_R^{\pi}(t)\, dt = \pi. \tag{12.3.29}$$

The effect of this is to invert the two-level system, so that

$$U_{\infty}^{\pi} = \begin{pmatrix} 0 & -i \\ -i & 0 \end{pmatrix}. \tag{12.3.30}$$

ii) A *2π-pulse* is similarly defined by a different choice of the function $\mathcal{E}(t)$ so that the total angle is 2π, that is

$$\tau^{2\pi}(\infty) = \int_{-\infty}^{\infty} \Omega_R^{2\pi}(t)\, dt = 2\pi. \tag{12.3.31}$$

This pulse inverts the two-level system twice, thus returning it to its ground state,

$$U_{\infty}^{2\pi} = \begin{pmatrix} -1 & 0 \\ 0 & -1 \end{pmatrix} \tag{12.3.32}$$

but with a negative sign for the amplitudes, because a rotation of a spin half system by 2π induces a sign change.

iii) A pulse can be chosen to give any rotation angle required. By allowing non-resonant pulses as well, we can also manipulate the relative phase of the excited and ground state amplitudes.

12.3.4 Effective Potential on a Ground State Atom

We consider the case where $A_g(0) = 1$, and $A_e(0) = 0$, that is an atom initially in its ground state. In this case we integrate (12.3.12) and substitute this into (12.3.13) to get

$$\dot{A}_g(t) = -\frac{\Omega_R^2}{4}\int_0^t dt'\, e^{i\Delta(t-t')} A_g(t') - \frac{\mathcal{E}}{\hbar} e^{i\Delta t} A_e(0). \tag{12.3.33}$$

Consider the case where $|\Omega_R|^2$ is sufficiently small that $A_g(t)$ varies slowly compared to the oscillating exponential; in this case:

i) The second term will oscillate very rapidly on this time scale, and average to zero—therefore we can omit this term.

ii) Because $A_g(t)$ varies slowly, we can write

$$\dot{A}_g(t) \approx -\frac{\Omega_R^2}{4} A_g(t) \int_0^t dt'\, e^{i\Delta(t-t')}, \tag{12.3.34}$$

$$= -i\frac{\Omega_R^2}{4} A_g(t) \frac{1 - e^{i\Delta t}}{\Delta}. \tag{12.3.35}$$

a) The Optical Potential: As above, on the time scale in which $A_g(t)$ changes, we can neglect the rapidly varying exponential term, to get a simple equation for the ground state amplitude

$$i\hbar \dot{A}_g(t) \approx \frac{\hbar\Omega_R^2}{4\Delta} A_g(t). \tag{12.3.36}$$

This is a very simple kind of Schrödinger equation for the ground state atom under the influence of an off-resonance light field, in which there is effectively an *optical potential* defined by

$$V_{\text{opt}} = \frac{\hbar\Omega_R^2}{4\Delta}. \tag{12.3.37}$$

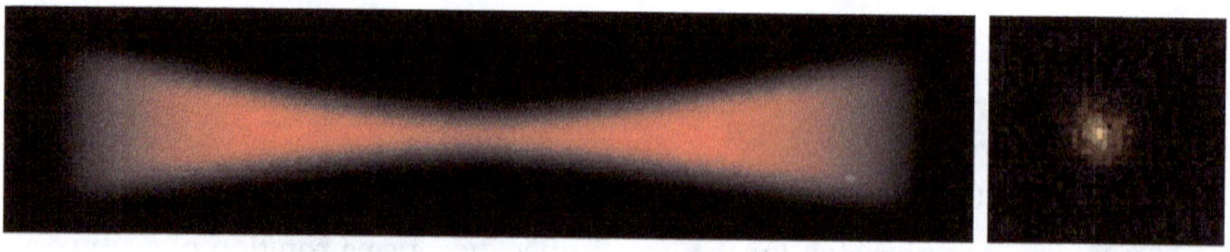

Fig. 12.3. Left: Schematic of an optical trap created by a tightly focused laser beam; Right: Individual atoms can be trapped on demand in a FORT [50], and the image shows a single ^{85}Rb atom trapped in a FORT in the laboratory of Dr Mikkel F. Andersen at the University of Otago (June 2010), where the technique was developed. (*Image kindly supplied by Dr Andersen.*)

b) Properties of the Optical Potential:

i) The result requires that $\Delta \gg \Omega_R$, since the time rate of change of A_g must be very much slower than that corresponding to Δ. From (12.3.37), this requires that $\Omega_R^2/\Delta \ll \Delta$, giving the result.

ii) *Red detuning* means that $\omega < \omega_{eg}$, that is the driving field has a frequency less than the transition frequency. In this case $V_{\rm opt} < 0$, and the potential attracts.

iii) *Blue detuning* means that $\omega > \omega_{eg}$, that is the driving field has a frequency greater than the transition frequency. In this case $V_{\rm opt} > 0$, and the potential repels.

c) Optical Trapping—the FORT: The acronym FORT means the *Far Off Resonance Trap.* If we take a spatially dependent optical field, in which the spatial dependence of the field is on a spatial scale much larger than that of the atomic wavefunction, the Rabi frequence becomes spatially dependent $\Omega_R(\boldsymbol{x})$, and the optical potential also becomes spatially dependent. The energies involved are in practice quite small, so that the potential is quite weak, but still strong enough to trap atoms at microKelvin temperatures—see Fig. 12.3.

Exercise 12.3 Direct Derivation from the Hamiltonian: Find the energy eigenvalues of the Hamiltonian matrix (12.3.19), and show that when $\Delta \gg \Omega_R$, they give the optical potential. What else can you deduce from these eigenvalues?

12.4 Interaction of a Two-Level Atom with a Single Mode

By using an optical cavity the interaction with a particular mode of the cavity can be so enhanced that the other modes can be neglected completely, leaving the *Jaynes–Cummings* [51] Hamiltonian

$$H = H_0 + H_1, \tag{12.4.1}$$

$$H_0 = \tfrac{1}{2}\hbar\omega\sigma_z + \hbar\omega a^\dagger a, \tag{12.4.2}$$

$$H_1 = \hbar g(a\sigma^+ + a^\dagger\sigma^-). \tag{12.4.3}$$

Notice that $[H_0, H_1] = 0$, so that we can move to the interaction picture without changing the interaction Hamiltonian.

If the initial state of the system is $|e,n\rangle$, a state of n photons with the atom excited, the only other state connected is $|g,n+1\rangle$. Thus, the state can be written

$$|\Psi,t\rangle = u(e,t)|e,n\rangle + u(g,t)|g,n+1\rangle, \tag{12.4.4}$$

$$\text{with} \quad u(e,0) = 1, \qquad u(g,0) = 0. \tag{12.4.5}$$

and the equations of motion in the interaction picture are

$$i\dot{u}(e,t) = g\sqrt{n+1}\,u(g,t), \tag{12.4.6}$$

$$i\dot{u}(g,t) = g\sqrt{n+1}\,u(e,t). \tag{12.4.7}$$

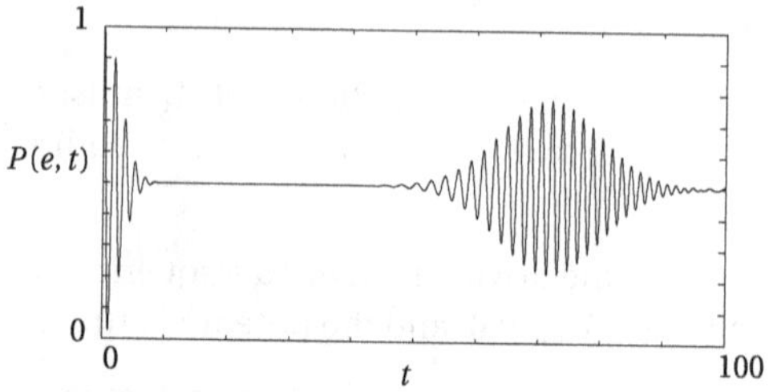

Fig. 12.4. Quantum revivals in the occupation probability of the excited state as given by (12.4.13).

Setting $\Omega_n = g\sqrt{n+1}$, these equations have the solution

$$u(e,t) = \cos(\Omega_n t), \quad (12.4.8)$$

$$u(g,t) = -\mathrm{i}\sin(\Omega_n t). \quad (12.4.9)$$

The system thus oscillates back and forth from excited to ground state of the atom, with the energy moving back and forth from the field to the atom. The probability of finding the atom in the excited state (and with n photons) is

$$P(e,t) = \cos^2 \Omega_n t. \quad (12.4.10)$$

12.4.1 Quantum Collapses and Revivals

Suppose we put the field in an initial coherent state, and the atom in the excited state

$$|\Psi,0\rangle = |e,\alpha\rangle = \mathrm{e}^{-\frac{1}{2}|\alpha|^2} \sum_n \frac{\alpha^n}{\sqrt{n!}} |e,n\rangle, \quad (12.4.11)$$

then this state evolves into

$$|\Psi,t\rangle = \mathrm{e}^{-|\alpha|^2} \sum_n \frac{\alpha^n}{\sqrt{n!}} \{\cos\Omega_n t |e,n\rangle + \sin\Omega_n t |g,n\rangle\}. \quad (12.4.12)$$

The probability of occupation of the excited state is

$$P(e,t) = \mathrm{e}^{-|\alpha|^2} \sum_n \frac{|\alpha|^n}{n!} \cos^2 \Omega_n t. \quad (12.4.13)$$

The different frequencies are incommensurate with each other, and soon get out of phase, so the probability of the excited state decays initially. However, since there are only a finite number which are effectively occupied, eventually the important ones can get in phase again, and the probability has a *quantum revival,* as illustrated in Fig. 12.4. The terminology "quantum revival" reflects the fact that this behaviour is a direct consequence of quantization.

IV Quantum Stochastic Processes

13. Quantum Markov Processes

The theory of quantum Markov processes is the modern theory of irreversible processes in quantum mechanics, and in particular the theory of the approach to equilibrium or to a non-equilibrium steady state. It was developed mainly in the field of quantum optics, and so retains a flavour of that field. However, such a framework is also highly relevant in studying the thermal processes in ultra-cold gases, and is well adapted to the development of a quantum kinetic theory for ultra-cold atoms, as is presented in *Book III.* This chapter gives a brief outline of the formulation of quantum Markov processes—a fuller development of all the techniques is given in *Book II.*

13.1 Two-Level Atom in a Finite-Temperature Electromagnetic Field

We shall introduce our description of quantum Markov processes with the very specific example of an atom in a radiation field, because this is much easier to grasp than a fully general treatment. It will be made clear, however, that the methods used are not specific to this problem. In particular, the finite temperature electromagnetic field is only one example of a range of possible environments, and does not permit the full range of behaviours available in more general environments.

13.1.1 The Quantum-Mechanical Master Equation

The treatment of atomic decay in Sect. 12.2 of the previous chapter is quite simple, but applies only in the case that the atom is initially in the excited state, and that the electromagnetic field is initially unoccupied. Generalization of that treatment to more general situations would be possible, but would yield a very complicated kind of formalism.

The idea of a quantum stochastic process in this context is to get a description of the quantum state of the *atom*, with the electromagnetic field being seen as both a source of noise, and as a reservoir into which energy can be dissipated. This, the essence of the procedure used in Sect. 12.2, is what we wish to generalize and formalize, and in particular, we want to include the Markov approximation of Sect. 12.2.1 b.

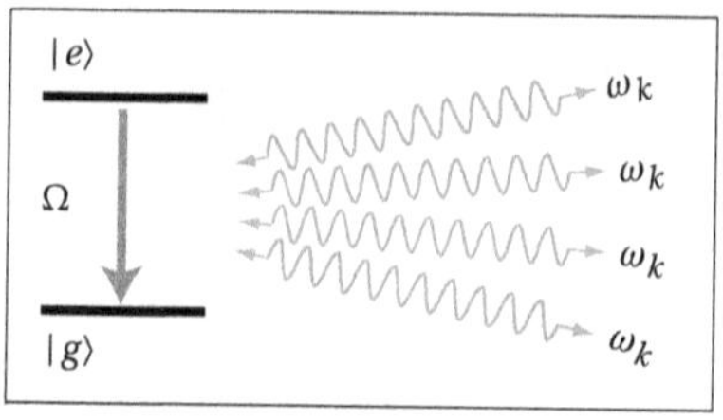

Fig. 13.1. Interaction of a two-level system with the electromagnetic field.

The density matrix formalism of quantum mechanics provides the most suitable framework for such a theory. For a two-level system the density matrix is a 2×2 matrix ρ_{sys}. This contains all the information about the state of the atom, and no direct information about the infinite numbers of modes which describe the electromagnetic field, or its coupling to the atom. This reduced description is very convenient and in practice nearly always provides all the information we need. However, in quantum-optical situations, information about the electromagnetic field can be important, and techniques for this are given in *Quantum Noise* and in *Book II.*

13.1.2 System and Heat Bath

Let us now be more specific in the framework we need. The kind of system we are considering is schematically shown in Fig. 13.1, and can described by a Hamiltonian consisting of three parts, which we write in the form

$$H = H_{sys} + H_B + H_{int}. \tag{13.1.1}$$

Here the three parts can be quite general, but for our example of a two-level atom, they take the following form:

i) *The System*: This has normally a few degrees of freedom, for example, in the case of a two-level atom, we need only the excited and ground states, namely

$$\text{Excited state} \quad |e\rangle, \tag{13.1.2}$$

$$\text{Ground state} \quad |g\rangle, \tag{13.1.3}$$

and the corresponding Hamiltonian can be written

$$H_{sys} = \tfrac{1}{2}\hbar\Omega\Big(|e\rangle\langle e| - |g\rangle\langle g|\Big). \tag{13.1.4}$$

ii) *The Heat Bath*: The Hamiltonian H_B describes the interaction with a large system with many degrees of freedom, conventionally known as a *heat bath.* It is a system which can exchange energy with the subsystem of interest, in this case the two-level atom. In this particular case, the large system will be the electromagnetic field, for which we take a simplified description by eliminating polarizations and restricting to one dimension. We introduce creation and destruction operators a_k, $a_k^\dagger$, with the commutation relations

$$[a_k, a_{k'}^\dagger] = \delta_{kk'}, \tag{13.1.5}$$

and the simplified electromagnetic field Hamiltonian is

$$H_B = \sum_k \hbar\omega_k a_k^\dagger a_k. \tag{13.1.6}$$

Neither the particular form of the bath Hamiltonian nor the commutation relations (13.1.5) are important, as long as the bath operators satisfy the commutation relations

$$[H_B, a_k] = -\omega_k a_k, \qquad [H_B, a_k^\dagger] = -\omega_k a_k^\dagger. \tag{13.1.7}$$

These can be derived from (13.1.5, 13.1.6), but can also be true quite independently of them. Examples are given in Sect. 13.3.1.

iii) *The Interaction between the Atom and the Heat Bath*: This is now put in the simplest form—we assume that it will allow the transitions

$$e \to g + \hbar\omega_k, \qquad \hbar\omega_k + g \to e, \tag{13.1.8}$$

corresponding to emission and absorption of a bath quantum. A Hamiltonian which permits this is

$$H_{\rm int} = \mathrm{i}\hbar \sum_k \kappa_k \left(A^\dagger a_k - A a_k^\dagger\right), \tag{13.1.9}$$

in which

$$A = |g\rangle\langle e|, \qquad A^\dagger = |e\rangle\langle g|, \tag{13.1.10}$$

$$[H_{\rm sys}, A] = -\hbar\Omega A, \qquad [H_{\rm sys}, A^\dagger] = \hbar\Omega A^\dagger. \tag{13.1.11}$$

Here κ_k is a coupling constant whose exact form is not in practice very important. For the case of an atom interacting with an electromagnetic field, it can be calculated exactly, but this is not universally the case. When the interaction Hamiltonian is written in the form (13.1.9) above it is a *real* quantity.

iv) *Density of States and Smoothness Requirements*: As in the case of the decay of the two-level atom, we replace the summation by an integral over frequency, and a density of states

$$\sum_k |\kappa_k|^2 \to \int d\omega_k\, g(\omega_k)|\kappa(\omega_k)|^2. \tag{13.1.12}$$

Generically, to derive an appropriate master equation, it is important only that $|\kappa(\omega_k)|^2$ be a *smooth function of* ω_k in the range of frequencies important for the dynamics.

13.1.3 The Master Equation

The *full density operator* $\rho_{\rm tot}$, which describes both the atom and the heat bath is related to the *system density operator* by summing over the bath degrees of freedom thus

$$\rho_{\rm sys} = \mathrm{Tr}_B\left\{\rho_{\rm tot}\right\}. \tag{13.1.13}$$

By making the same kinds of approximations as in Sect. 12.2, we will be able to show that the equation of motion for the system density operator takes the form

$$\begin{aligned}\frac{\partial\rho_{\rm sys}}{\partial t} &= -\frac{\rm i}{\hbar}\left[H_{\rm sys},\rho_{\rm sys}\right]-{\rm i}\left[\delta_1 A^\dagger A-\delta_2 AA^\dagger,\rho_{\rm sys}\right]\\ &+\frac{\gamma}{2}(\bar N(\Omega)+1)\left(2A\rho_{\rm sys}A^\dagger-\rho_{\rm sys}A^\dagger A-A^\dagger A\rho_{\rm sys}\right)\\ &+\frac{\gamma}{2}\bar N(\Omega)\left(2A^\dagger\rho_{\rm sys}A-\rho_{\rm sys}AA^\dagger-AA^\dagger\rho_{\rm sys}\right). \end{aligned} \tag{13.1.14}$$

This is an equation for the reduced density operator in which the influence of the heat bath is encapsulated in the four quantities $\bar N(\Omega)$, δ_1, δ_2, and γ, all of which represent properties of the heat bath. They are given by

$$\gamma = 2\pi g(\Omega)|\kappa(\Omega)|^2, \tag{13.1.15}$$

$$\delta_1 = {\rm P}\int d\omega\,\frac{g(\omega)|\kappa(\omega)|^2(\bar N(\omega)+1)}{\Omega-\omega}, \tag{13.1.16}$$

$$\delta_2 = {\rm P}\int d\omega\,\frac{g(\omega)|\kappa(\omega)|^2\bar N(\omega)}{\Omega-\omega}. \tag{13.1.17}$$

Here we use the *principal value integral* ${\rm P}\int d\omega$, as defined in (12.2.21). The use here of the principal value integral has a close relationship to its use in Chap. 12. In fact, when $\bar N(\omega)=0$ (which corresponds to a heat bath at absolute zero) the constants δ_1,δ_2,γ correspond precisely to those found in the derivation of atomic decay in Chap. 12.

a) State of the Heat Bath: The quantity $\bar N(\Omega)$ is determined by the state of the heat bath, in this case the electromagnetic field. Here it has been assumed to be thermal at temperature T; specifically

$$\mathrm{Tr}_B\left\{a_k^\dagger a_{k'}\rho_B\right\} = \delta_{k,k'}\bar N(\omega_k), \tag{13.1.18}$$

$$\mathrm{Tr}_B\left\{a_k a_{k'}^\dagger\rho_B\right\} = \delta_{k,k'}\left(\bar N(\omega_k)+1\right), \tag{13.1.19}$$

$$\bar N(\omega_k) \equiv \frac{1}{e^{\hbar\omega_k/k_B T}-1}. \tag{13.1.20}$$

However, the conditions necessary for a master equation of this kind to be valid are more general, as will be shown in Sect. 13.3, where we consider different heat baths, yielding different values for the bath correlation functions (13.1.18, 13.1.19).

b) The Lindblad Form: The form of the master equation above, that is, a combinations of Hamiltonian terms, like that in the first line, and dissipative terms as in the next two lines, is known as the *Lindblad form*—for a general discussion of this form, see *Quantum Noise*.

Explicitly, a master equation of the Lindblad form is one which can be written in the form

$$\frac{\partial\rho}{\partial t} = L\rho \equiv -\frac{i}{\hbar}[H,\rho] + \sum_i \tfrac{1}{2}\gamma_i \mathcal{M}_i \left\{2A_i\rho A_i^\dagger - \rho A_i^\dagger A_i - A_i^\dagger A_i \rho\right\} + \sum_i \tfrac{1}{2}\gamma_i \mathcal{N}_i \left\{2A_i^\dagger\rho A_i - \rho A_i A_i^\dagger - A_i A_i^\dagger \rho\right\}. \quad (13.1.21)$$

In this master equation:

i) The Hamiltonian H is Hermitian, and in principle quite arbitrary.

ii) The A_i are arbitrary operators, which in practice are usually operators like raising and lowering operators, with properties similar to (13.1.10, 13.1.11) with respect to some Hamiltonian, which we can call a *basis Hamiltonian* H_{basis}. This is normally closely related to the Hamiltonian H in the Lindblad definition (13.1.21), but need not be identical with it. When this is the case, it is usual to take the A_i as lowering operators, and the $A_i^\dagger$ as raising operators.

iii) The coefficients $\mathcal{M}_i, \mathcal{N}_i$ are real non-negative c-numbers (either of which can be the larger). They normally represent the properties of the heat bath, derived from appropriate bath correlation functions.

iv) The coefficients γ_i (which are positive) are introduced to represent the coupling between the system and the heat bath. From a purely mathematical point of view, they represent only notational convenience, since they could be absorbed into the definition of the $\mathcal{M}_i, \mathcal{N}_i$.

If a master equation is of the Lindblad form, the time evolution is completely consistent with quantum mechanics, and in particular, this means that the solution for the density operator is always a positive definite operator, so that no negative probabilities occur. Master equations which are not of the Lindblad form do occur, and the lack of any guarantee that their solutions give positive probabilities is often not a problem, since the deviations are often small, and transient—see [52].

Exercise 13.1 Equations of Motion for Mean Values: If G is an arbitrary operator and $\rho(t)$ is a solution of the master equation (13.1.21), show that the equation of motion for $\langle G(t)\rangle \equiv \mathrm{Tr}\left\{G\rho(t)\right\}$ is

$$\frac{d\langle G(t)\rangle}{dt} = \frac{i}{\hbar}\langle [H,G]\rangle + \sum_i \tfrac{1}{2}\gamma_i \mathcal{M}_i \left\langle A_i^\dagger[G,A_i] + [A_i^\dagger,G]A_i\right\rangle + \sum_i \tfrac{1}{2}\gamma_i \mathcal{N}_i \left\langle A_i[G,A_i^\dagger] + [A_i,G]A_i^\dagger\right\rangle. \quad (13.1.22)$$

Very often the commutation relations between G and the operators H, A_i and $A_i^\dagger$ are known, so that they can be directly inserted in this form to derive usable equations of motion.

c) Gain and Loss: The second line of the master equation (13.1.14) represents the process $|g\rangle \to |e\rangle$, a fact which follows from the definition (13.1.10) of the operator A, and this is viewed as a *loss process*, since energy is lost by this transition. The first line corresponds to the process $|e\rangle \to |g\rangle$, and is correspondingly viewed as a *gain process*. In this case, since the coefficient of the gain is larger than that of the loss, the overall result is loss.

However, heat baths which give rise to net gain are quite possible, and indeed are used to make optical amplifiers and lasers. The generic Lindblad form given above explicitly includes the possibility that $\mathcal{N}_i$ may be larger than $\mathcal{M}_i$, which gives rise to *gain*. In Sect. 13.3 we will give examples of how this can be done, while in Sect. 17.3, we will show how gain can be exploited to make a laser.

13.2 Derivation of the Master Equation

The derivation of the master equation is complex and technical, unlike the master equation itself, which is an elegant and simple encapsulation of the concept of a quantum Markov process. There are many ways to do the derivation, and the derivation we present here is based on the use of projectors, and has been chosen because technically it is very similar to the derivation of the white noise limit for classical stochastic processes, as presented in Chap. 8. The actual limits used, however, are of a different kind, as will be explained in Sect. 13.2.1g. Other derivations can be found in *Quantum Noise*, and in *Book II*.

The physics involved in the derivation is rather similar to that used for the decay of an excited atom in Sect. 12.2, but is not limited by the need to have no photons in the initial state. The approximations involved are essentially identical.

13.2.1 Description of Projection Method

The equation for the total density operator is

$$\dot{\rho}_{\text{tot}} = -\frac{i}{\hbar}[H, \rho_{\text{tot}}] = (\mathcal{L}_{\text{sys}} + \mathcal{L}_{\text{int}} + \mathcal{L}_B)\rho_{\text{tot}}\,, \tag{13.2.1}$$

in which we introduce the *Liouvillian operators*, defined by

$$\mathcal{L}_{\text{sys}}\,\rho_{\text{tot}} = -\frac{i}{\hbar}[H_{\text{sys}}, \rho_{\text{tot}}]\,, \tag{13.2.2}$$

$$\mathcal{L}_{\text{int}}\,\rho_{\text{tot}} = -\frac{i}{\hbar}[H_{\text{int}}, \rho_{\text{tot}}]\,, \tag{13.2.3}$$

$$\mathcal{L}_B\,\rho_{\text{tot}} = -\frac{i}{\hbar}[H_B, \rho_{\text{tot}}]\,. \tag{13.2.4}$$

a) Heat Bath Density Operator: We will assume that the heat bath is, to a first approximation, described by a stationary solution of $\dot{\rho}_B = \mathcal{L}_B \rho_B$, and for this case the canonical ensemble gives the thermal stationary solution

$$\rho_B = \frac{\exp(-H_B/k_B T)}{\mathrm{Tr}_B\{\exp(-H_B/k_B T)\}}. \tag{13.2.5}$$

The specific form of the density operator is not very important in the derivation, and the results are easily generalized to different choices for ρ_B.

The projector we shall use will be defined by

$$\mathcal{P}\rho_{\mathrm{tot}} = \rho_B \otimes \mathrm{Tr}_B\{\rho_{\mathrm{tot}}\}. \tag{13.2.6}$$

Thus, the projector turns the density operator for the coupled system of atom and heat bath into a direct product of a density operator ρ_B (given by (13.2.5)) which is an operator only in the heat bath degrees of freedom, and the density operator $\mathrm{Tr}_B\{\rho_{\mathrm{tot}}\}$, which is traced out over the heat bath degrees of freedom. In the case that we have two-level system, we can therefore write

$$\mathcal{P}\rho_{\mathrm{tot}} = \begin{pmatrix} a\rho_B & \vdots & b\rho_B \\ \cdots & \cdots & \cdots \\ b^*\rho_B & \vdots & c\rho_B \end{pmatrix}. \tag{13.2.7}$$

The operation $\mathrm{Tr}_B\{\}$ sums over heat bath indices only—therefore it is obvious that $\mathcal{P}^2 = \mathcal{P}$, and $\mathcal{P}$ is indeed a projector.

b) Laplace Transform and Projectors: We use the notation (as in Sect. 8.1.2) for the Laplace transform of a function $v(t)$

$$\tilde{v}(s) = \int_0^\infty \mathrm{e}^{-st} v(t)\, dt. \tag{13.2.8}$$

The Laplace transform of (13.2.1) is

$$s\tilde{\rho}_{\mathrm{tot}}(s) - \rho_{\mathrm{tot}}(0) = (\mathcal{L}_B + \mathcal{L}_{\mathrm{int}} + \mathcal{L}_{\mathrm{sys}})\rho_{\mathrm{tot}}(s). \tag{13.2.9}$$

We now introduce the projection operator $\mathcal{P}$ defined by (13.2.6), which has the following properties:

$$\mathcal{P}^2 = \mathcal{P}, \tag{13.2.10a}$$

$$\mathcal{P}\mathcal{L}_B = 0, \tag{13.2.10b}$$

$$\mathcal{L}_B\mathcal{P} = 0, \tag{13.2.10c}$$

$$\mathcal{P}\mathcal{L}_{\mathrm{int}}\mathcal{P} = 0, \tag{13.2.10d}$$

$$\mathcal{P}\mathcal{L}_{\mathrm{sys}} = \mathcal{L}_{\mathrm{sys}}\mathcal{P}. \tag{13.2.10e}$$

For convenience we also define the complementary projection operator

$$\mathcal{Q} = 1 - \mathcal{P}. \tag{13.2.11}$$

These are very similar to the corresponding identities for the projectors used in Sect. 8.1.2 for the classical white noise limit.

c) Exact Equation for the System Density Operator: We now set

$$\tilde{v}(s) = \mathcal{P}\tilde{\rho}_{\text{tot}}(s), \qquad \tilde{w}(s) = \mathcal{Q}\tilde{\rho}_{\text{tot}}(s). \tag{13.2.12}$$

The quantity

$$\tilde{v}(s) = \rho_B \otimes \mathrm{Tr}_B\left\{\tilde{\rho}_{\text{tot}}(s)\right\}, \tag{13.2.13}$$

is proportional to the Laplace transform of the density operator for the atom alone, thus we want to derive an equation for $\tilde{v}(s)$. We can straightforwardly derive from (13.2.9)

$$s\tilde{v}(s) - v(0) = \mathcal{L}_{\text{sys}}\tilde{v}(s) + \mathcal{P}\mathcal{L}_{\text{int}}\tilde{w}(s), \tag{13.2.14}$$

$$s\tilde{w}(s) - w(0) = \left(\mathcal{L}_{\text{sys}} + \mathcal{L}_B + \mathcal{Q}\mathcal{L}_{\text{int}}\right)\tilde{w}(s) + \mathcal{Q}\mathcal{L}_{\text{int}}\tilde{v}(s), \tag{13.2.15}$$

and we can solve for $\tilde{w}(s)$, using the second equation, and substitute into the first equation, to get

$$\begin{aligned} s\tilde{v}(s) - v(0) - \mathcal{P}\mathcal{L}_{\text{int}}\left(s - \mathcal{L}_{\text{sys}} - \mathcal{L}_B - \mathcal{Q}\mathcal{L}_{\text{int}}\right)^{-1} w(0), \\ = \left\{\mathcal{L}_{\text{sys}} + \mathcal{P}\mathcal{L}_{\text{int}}\left(s - \mathcal{L}_{\text{sys}} - \mathcal{L}_B - \mathcal{Q}\mathcal{L}_{\text{int}}\right)^{-1}\mathcal{Q}\mathcal{L}_{\text{int}}\right\}\tilde{v}(s). \end{aligned} \tag{13.2.16}$$

The term proportional to $w(0)$ in the first line of this equation carries all the information about the initial state of the heat bath. In practice it is very small, giving rise to a transient behaviour which rapidly damps out. We will therefore set $w(0) \to 0$ from now on.

d) The Born Approximation: We want to consider the situation in which the $\mathcal{L}_{\text{sys}}$ and $\mathcal{L}_B$ are the dominant terms, and $\mathcal{L}_{\text{int}}$ is a weak coupling. For example, in the case of a single atom in a radiation field, we know that in practice the atom interacts very little—quite accurate descriptions of atoms can be given without even including the interaction with the radiation field.

Thus we neglect $\mathcal{L}_{\text{int}}$ in the $(\ )^{-1}$ terms in (13.2.16). We then get the Laplace transformed equation

$$s\tilde{v}(s) - v(0) = \left\{\mathcal{L}_{\text{sys}} + \mathcal{P}\mathcal{L}_{\text{int}}\left(s - \mathcal{L}_{\text{sys}} - \mathcal{L}_B\right)^{-1}\mathcal{Q}\mathcal{L}_{\text{int}}\right\}\tilde{v}(s). \tag{13.2.17}$$

The neglect of such terms is essentially the same as that used in the Born approximation of scattering theory. Without this approximation, the procedure yields intractable equations.

In *Book III* these methods are adapted to the quantum gas, and in that case the Born approximation is an essential part of the development of a quantum kinetic theory.

e) Inversion of the Laplace Transform: The convolution theorem for the Laplace transform is: If $\tilde{f}(s)$, $\tilde{g}(s)$ are Laplace transforms of $f(t)$, $g(t)$, and if we define their *convolution* by

$$h(t) = \int_0^t d\tau\, f(\tau)g(t-\tau), \tag{13.2.18}$$

then the Laplace transform of $h(t)$ is

$$\tilde{h}(s) = \tilde{f}(s)\tilde{g}(s). \tag{13.2.19}$$

Now note that

$$\int_0^\infty e^{-st} \exp\{(\mathcal{L}_{\text{sys}} + \mathcal{L}_B)t\}\, dt = (s - \mathcal{L}_{\text{sys}} - \mathcal{L}_B)^{-1}, \tag{13.2.20}$$

so that inverting (13.2.17), and using the convolution theorem

$$\frac{dv}{dt} = \mathcal{L}_{\text{sys}} v(t) + \mathcal{P}\mathcal{L}_{\text{int}} \int_0^t d\tau \exp\{(\mathcal{L}_{\text{sys}} + \mathcal{L}_B)\tau\} \mathcal{Q}\mathcal{L}_{\text{int}} v(t-\tau). \tag{13.2.21}$$

f) The Markov Approximation: By assumption $\mathcal{L}_{\text{int}}$ is small; the second part of (13.2.21) is therefore small, so that we can consistently make the approximation that, in this second part, $v(t-\tau)$ can be written as

$$v(t-\tau) \approx \exp\{-\mathcal{L}_{\text{sys}}\tau\}\, v(t). \tag{13.2.22}$$

The error thus introduced will be of the second order in $\mathcal{L}_{\text{int}}$ in (13.2.21), and such terms have already been neglected. This is essentially an extension of the Born approximation, but since it yields an equation of motion which involves $v(t)$ only at a single time, it is known as the *Markov approximation*, and is equivalent to that used in Sect. 12.2.1 b for the decay of an atom from its excited state.

g) Comparison with the Classical Markov Limit: In the situation being treated here, $\mathcal{L}_{\text{sys}}$ and $\mathcal{L}_B$ are of similar size, while $\mathcal{L}_{\text{int}}$ is a small perturbation. In the classical Markov limit derivation of Chap. 8, the overall Liouvillian operator had the form $\gamma^2 L_1 + \gamma L_2 + L_3$ and in terms of the *algebra*, there is the correspondence

$$\left.\begin{array}{lcl} \gamma^2 L_1 & \longleftrightarrow & \mathcal{L}_B, \\ \gamma L_2 & \longleftrightarrow & \mathcal{L}_{\text{int}}, \\ L_3 & \longleftrightarrow & \mathcal{L}_{\text{sys}}. \end{array}\right\} \tag{13.2.23}$$

The major difference is in the orders of magnitude. In the classical case, γ is considered to be large, so that the terms listed above are listed in order of magnitude. In the case under consideration here, $\mathcal{L}_B$ and $\mathcal{L}_{\text{sys}}$ are of a comparable size, and $\mathcal{L}_{\text{int}}$ is to be considered as a small perturbation. This means that the perturbation theory will be different, but the methodology used here is nevertheless very similar to that used for the classical case in Chap. 8.

In terms of the *dynamics*, the fact that $\mathcal{L}_B$ and $\mathcal{L}_{\text{sys}}$ are of a comparable size leads to the inclusion of $\mathcal{L}_{\text{sys}}$ in the second term of the master equation (13.2.21), which we will see leads to a somewhat different kind equation of motion from that used classically. This is known as the *quantum Brownian motion master equations*, and is developed in full in *Quantum Noise.*

Exercise 13.2 Hamiltonian Evolution Solutions: Using the Liouvillian operator notation

$$L\rho = -\frac{\mathrm{i}}{\hbar}[H,\rho], \tag{13.2.24}$$

for any Hamiltonian H, show that

$$\exp(Lt)\rho = \exp\left(-\frac{\mathrm{i}Ht}{\hbar}\right)\rho\exp\left(\frac{\mathrm{i}Ht}{\hbar}\right). \tag{13.2.25}$$

Exercise 13.3 Mean Values of the Bath Operators: Show that $\mathcal{P}\mathcal{L}_{\mathrm{int}}\mathcal{P} = 0$, where from (13.2.1), $\mathcal{L}_{\mathrm{int}}\rho = -(\mathrm{i}/\hbar)[H_{\mathrm{int}},\rho]$, and that this is equivalent to saying that

$$\mathrm{Tr}_B\{a(\omega)\rho_B\} = \mathrm{Tr}_B\{a^\dagger(\omega)\rho_B\} = 0. \tag{13.2.26}$$

In situations where $\mathrm{Tr}_B\{a(\omega)\rho_B\} \neq 0$, one can usually separate the interaction into a term arising from the mean value and a remainder $\mathcal{L}'_{\mathrm{int}}$, which does satisfy $\mathcal{P}\mathcal{L}'_{\mathrm{int}}\mathcal{P} = 0$ and include the mean value term in H_{sys}.

This procedure is used in Sect. 14.2 to treat the case of an atom under the influence of both a thermal electromagnetic field and a coherent driving electromagnetic field, such as the light from a laser.

13.2.2 Explicit Formulation as a Quantum Master Equation

Using the results of Ex. 13.2 and Ex. 13.3, we can now write (13.2.21) as

$$\frac{dv}{dt} = \mathcal{L}_{\mathrm{sys}}v - \frac{1}{\hbar^2}\mathcal{P}\left\{\left[H_{\mathrm{int}}, \int_0^t d\tau\, \mathrm{e}^{-\frac{\mathrm{i}(H_{\mathrm{sys}}+H_B)\tau}{\hbar}}\left[H_{\mathrm{int}}, \mathrm{e}^{\frac{\mathrm{i}H_{\mathrm{sys}}\tau}{\hbar}}v(t)\mathrm{e}^{-\frac{\mathrm{i}H_{\mathrm{sys}}\tau}{\hbar}}\right]\mathrm{e}^{\frac{\mathrm{i}(H_{\mathrm{sys}}+H_B)\tau}{\hbar}}\right]\right\}. \tag{13.2.27}$$

This equation can still be further simplified by taking the specific form of H_{int} in (13.1.9), as follows.

a) Reduced Density Operator: The equation (13.2.27) involves only $v(t)$, which is explicitly of the form of a direct product with ρ_B, as in (13.2.5). It is more convenient to remove the dependence on ρ_B by writing

$$v(t) \equiv \mathcal{P}\rho_{\mathrm{tot}}(t) = \rho_B \otimes \mathrm{Tr}_B\{\rho_{\mathrm{tot}}\} \equiv \rho_B \otimes \rho_{\mathrm{sys}}. \tag{13.2.28}$$

This defines ρ_{sys} as the *reduced density operator* for the atom, alternatively known as the *atomic density operator*. In this case, in which we are dealing with a two-level atom, it is a 2×2 Hermitian matrix.

b) Simplification of Terms: There are two terms in this, each integrated over ω. In the double commutator in (13.2.27) these will give rise to 16 different kinds of terms. We will find that eight of these are zero, and there are only six different non-zero terms in the remaining eight. Let us consider the term which involves the parts:

i) $i\hbar \sum_k \kappa_k A^\dagger a_k$ in the first H_{int},

ii) $-i\hbar \sum \kappa_{k'} A a_{k'}^\dagger$ in the second H_{int},

iii) and consider the term $H_{\text{int}} H_{\text{int}}$ arising from $[H_{\text{int}}, [H_{\text{int}}, \ldots]]$.

This term takes the form

$$-\frac{1}{\hbar^2}\rho_B \otimes \int_0^t \mathrm{Tr}_B\Big\{\hbar^2 \sum_{kk'} \kappa_k \kappa_{k'} A^\dagger a_k \mathrm{e}^{-\frac{i(H_{\text{sys}}+H_B)\tau}{\hbar}} A \\ \times a_{k'}^\dagger \mathrm{e}^{\frac{iH_{\text{sys}}\tau}{\hbar}} \rho_B \otimes \rho_{\text{sys}} \mathrm{e}^{-\frac{iH_{\text{sys}}\tau}{\hbar}} \mathrm{e}^{\frac{i(H_{\text{sys}}+H_B)\tau}{\hbar}}\Big\} d\tau . \tag{13.2.29}$$

Exercise 13.4 Explicit Forms of Time-Dependent Operators: By using (13.1.7) and (13.1.11), show that

$$\mathrm{e}^{-\frac{iH_{\text{sys}}\tau}{\hbar}} A \mathrm{e}^{\frac{iH_{\text{sys}}\tau}{\hbar}} = \mathrm{e}^{i\Omega\tau} A, \tag{13.2.30}$$

$$\mathrm{e}^{-\frac{iH_B\tau}{\hbar}} a_k^\dagger \mathrm{e}^{\frac{iH_B\tau}{\hbar}} = \mathrm{e}^{-i\omega_k\tau} a_k^\dagger . \tag{13.2.31}$$

c) Bath Correlations: Note that all system operators, $H_{\text{sys}}, A^\dagger, A$, commute with all field operators, a_k, $a_k^\dagger$, H_B, so that (13.2.29) simplifies to

$$-\rho_B \otimes \int_0^t d\tau \sum_{kk'} \kappa_k \kappa_{k'} \mathrm{e}^{i(\Omega-\omega_{k'})\tau} A^\dagger A \rho_{\text{sys}} \mathrm{Tr}_B \left\{ a_k a_{k'}^\dagger \rho_B \right\}. \tag{13.2.32}$$

Traces over the bath operators are evaluated in this case by using (13.1.18–13.1.20), so that the end result is determined entirely by the second-order correlation functions of the heat bath. In fact, this is the *only* way in which the bath density operator enters this final result.

d) Final Simplification: The time t is arbitrary, and will increase indefinitely. On a time scale which is much larger than the typical periods $\Omega^{-1}, \omega_k^{-1}$, we can set $t \to \infty$ in the upper limit of the τ integral. The τ integral is then

$$\int_0^\infty d\tau\, \mathrm{e}^{i(\Omega-\omega_{k'})\tau} = \pi\delta(\Omega-\omega_{k'}) + i\frac{\mathrm{P}}{\Omega-\omega_{k'}}, \tag{13.2.33}$$

as was shown in (12.2.20). We can now:

i) Use (13.2.33) to substitute in (13.2.32).

ii) As in Sect. 12.2.1b, transform the sum over the index k to an integral over frequencies ω_k, and a density of states $g(\omega_k)$ to get

$$-\rho_B \otimes \left\{\pi g(\Omega)\kappa(\Omega)^2(\bar{N}(\Omega)+1) + \mathrm{i}\mathrm{P}\int d\omega\, \frac{g(\omega)|\kappa(\omega)|^2(\bar{N}(\omega)+1)}{\Omega-\omega}\right\} A^\dagger A\rho_{\rm sys}$$
$$= -\rho_B \otimes \left\{\frac{\gamma}{2}(\bar{N}(\Omega)+1) + \mathrm{i}\delta_1\right\} A^\dagger A\rho_{\rm sys}\,. \qquad (13.2.34)$$

The only operator part (apart from the factor ρ_B, which drops out eventually) is the very simple expression $A^\dagger A\rho_{\rm sys}$.

e) The Other Terms: These come from other parts of $H_{\rm int}$, and from other orderings of the terms in the double commutator, but are all evaluated in the same way. Coefficients which turn up are

$$\mathrm{Tr}_B\{a_k\rho_B a_{k'}^\dagger\} = \mathrm{Tr}_B\{a_{k'}^\dagger a_k\rho_B\} = \mathrm{Tr}_B\{\rho_B a_{k}^\dagger a_k\} = \delta_{kk'}\bar{N}(\omega_k)\,, \qquad (13.2.35)$$
$$\mathrm{Tr}_B\{a_k^\dagger\rho_B a_{k'}\} = \mathrm{Tr}_B\{a_{k'} a_k^\dagger\rho_B\} = \mathrm{Tr}_B\{\rho_B a_{k'} a_k^\dagger\} = \delta_{kk'}(\bar{N}(\omega_k)+1)\,. \qquad (13.2.36)$$

There are also traces involving $a_{k'}$ and a_k, or $a_{k'}^\dagger$ and $a_k^\dagger$ and these are all zero. Because the $A^\dagger$, A are attached to $a_{k'}$, $a_k^\dagger$, the different orderings of $A^\dagger$, A will match with corresponding orderings of $a_{k'}$, $a_k^\dagger$.

Exercise 13.5 Bath Correlation Functions: Derive all of the identities (13.2.35) and (13.2.36). Show also that other traces involving two a operators or two $a^\dagger$ operators vanish. Use the cyclic property of the trace, $\mathrm{Tr}(ABC) = \mathrm{Tr}(BCA) = \mathrm{Tr}(CAB)$.

13.2.3 Final Form of the Master Equation

Putting everything together, we find all terms involve a factor $\rho_B\otimes$, which can be cancelled, leaving a *generalized master equation* in the form (13.1.14–13.1.17), given in Sect. 13.1.3.

13.3 More General Heat Baths

The derivation of the master equation we have given here has been made using a *thermal heat bath of harmonic oscillators.* However, the derivation is really much more generally applicable, and in fact, in the derivation, the only properties of H_B and the operators $a_{k'}$, $a^\dagger(\omega_k)$ which were used were:

i) The commutation relations with H_B are

$$[H_B, a_k] = -\hbar\omega_k a(\omega_k)\,, \qquad (13.3.1)$$
$$[H_B, a_k^\dagger] = \hbar\omega_k a^\dagger(\omega_k)\,. \qquad (13.3.2)$$

ii) We have attributed values to the averages

$$\mathrm{Tr}_B\left\{a^\dagger(\omega_k)a_{k'}\rho_B\right\} = \delta_{k,k'}\bar{N}(\omega_k), \tag{13.3.3}$$

$$\mathrm{Tr}_B\left\{a(\omega_k)a^\dagger(\omega_k)\rho_B\right\} = \delta_{k,k'}\left(\bar{N}(\omega_k)+1\right), \tag{13.3.4}$$

$$\mathrm{Tr}_B\left\{a_{k'}a(\omega_k)\rho_B\right\} = \mathrm{Tr}_B\left\{a^\dagger_{k'}a^\dagger(\omega_k)\rho_B\right\} = 0. \tag{13.3.5}$$

As noted in Sect. 13.1.3c the master equation of the form (13.1.14) which results leads to *loss* of energy from the system at any temperature. This is a characteristic of this particular kind of heat bath—for other kinds of heat bath *gain* of energy is possible, as we shall now explain.

13.3.1 Generalized Bath Operators

Let us introduce *generic* bath operators $\Gamma_k, \Gamma^\dagger_{k'}$, which have no particular commutation relations with each other, but have the properties

$$[H_B, \Gamma_k] = -\hbar\omega_k\Gamma_k, \tag{13.3.6}$$

$$\left[H_B, \Gamma^\dagger_k\right] = \hbar\omega_k\Gamma^\dagger_k, \tag{13.3.7}$$

$$\mathrm{Tr}_B\{\Gamma_k\rho_B\} = 0, \tag{13.3.8}$$

$$\mathrm{Tr}_B\{\Gamma^\dagger_k\rho_B\} = 0, \tag{13.3.9}$$

$$\mathrm{Tr}_B\{\Gamma^\dagger_k\Gamma_{k'}\rho_B\} = \delta_{kk'}\mathcal{N}(\omega_k), \tag{13.3.10}$$

$$\mathrm{Tr}_B\{\Gamma_k\Gamma^\dagger_{k'}\rho_B\} = \delta_{kk'}\mathcal{M}(\omega_k). \tag{13.3.11}$$

The interaction with the system operators is then given by the form analogous to (13.1.9)

$$H_{\mathrm{int}} = \mathrm{i}\hbar\sum_k \kappa_k\left(A^\dagger\Gamma_k - A\Gamma^\dagger_k\right). \tag{13.3.12}$$

These are the only properties of the bath operators necessary for the derivation of a master equation.

a) An Assembly of Two-Level Atoms: This provides the most natural alternative choice for $\Gamma_k, \Gamma^\dagger_k$ other than harmonic oscillator operators.

In this case $\Gamma_k \to \sigma^-_k$, $\Gamma^\dagger_k \to \sigma^+_k$, where the bath is characterized by independent sets of the three operators $\sigma^-_k, \sigma^+_k, \sigma^k_z$, with the two-level atom commutation relations

$$[\sigma^-_k, \sigma^+_{k'}] = \delta_{kk'}\sigma^z_{k'}, \tag{13.3.13}$$

and so on. In this case we can deduce that

$$\langle\sigma^+_k\sigma^-_k\rangle = \left\langle\frac{1+\sigma^k_z}{2}\right\rangle \equiv \mathcal{M}(\omega_k), \tag{13.3.14}$$

$$\langle\sigma^-_k\sigma^+_k\rangle = \left\langle\frac{1-\sigma^z_k}{2}\right\rangle \equiv \mathcal{N}(\omega_k). \tag{13.3.15}$$

Thus, $\mathcal{M}(\omega_k)$ is the mean occupation of the upper level, and $\mathcal{N}(\omega_k)$ is the mean occupation of the lower level, and $\mathcal{M}(\omega_k)$ can be chosen greater than, equal to, or less than $\mathcal{N}(\omega_k)$—a bath of two-level atoms can be maintained so that the population of the upper level is greater than that of the lower level by a process of optical pumping, or indeed by many other methods.

b) Gain and Amplification: Using the generic bath operators (or the particular choice of the two-level atom bath) the master equation takes the same form as (13.1.14), with the substitution throughout of

$$\bar{N}(\omega_k)+1 \to \mathcal{M}(\omega_k), \qquad \bar{N}(\omega_k) \to \mathcal{N}(\omega_k). \tag{13.3.16}$$

In neither case is it necessary to require that $\mathcal{M}(\omega_k) > \mathcal{N}(\omega_k)$—thus, in the case that $\mathcal{M}(\omega_k) < \mathcal{N}(\omega_k)$, *gain* will be present.

c) Master Equation Parameters: The master equation and its parameters take the form (13.1.14–13.1.17) with the substitutions

$$\bar{N}(\omega) \to \mathcal{N}(\omega), \qquad \bar{N}(\omega)+1 \to \mathcal{M}(\omega). \tag{13.3.17}$$

In particular, the expression for γ is unchanged, and the frequency shifts δ_1, δ_2 have the same form apart from these substitutions.

Exercise 13.6 Thermal Heat Baths: If a heat bath of any kind is *thermal*, its density operator is proportional to $\exp(-H_B/k_B T)$. Using the commutation relations (13.3.6, 13.3.7), show that this means that

$$\mathcal{M}_{\text{Thermal}}(\omega) = e^{\hbar\omega/k_B T}\mathcal{N}_{\text{Thermal}}(\omega). \tag{13.3.18}$$

Exercise 13.7 Stationary Solution of the Master Equation: Using the commutation relations (13.1.11) and the relations (13.3.18), show that

$$\rho_{\text{Thermal}} \equiv \frac{\exp(-H_{\text{sys}}/k_B T)}{\text{Tr}\left\{\exp(-H_{\text{sys}}/k_B T)\right\}}, \tag{13.3.19}$$

is an *exact* solution of the master equation with a thermal heat bath. Thus, a system in interaction with a thermal heat bath has a stationary solution which is also in thermal equilibrium, as expected from statistical mechanics.

The thermal stationary solution involves only the system Hamiltonian H_{sys}, and therefore does not take account of the Hamiltonian corrections involving δ_1 and δ_2, which appear in (13.1.14–13.1.17). Show that the effect of these corrections in the procedure used to derive the master equation is of fourth order in the interaction Hamiltonian, and thus involves terms neglected in the derivation.

It is important to note that the thermal stationary solution does not pertain if extra driving terms are added, such as will be done, for example, in the treatment of the driven two-level atom in Sect. 14.2. In such a case, the heat bath is no longer strictly thermal.

13.4 Quantum Correlation Functions and Spectra

The concepts of *correlation function* and *spectrum* play as important role in all systems, whether classical or quantum-mechanical. In order to define them in the context of a quantum Markov process, we need a little more formalism, since the classical concepts of joint and conditional probabilities are not straightforward in quantum mechanics.

13.4.1 The Evolution Operator

We need a way to express the time evolution of a quantum stochastic variable within the context of the master equation. We will introduce the evolution operator as the operator which generates, from the density operator at a given time, the density operator at a later time. We note that the solutions to a master equation

$$\frac{\partial \rho(t)}{\partial t} = L\rho(t), \tag{13.4.1}$$

where $L\rho$ is the linear operation such as in the right-hand side of the master equation (13.1.21), can be written as

$$\rho(t) = V(t, t_0)\rho(t_0), \tag{13.4.2}$$

where $V(t, t_0)$ is the *evolution operator*, which satisfies the equation of motion

$$\frac{d}{dt}V(t, t_0) = LV(t, t_0). \tag{13.4.3}$$

Clearly, $V(t, t_0)$ obeys the *semigroup property*

$$V(t, t_1)V(t_1, t_0) = V(t, t_0), \qquad \text{where } t \geq t_1 \geq t_0, \tag{13.4.4}$$

and $V(t_0, t_0) = 1$.

a) Two-Sided Operators: Both L and $V(t, t_0)$ are *two-sided* operators—that is, if ρ can be represented as an $n \times n$ matrix then L and $V(t, t_0)$ are to be understood as matrices of size $n^2 \times n^2$, acting on a column vector of dimension n^2, composed of all the elements of ρ written in a chosen order.

b) Time Dependence of the Operator L: It is possible, and quite common, to have a time-dependent evolution operator $L(t)$. This does not change any of the results of this section.

13.4.2 Multitime Averages

If F, G, are two system operators, then the correlation function $\langle F(t+\tau)G(t)\rangle$ can be expressed in the Heisenberg picture as

$$\langle F(t+\tau)G(t)\rangle = \mathrm{Tr}_{\mathrm{sys}}\mathrm{Tr}_B\{F(t+\tau)G(t)\;\rho_s \otimes \rho_B\} \tag{13.4.5}$$

and this can be rewritten

$$\langle F(t+\tau)G(t)\rangle = \mathrm{Tr}_{\mathrm{sys}}\Big\{\mathrm{Tr}_B\{e^{iH(t+\tau)/\hbar}Fe^{-iH(t+\tau)/\hbar}e^{iHt/\hbar}Ge^{-iHt/\hbar}\rho_s\otimes\rho_B\}\Big\}. \tag{13.4.6}$$

By using the cyclic property of the trace, we can rewrite (13.4.6) as

$$\langle F(t+\tau)G(t)\rangle = \mathrm{Tr}_{\mathrm{sys}}\Big\{F\,\mathrm{Tr}_B\underbrace{\{e^{-iH\tau/\hbar}G\rho_{\mathrm{tot}}(t)e^{iH\tau/\hbar}\}}_{X(\tau,t)}\Big\}. \tag{13.4.7}$$

The equation of motion for $X(\tau,t)$ in terms of τ is clearly

$$i\hbar\frac{\partial}{\partial\tau}X(\tau,t) = [H, X(\tau,t)] \tag{13.4.8}$$

and in exactly the same way as we derived the master equation for $\rho = \mathrm{Tr}_B\{\rho_{\mathrm{tot}}\}$, we can derive an equation of motion for $\mathrm{Tr}_B\{X(t,\tau)\}$, and this will be the same master equation.

This means that we can write

$$\mathrm{Tr}_B\{X(t,\tau)\} = V(t+\tau,t)\mathrm{Tr}_B\{X(0,t)\} = V(t+\tau,t)G\rho(t). \tag{13.4.9}$$

We now put this back into the equation (13.4.6), to get

$$\langle F(t+\tau)G(t)\rangle = \mathrm{Tr}_{\mathrm{sys}}\Big\{F\,V(t+\tau,t)\{G\rho(t)\}\Big\}. \tag{13.4.10}$$

Exercise 13.8 Alternative Time Ordering: Show that

$$\langle F(t)G(t+\tau)\rangle = \mathrm{Tr}_{\mathrm{sys}}\Big\{GV(t+\tau,t)\{\rho(t)F\}\Big\}. \tag{13.4.11}$$

It is not difficult to see that this process can be extended to yield general time-ordered correlation functions, in the sense that we can evaluate any correlation function of the form

$$\langle F_0(s_0)F_1(s_1)\ldots F_m(s_m)G_n(t_n)G_{n-1}(t_{n-1})\ldots G_0(t_0)\rangle \tag{13.4.12}$$

where the terms are ordered

$$t_n \ge t_{n-1} \ge \ldots t_0, \qquad s_m \ge s_{m-1} \ge \ldots s_0. \tag{13.4.13}$$

This kind of *time-ordered correlation function* corresponds to that which would be produced by a sequence of quantum measurements, as is demonstrated in Sect. 18.4.3. A somewhat more extensive formulation is given in *Quantum Noise.*

13.4.3 Quantum Regression Theorem

The arguments we used to derive the equation of motion for the mean in the form (13.1.22) will be exactly the same, so that we can say the *time correlation function is the solution of the mean value equation with an initial system density operator set equal to* $G\rho_{\mathrm{sys}}$. This initial condition density operator is not a genuine density operator, for example it need not be positive definite or even Hermitian, but this does not invalidate the argument.

a) Derivation of the Quantum Regression Theorem: The previous result is not in itself very powerful, since the equation of motion for the mean (13.1.22) is in general not a closed equation, as the left-hand side cannot necessarily be evaluated in terms of the mean $\langle G(t)\rangle$. However it is quite often the case that we deal with a space in which the operators of interest form a complete set in terms of which any operator can be expressed. Alternatively, the equations of motion (13.1.22) may turn out to be linear in terms of a set of operators, such as often happens in systems of linear harmonic oscillators.

To formulate the quantum regression theorem, let us suppose that there is a set of operators G_i in terms of which the mean value equation of motion becomes

$$\frac{d\langle G_i(\tau)\rangle}{dt} = \sum_l R_{i,l}(\tau)\langle G_l(\tau)\rangle, \qquad \tau > 0, \tag{13.4.14}$$

where the $R_{i,l}(\tau)$ are certain numerical coefficients.

Then, since the time correlation function $\langle G_i(t+\tau)G_j(t)\rangle$ is simply a mean value evaluated with respect to a modified initial density operator $G_j\rho_{\text{sys}}(t)$, it also obeys the same equation, so we can say that

$$\frac{d\langle G_i(t+\tau)G_j(t)\rangle}{d\tau} = \sum_l R_{i,l}(t+\tau)\langle G_l(t+\tau)G_j(t)\rangle, \qquad \tau > 0. \tag{13.4.15}$$

The case that $\tau < 0$ is equivalent to $\tau > 0$ and $j \leftrightarrow i$, for which a similar argument gives

$$\frac{d\langle G_j(t)G_j(t+\tau)\rangle}{d\tau} = \sum_l R_{i,l}(t+\tau)\langle G_j(t)G_l(t+\tau)\rangle, \qquad \tau < 0. \tag{13.4.16}$$

b) Using the Quantum Regression Theorem: To use this theorem, we need to find the initial values of the time correlation functions for $\tau \to 0$, that is, we need to know $\langle G_i(t)G_j(t)\rangle$. The most useful case is when the system is in a stationary state, so that the dependence on t disappears, and often (but definitely not always) this will also be a state of thermodynamic equilibrium.

c) Most General Form: The same kind of argument can be also used to show that

$$\frac{d\langle G_j(t)G_j(t+\tau)G_k(t)\rangle}{d\tau} = \sum_l R_{i,l}(t+\tau)\langle G_j(t)G_l(t+\tau)G_k(t)\rangle, \qquad \tau > 0. \tag{13.4.17}$$

This is in fact quite a common type of correlation function, particularly relevant to the theory of photon counting, which is covered in *Book II.*

13.4.4 Spectrum and Quantum Correlation Functions

Because quantum operators do not commute, there are many correlation functions for any give set of physical observables. In fact, since operators which commute at equal times do not commute at different times, the correlation functions

$$g_l(\tau,t) \equiv \langle G(t+\tau)G(t)\rangle, \tag{13.4.18}$$

$$g_r(\tau,t) \equiv \langle G(t)G(t+\tau)\rangle, \tag{13.4.19}$$

are in general different from each other. Normally, we consider correlation functions in the stationary state, for which these correlation functions become independent of the absolute time t. In this case, the relationship between these *stationary correlation functions* can be written

$$g_l(\tau)_{\text{stationary}}) = g_r(-\tau)_{\text{stationary}}. \tag{13.4.20}$$

The spectrum in quantum mechanics is therefore not quite so straightforward to define in quantum mechanics as in classical mechanics. The kind of reasoning used to yield equation (3.3.16) of Sect. 3.3.4 does not take account of the measurement process required to produce the measurements needed. In quantum mechanics this must be accounted for. A dramatic example is given in *Quantum Noise*, Chap. 1, where the spectrum of black body radiation measured by direct photodetection is compared with that measured by means of heterodyne amplification; the first does not exhibit zero point fluctuations, while the second displays them dramatically. This effect was predicted and experimentally verified by *Koch, van Harlingen* and *Clarke* [53], and provided probably the first demonstration that vacuum fluctuations in the electromagnetic field were real and measurable.

We will leave more details on how to calculate quantum-mechanical spectra to specific cases, since the method of measurement is best explained in context.

14. Applications of the Master Equation

The basic building blocks used to model a range of interesting phenomena are the *harmonic oscillator* and the *two-level system.* In both of these there is a single transition frequency between the available system states. In the two-level system, this arises because there is only one transition possible, and in the harmonic oscillator, there are many transitions, but all have the same frequency. This chapter gives the elements of the application of the master equation to these kinds of systems.

The treatment of systems of harmonic oscillators will be further elaborated in Part V, where we will show how harmonic oscillator systems can in many cases be described by classical stochastic processes. The techniques thus developed are very powerful. Further development based on the two-level system can be found in *Book II,* where more realistic descriptions of atoms and the kinds of modelling required will be developed.

14.1 A Two-Level Atom Interacting with a Thermal Heat Bath

For a two-level atom the density operator ρ is a 2×2 matrix, as are H_{sys}, A and $A^\dagger$. We can write these in terms of the Pauli matrices (see Sect. 10.7) thus

$$A \to \sigma^-, \qquad A^\dagger \to \sigma^+, \qquad H_{\text{sys}} \to \tfrac{1}{2}\hbar\omega_{eg}\sigma_z. \tag{14.1.1}$$

Using the Pauli matrix identities (10.7.14, 10.7.15), we can write the frequency shift terms in the master equation (13.1.14) as

$$[\delta_1\sigma^+\sigma^- - \delta_2\sigma^-\sigma^+, \rho] = \tfrac{1}{2}\left[(\delta_1+\delta_2)\sigma_z, \rho\right], \tag{14.1.2}$$

which is exactly of the same form as the main Hamiltonian part of the master equation, which then takes the form

$$\begin{aligned}\frac{\partial \rho_{\text{sys}}}{\partial t} &= -\tfrac{1}{2}\mathrm{i}(\omega_{eg}+\delta_1+\delta_2)\left[\sigma_z, \rho_{\text{sys}}\right] \\ &\quad + \frac{\Gamma}{2}(\bar{N}(\omega_{eg})+1)\left(2\sigma^-\rho_{\text{sys}}\sigma^+ - \rho_{\text{sys}}\sigma^+\sigma^- - \sigma^+\sigma^-\rho_{\text{sys}}\right) \\ &\quad + \frac{\Gamma}{2}\bar{N}(\omega_{eg})\left(2\sigma^+\rho_{\text{sys}}\sigma^- - \rho_{\text{sys}}\sigma^-\sigma^+ - \sigma^-\sigma^+\rho_{\text{sys}}\right).\end{aligned} \tag{14.1.3}$$

In this equation, the decay constant takes the form

$$\Gamma = 2\pi g(\omega_{eg})|\kappa(\omega_{eg})|^2, \tag{14.1.4}$$

which was derived as (13.1.15) in Sect. 13.1.3.

14.1.1 Frequency Shifts

In practice $\delta_1 + \delta_2 \ll \omega_{eg}$, so the basic effect of these terms is an almost negligible shift in the transition frequency ω_{eg}. Nevertheless this shift does have a fundamental significance. It is divided into two parts:

a) Lamb Shift: This is the part *not* proportional to $\bar{N}(\omega)$; i.e. that part obtained by setting $\bar{N}(\omega) = 0$; it is called the Lamb shift, and is given explicitly by

$$\delta_{\text{Lamb}} = \text{P}\int d\omega\, \frac{g(\omega)|\kappa(\omega)|^2}{\omega_{eg} - \omega}. \tag{14.1.5}$$

This is a particular model-dependent calculation of this shift, which can only be calculated accurately by using relativistic quantum electrodynamics and renormalization theory. However the basic principle is clear—the energy levels are moved a little because of the interaction with the electromagnetic field, and this happens at $T = 0$, that is *even if this field is a vacuum.*

The Lamb shift can be exaggerated by modifying the quantity $g(\omega)|\kappa(\omega)|^2$, by changing the density of states with mirrors or similar techniques, as illustrated in Fig. 14.1. By choosing particular configurations, it can be arranged that $g(\omega)|\kappa(\omega)|^2$ for ω just less than ω_{eg} is very different from $g(\omega)|\kappa(\omega)|^2$ for ω a little greater than ω_{eg}. This means that the amount of the Lamb shift can be considerably enhanced, and this effect can be accurately calculated by the methods above.

b) Stark Shift: The terms in δ_1 and δ_2 which depend on $\bar{N}(\omega)$ are temperature dependent, and give a level shift which increases with temperature, namely

$$\delta_{\text{Stark}} = \text{P}\int d\omega\, \frac{g(\omega)|\kappa(\omega)|^2 \bar{N}(\omega)}{\omega_{eg} - \omega}. \tag{14.1.6}$$

Obviously, since $\bar{N}(\omega)$ is only significant if $\hbar\omega \ll kT$, and since the principal effect is at $\omega \approx \omega_{eg}$, an optical frequency, this effect will not be large in the optical

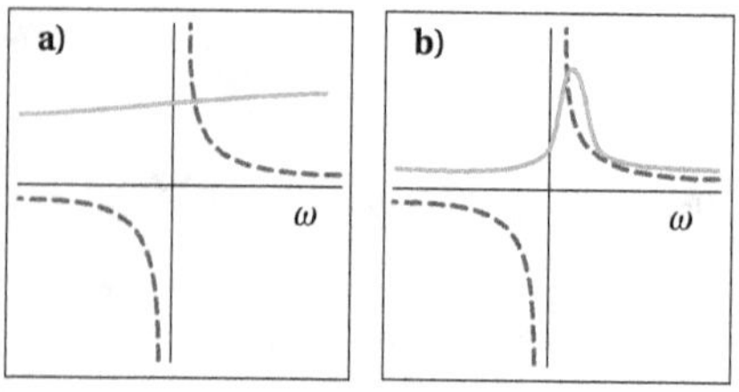

Fig. 14.1. The green curves represent $g(\omega)|\kappa(\omega)|^2$, the brown dashed curves represent $1/(\omega_{eg} - \omega)$. The integral of the product of these is very small in case **a)**, and quite large in case **b)**, where the density of states $g(\omega)$ has been modified by using an appropriate cavity.

region. However using very highly excited atoms, the energy levels become more densely spaced, so that the transition frequency is in the microwave region, and these effects are then so strong that liquid helium temperatures are necessary to observe energy levels.

14.1.2 Equations of Motion

Since the density matrix is a 2 × 2 matrix with unit trace the mean values

$$\bar{\sigma}_z(t) \equiv \mathrm{Tr}\left\{\sigma_z \rho(t)\right\}, \tag{14.1.7a}$$

$$\bar{\sigma}^+(t) \equiv \mathrm{Tr}\left\{\sigma^+ \rho(t)\right\}, \tag{14.1.7b}$$

$$\bar{\sigma}^-(t) \equiv \mathrm{Tr}\left\{\sigma^- \rho(t)\right\}, \tag{14.1.7c}$$

provide a complete specification of $\rho(t)$ through

$$\rho(t) = \tfrac{1}{2} + \tfrac{1}{2}\bar{\sigma}_z(t)\sigma_z + \bar{\sigma}^+(t)\sigma^- + \bar{\sigma}^-(t)\sigma^+. \tag{14.1.8}$$

Using the mean value equation of motion (13.1.22) in this master equation, the equations of motion for the mean values are

$$\frac{d\bar{\sigma}_z}{dt} = -\Gamma(2\bar{N}+1)\bar{\sigma}_z - \Gamma, \tag{14.1.9}$$

$$\frac{d\bar{\sigma}^+}{dt} = \left(-\tfrac{1}{2}\Gamma(2\bar{N}+1) + \mathrm{i}\omega_{eg}\right)\bar{\sigma}^+, \tag{14.1.10}$$

$$\frac{d\bar{\sigma}^-}{dt} = \left(-\tfrac{1}{2}\Gamma(2\bar{N}+1) - \mathrm{i}\omega_{eg}\right)\bar{\sigma}^-. \tag{14.1.11}$$

a) Solutions of the Equations of Motion: These are damped equations of motion with the solutions

$$\bar{\sigma}_z(t) = \mathrm{e}^{-\Gamma(2\bar{N}+1)t}\bar{\sigma}_z(0) - \frac{1-\mathrm{e}^{-\Gamma(2\bar{N}+1)t}}{2\bar{N}+1}, \tag{14.1.12}$$

$$\bar{\sigma}^{\pm}(t) = \mathrm{e}^{(\pm\mathrm{i}\omega_{eg}-\frac{1}{2}\Gamma(2\bar{N}+1))t}\bar{\sigma}^{\pm}(0). \tag{14.1.13}$$

b) Density Matrix Elements: Using (14.1.8) we find

$$\langle e|\rho(t)|e\rangle = \left(1-\mathrm{e}^{-\Gamma(2\bar{N}+1)t}\right)\frac{\bar{N}}{2\bar{N}+1} + \frac{\mathrm{e}^{-\Gamma(2\bar{N}+1)t}}{2\bar{N}+1}\langle e|\rho(0)|e\rangle, \tag{14.1.14}$$

$$\langle g|\rho(t)|g\rangle = \left(1-\mathrm{e}^{-\Gamma(2\bar{N}+1)t}\right)\frac{\bar{N}+1}{2\bar{N}+1} + \frac{\mathrm{e}^{-\Gamma(2\bar{N}+1)t}}{2\bar{N}+1}\langle g|\rho(0)|g\rangle, \tag{14.1.15}$$

$$\langle e|\rho(t)|g\rangle = \mathrm{e}^{\left(\mathrm{i}\omega_{eg}-\frac{1}{2}\Gamma(2\bar{N}+1)\right)t}\langle e|\rho(0)|g\rangle, \tag{14.1.16}$$

$$\langle g|\rho(t)|e\rangle = \mathrm{e}^{\left(-\mathrm{i}\omega_{eg}-\frac{1}{2}\Gamma(2\bar{N}+1)\right)t}\langle g|\rho(0)|e\rangle. \tag{14.1.17}$$

i) Thus the *populations* of the two levels decay to their stationary values on a time scale $\tau = 1/(2\bar{N}+1)\Gamma$. Since $\bar{N}$ is normally negligible, the quantity $\tau_A \equiv \Gamma^{-1}$ is the quantity which one knows as the *atomic lifetime.*

ii) However, in principle the decay constant does depend on the temperature, in contrast to the case of a damped harmonic oscillator, which will be treated in Sect. 14.3.1.

iii) The off-diagonal elements are known as *coherences,* and display an oscillation at the transition frequency. Their decay time scale is half that of the populations.

c) Consistency with Statistical Mechanics: Noting that the thermal occupation $\bar{N}(\omega_{eg}) = 1/(e^{\hbar\omega_{eg}/k_B T} - 1)$, we can see that the stationary populations are given by the stationary density operator ρ_s, and are

$$\langle e|\rho_s|e\rangle = \frac{\bar{N}}{2\bar{N}+1}, \tag{14.1.18}$$

$$\langle g|\rho_s|g\rangle = \frac{\bar{N}+1}{2\bar{N}+1}, \tag{14.1.19}$$

$$\langle e|\rho_s|g\rangle = 0, \tag{14.1.20}$$

and

$$\frac{\langle e|\rho_s|e\rangle}{\langle g|\rho_s|g\rangle} = \frac{\bar{N}}{\bar{N}+1} = e^{-\hbar\omega_{eg}/k_B T}, \tag{14.1.21}$$

as predicted by the canonical distribution. In fact, we can see that

$$\rho_s = \begin{pmatrix} \frac{\bar{N}}{2\bar{N}+1} & 0 \\ 0 & \frac{\bar{N}+1}{2\bar{N}+1} \end{pmatrix} = \frac{\exp(-H_{sys}/k_B T)}{\mathrm{Tr}\{\exp(-H_{sys}/k_B T)\}}. \tag{14.1.22}$$

As pointed out in Ex. 13.7, the correction terms, (i.e., the Lamb and Stark shifts) do *not* appear in this thermal stationary solution, because they are higher-order effects.

14.1.3 Master Equation for the Occupation Numbers

We note that the occupation probabilities for excited and ground states are

$$p(e) \equiv \langle e|\rho|e\rangle = \tfrac{1}{2}\langle 1+\sigma_z(t)\rangle, \qquad p(g) \equiv \langle g|\rho|g\rangle = \tfrac{1}{2}\langle 1-\sigma_z(t)\rangle, \tag{14.1.23}$$

and that the equation of motion (14.1.9) for $\langle\sigma_z(t)\rangle$ is equivalent to the stochastic master equation for these probabilities (and independent of the off-diagonal elements)

$$\dot{p}(e) = -\Gamma(\bar{N}+1)p(e) + \Gamma\bar{N}p(g), \tag{14.1.24}$$

$$\dot{p}(g) = \Gamma(\bar{N}+1)p(e) - \Gamma\bar{N}p(g). \tag{14.1.25}$$

This is the simplest stochastic master equation imaginable—that of a random telegraph process, corresponding to random transitions back and forth between the two levels. It has been explicitly simulated in Sect. 6.1.4. It provides a partial model of *quantum jumps*, since we can see that the system jumps back and forth between the two available levels, but gives no interpretation of the off-diagonal elements of the density operator. The extension of the picture of quantum jumps to include quantum-mechanical coherences is formulated in Sect. 18.3.2, and more extensively in *Book II.*

14.1.4 Comparison with Classical Damping

The equations (14.1.9–14.1.11) were first introduced in the theory of nuclear magnetic resonance, where the basic Hamiltonian is that of a magnetic moment interacting with a magnetic field; i.e.

$$H = \boldsymbol{\mu}\cdot\boldsymbol{B}, \qquad \boldsymbol{\mu} = \tfrac{1}{2}g\hbar\boldsymbol{\sigma}. \tag{14.1.26}$$

The typical frequencies in nuclear magnetic resonance are in the radio frequency region, so that at room temperature, $\bar{N}(\omega_{eg})$ is enormous, and the behaviour therefore almost classical.

We have already treated the classical problem in Sect. 9.2.2, where it was posed in terms of a stochastic differential equation as follows. We suppose a magnetic moment is in a static magnetic field, and couples to a fluctuating field along an orthogonal direction. We thus set up a Hamiltonian

$$H = \tfrac{1}{2}\hbar\big(\omega_{eg}\sigma_z + g\xi(t)\sigma_x\big), \tag{14.1.27}$$

where $\xi(t)$ is a Langevin white noise, so that the equations of motion for σ are the Heisenberg equations of motion

$$\left.\begin{aligned} \dot\sigma_x &= -\omega_{eg}\sigma_y, \\ \dot\sigma_y &= \omega_{eg}\sigma_x - g\xi(t)\sigma_z, \\ \dot\sigma_z &= g\xi(t)\sigma_y, \end{aligned}\right\} \tag{14.1.28}$$

and correspond to Stratonovich stochastic differential equations.

Exercise 14.1 Equivalent Ito Stochastic Differential Equations: Show that the corresponding Ito stochastic differential equations are

$$\left.\begin{aligned} (\mathrm{I})\, d\sigma_x &= -\omega_{eg}\sigma_y\, dt, \\ (\mathrm{I})\, d\sigma_y &= (\omega_{eg}\sigma_x - g^2\sigma_y)\, dt - g\sigma_z\, dW(t), \\ (\mathrm{I})\, d\sigma_z &= -g^2\sigma_z\, dt + g\sigma_y\, dW(t). \end{aligned}\right\} \tag{14.1.29}$$

Since, in taking averages, we put equal to zero all terms involving $dW(t)$ in the Ito formalism, we can write down the equations of motion for the averages, which are

$$\left.\begin{aligned}\langle\dot{\sigma}_x\rangle &= -\omega_{eg}\langle\sigma_y\rangle,\\ \langle\dot{\sigma}_y\rangle &= \omega_{eg}\langle\sigma_x\rangle - g^2\langle\sigma_y\rangle,\\ \langle\dot{\sigma}_z\rangle &= -g^2\langle\sigma_z\rangle.\end{aligned}\right\} \tag{14.1.30}$$

These are in fact very similar to equations (14.1.9–14.1.11), a fact which we can demonstrate by writing the equation for $\langle\dot{\sigma}^+\rangle$, which is

$$\langle\dot{\sigma}^+\rangle = \mathrm{i}\omega_{eg}\langle\sigma^+\rangle - \tfrac{1}{2}g^2\langle\sigma^+\rangle + \tfrac{1}{2}g^2\langle\sigma^-\rangle. \tag{14.1.31}$$

This differs from (14.1.10) by the last term—but if $\omega_{eg} \gg \frac{1}{2}g^2$, i.e., there is only a weak interaction with the fluctuating field, then essentially $\langle\sigma^+\rangle \sim \mathrm{e}^{\mathrm{i}\omega_{eg}t}$, and $\langle\sigma^-\rangle \sim \mathrm{e}^{-\mathrm{i}\omega_{eg}t}$. Thus the $\langle\sigma^-\rangle$ term is out of resonance with all the other factors, and can be neglected.

The similarity is clear, but the equations are not identical. We see that the relaxation decay constants are $\frac{1}{2}g^2$ and g^2; corresponding to $\Gamma(2\bar{N}+1)/2$ and $\Gamma(2\bar{N}+1)$. These are called the *longitudinal* and *transverse* relaxation decay constants. The equation has no $-\Gamma$ term, so that $\langle\sigma_z\rangle = 0$ in the stationary state.

Essentially, the classical equations give only those damping terms proportional to $\bar{N}$ in the quantum equations. The other terms, which are responsible for atomic decay, are absent. This means that no classical magnetic field can give damping which leads to any stationary solution other than $\boldsymbol{\sigma} = 0$. The damping terms which lead to non-zero values of $\boldsymbol{\sigma}$, such as occur in (14.1.9), must be added "by hand", as indeed was done by *Bloch* himself [54], when he introduced these equations which now bear his name.

In contrast, in the quantum-mechanical situation, even if there is no imposed noisy electromagnetic field, damping occurs, and eventually $\langle\sigma_z\rangle \to -1$, corresponding to the radiation of all electromagnetic energy. Correspondingly, at finite temperature, $\langle\sigma_z\rangle \to -1/(2\bar{N}+1)$, corresponding to equilibrium with the heat bath.

14.2 The Two-Level Atom Driven by a Coherent Light Field

The two-level atom may interact with an electromagnetic field which has a coherent part as well as the thermal and vacuum noise parts. As noted in Ex. 13.3, we could model this by choosing more general mean values for the field operators, in which the mean of the operators $a(\omega)$ is not zero. However, it is simpler to add a driving Hamiltonian

$$H_{\text{driving}} \equiv \mathrm{i}g(\sigma^+E(t) - \sigma^-E^*(t)), \tag{14.2.1}$$

and this is an equivalent procedure. We will consider here the configuration illustrated in Fig. 14.2, that is, the particular case of monochromatic driving, in which

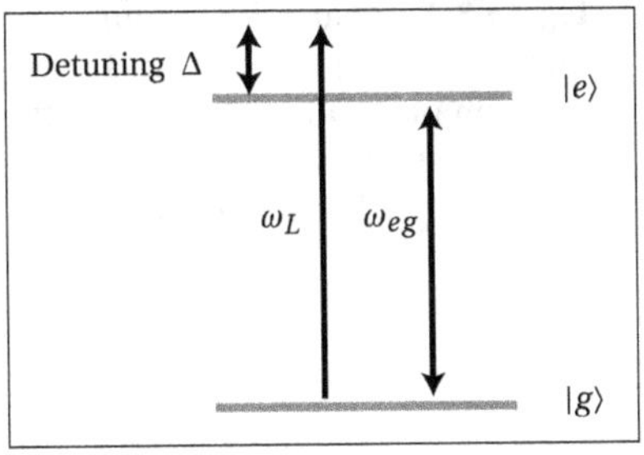

Fig. 14.2. Coherent driving of a two-level system, in which the driving frequency is ω_L, a frequency detuned by Δ from the transition frequency ω_{eg}.

the driving frequency ω_L is close to ω_{eg}, the natural frequency of the atom; thus we set

$$gE(t) = \mathcal{E}e^{-i\omega_L t}. \tag{14.2.2}$$

To derive the master equation when the Hamiltonian is time dependent we need to go into an interaction picture first, by setting

$$\rho_{\text{Int}} = \exp(iH_A t/\hbar)\,\rho(t)\exp(-iH_A t/\hbar), \tag{14.2.3}$$

where

$$H_A \equiv \hbar\omega_L\left(\sigma_z + \sum_k a_k^\dagger a_k\right). \tag{14.2.4}$$

We find the correct equations of motion are obtained from the Hamiltonian

$$\tilde{H} = \tilde{H}_{\text{sys}} + \tilde{H}_B + \tilde{H}_{\text{int}} + i(\sigma^+\mathcal{E} - \sigma^-\mathcal{E}^*), \tag{14.2.5}$$

in which

$$\tilde{H}_{\text{sys}} = \tfrac{1}{2}\hbar(\omega_{eg} - \omega_L)\sigma_z, \tag{14.2.6}$$

$$\tilde{H}_B = \sum_k \hbar(\omega_k - \omega_L)a_k^\dagger a_k, \tag{14.2.7}$$

$$\tilde{H}_{\text{int}} = i\hbar\sum_k \kappa(\omega_k)\left\{\sigma^+ a_k - \sigma^- a_k^\dagger\right\}. \tag{14.2.8}$$

The master equation can be derived as before, and it takes the same form as (14.1.3), apart from the replacement

$$H_{\text{sys}} \to \tfrac{1}{2}\hbar(\omega_{eg} - \omega_L)\sigma_z + i(\sigma^+\mathcal{E} - \sigma^-\mathcal{E}^*). \tag{14.2.9}$$

14.2.1 The Resonant Optical Bloch Equations

The simplest case occurs when the incoming light field is resonant, that is $\omega_{eg} = \omega_L$. In that case, from this master equation we derive the equations of motion for the mean values, known as the *optical Bloch equations*:

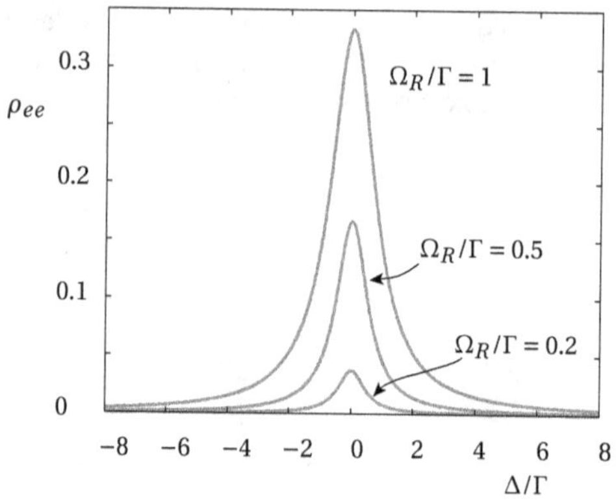

Fig. 14.3. Resonance behaviour of the excited state population (14.2.15) as a function of the detuning Δ from the central frequency, for a range of values of the Rabi frequency Ω_R.

$$\left.\begin{aligned} \frac{d}{dt}\langle\sigma_z\rangle &= -\Gamma(2\bar{N}+1)\langle\sigma_z\rangle + \frac{2\mathcal{E}}{\hbar}\langle\sigma^+\rangle + \frac{2\mathcal{E}^*}{\hbar}\langle\sigma^-\rangle - \Gamma, \\ \frac{d}{dt}\langle\sigma^+\rangle &= -\tfrac{1}{2}\Gamma(2\bar{N}+1)\langle\sigma^+\rangle - \frac{\mathcal{E}^*}{\hbar}\langle\sigma_z\rangle, \\ \frac{d}{dt}\langle\sigma^-\rangle &= -\tfrac{1}{2}\Gamma(2\bar{N}+1)\langle\sigma^-\rangle - \frac{\mathcal{E}}{\hbar}\langle\sigma_z\rangle. \end{aligned}\right\} \qquad (14.2.10)$$

a) Stationary Solutions: These equations can be solved in the stationary state. The expression is best written in terms of the Rabi frequency $\Omega_R \equiv |2\mathcal{E}/\hbar|$ and the phase ϕ of $\mathcal{E}$ as introduced in Sect. 12.3.2, equations (12.3.15, 12.3.16) to produce

$$\left.\begin{aligned} \langle\sigma_z\rangle &= \frac{-\Gamma\gamma_T}{2\gamma_T{}^2+\Omega_R{}^2}, \\ \langle\sigma^+\rangle &= \frac{\frac{1}{2}\Gamma\Omega_R e^{-i\phi}}{2\gamma_T{}^2+\Omega_R{}^2}, \\ \langle\sigma^-\rangle &= \frac{\frac{1}{2}\Gamma\Omega_R e^{i\phi}}{2\gamma_T{}^2+\Omega_R{}^2}, \end{aligned}\right\} \qquad (14.2.11)$$

where

$$\gamma_T = \tfrac{1}{2}\Gamma(2\bar{N}+1) \to \tfrac{1}{2}\Gamma \quad \text{when } T \to 0. \qquad (14.2.12)$$

Exercise 14.2 Driving with a Detuned Field: Suppose we choose a slightly non-resonant coherent driving field in which

$$\omega_L - \omega_{eg} = \Delta, \qquad (14.2.13)$$

and Δ is known as a *detuning*. Carry out the procedure as above, and show that the Bloch equations (14.2.10) are modified by the addition to the equation of motion for $\langle\sigma^+\rangle$ of a term $-i\Delta\langle\sigma^+\rangle$ on the right-hand side, with a conjugate modification to the equation of motion for $\langle\sigma^-\rangle$.

Solve these equations in the stationary state; in particular show

$$\langle\sigma_z\rangle = \frac{-\Gamma({\gamma_T}^2+\Delta^2)}{2\gamma_T\left({\gamma_T}^2+\Delta^2+\frac{1}{2}\Omega_R^2\right)} \longrightarrow -\frac{\frac{1}{4}\Gamma^2+\Delta^2}{\frac{1}{4}\Gamma^2+\Delta^2+\frac{1}{2}\Omega_R^2} \quad \text{when } T=0. \tag{14.2.14}$$

This shows a resonance phenomenon, as shown in Fig. 14.3. As $\Delta \to \pm\infty$, we recover the value of $\langle\sigma_z\rangle$ for a purely thermal field, while for $\Delta \to 0$, we obtain the resonant result.

For convenience, let us note that the excited state population corresponding to the $T = 0$ result is

$$\rho_{ee} = \frac{\frac{1}{4}\Omega_R^2}{\frac{1}{4}\Gamma^2+\Delta^2+\frac{1}{2}\Omega_R^2}. \tag{14.2.15}$$

This is illustrated in Fig. 14.3.

b) Optical Bloch Equations in Cartesian Form: The phase of the driving field can be chosen arbitrarily, so for simplicity let us choose $\mathcal{E} = -\mathrm{i}|\mathcal{E}|$, in which case the mean value equations for $\boldsymbol{S} \equiv \langle\boldsymbol{\sigma}\rangle$ corresponding to (14.2.10) become, in Cartesian form,

$$\dot{S}_x = -\gamma_T S_x, \tag{14.2.16}$$

$$\dot{S}_y = -\gamma_T S_y + \Omega_R S_z, \tag{14.2.17}$$

$$\dot{S}_z = -2\gamma_T S_z - \Omega_R S_y - \Gamma. \tag{14.2.18}$$

c) Stationary Solutions: The stationary solutions of (14.2.16–14.2.18) can be written as the vector $\bar{\boldsymbol{S}} \equiv (\bar{S}_x, \bar{S}_y, \bar{S}_z)$, given by

$$\bar{S}_x = 0, \tag{14.2.19}$$

$$\bar{S}_y = -\frac{\Gamma\Omega_R}{2{\gamma_T}^2+\Omega_R^2}, \tag{14.2.20}$$

$$\bar{S}_z = -\frac{\Gamma\gamma_T}{2{\gamma_T}^2+\Omega_R^2}. \tag{14.2.21}$$

d) Stationary Density Matrix: This then takes the form

$$\rho_s = \tfrac{1}{2}\left(1+\sigma_z\bar{S}_z\right)+\sigma^+\langle\sigma^-\rangle+\sigma^-\langle\sigma^+\rangle = \tfrac{1}{2}\left(1+\bar{\boldsymbol{S}}\cdot\boldsymbol{\sigma}\right). \tag{14.2.22}$$

e) Time-Dependent Solutions: The equations of motion (14.2.16) can be written symbolically as

$$\dot{\boldsymbol{S}}(t) = A(\boldsymbol{S}(t)-\bar{\boldsymbol{S}}), \tag{14.2.23}$$

with

$$A = \begin{pmatrix} -\gamma_T & 0 & 0 \\ 0 & -\gamma_T & \Omega_R \\ 0 & -\Omega_R & -2\gamma_T \end{pmatrix}, \tag{14.2.24}$$

and solutions given by

$$\mathbf{S}(t) = \mathrm{e}^{At}\left(\mathbf{S}(0) - \bar{\mathbf{S}}\right) + \bar{\mathbf{S}}. \tag{14.2.25}$$

Using the property (10.7.10) for exponentiating the 2×2 submatrix, one can write

$$\mathrm{e}^{At} = \begin{pmatrix} \mathrm{e}^{-\gamma_T t} & 0 & 0 \\ 0 & f_1(t) & f_3(t) \\ 0 & -f_3(t) & f_2(t) \end{pmatrix}, \tag{14.2.26}$$

where

$$f_1(t) = \mathrm{e}^{-3\gamma_T t/2}\left(\cos\Omega_\Gamma t + \frac{\gamma_T}{2\Omega_\Gamma}\sin\Omega_\Gamma t\right), \tag{14.2.27}$$

$$f_2(t) = \mathrm{e}^{-3\gamma_T t/2}\left(\cos\Omega_\Gamma t - \frac{\gamma_T}{2\Omega_\Gamma}\sin\Omega_\Gamma t\right), \tag{14.2.28}$$

$$f_3(t) = \mathrm{e}^{-3\gamma_T t/2}\frac{\Omega_R}{\Omega_\Gamma}\sin\Omega_\Gamma t, \tag{14.2.29}$$

and

$$\Omega_\Gamma = \sqrt{\Omega_R^2 - \frac{\gamma_T{}^2}{4}}\,. \tag{14.2.30}$$

The frequency Ω_Γ is a *modified Rabi frequency*, and this represents the frequency of the Rabi oscillations, as modified by the damping γ_T.

The solutions change from circular to hyperbolic functions when $\Omega_R < \gamma_T/2$; for smaller Ω_R one makes the replacement

$$\Omega_\Gamma \to \mathrm{i}\Lambda_\Gamma = \mathrm{i}\sqrt{\frac{\gamma_T{}^2}{4} - \Omega_R^2}\,. \tag{14.2.31}$$

14.2.2 Correlation Functions and Spectrum

The equations of motion for the mean values (14.2.10) are linear, and therefore the quantum regression theorem of Sect. 13.4.3 can be applied to compute the correlation functions and corresponding spectra. Algebraically, this is straightforward, but tedious. A detailed derivation is given in *Quantum Noise*, Chap. 9, in which it is shown that the stationary correlation function $\langle\sigma^+(t)\sigma^-(0)\rangle_s$ is given—in terms of $\bar{\mathbf{S}}$, as defined in (14.2.19–14.2.21)—by

$$\langle\sigma^+(t)\sigma^-(0)\rangle_s = \frac{1}{4}\left(\bar{S}_y^2 + (1+\bar{S}_z)\mathrm{e}^{-\gamma_T|t|} + (1-\bar{S}_z-\bar{S}_y^2)f_1(|t|) - \bar{S}_y(1-\bar{S}_z)f_3(|t|)\right). \tag{14.2.32}$$

This is the correlation function which determines the spectrum of the fluorescent light from the atom—methods for treating this are given in *Book II*, and in *Quantum Noise*.

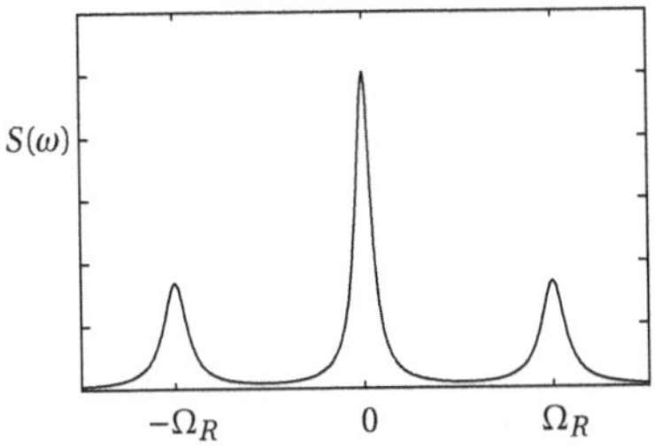

Fig. 14.4. The characteristic three-peaked spectrum corresponding to (14.2.34).

The spectrum corresponding to this correlation function is in general a complicated expression, but it has a simple form in the limit of very large Ω_R, which takes the form

$$S(\omega) \equiv \frac{1}{\pi}\,\mathrm{Re}\left(\int_0^\infty \langle\sigma^+(t)\sigma^-(0)\rangle_s e^{-i\omega t}\,dt\right), \tag{14.2.33}$$

$$\longrightarrow \frac{\gamma_T}{16\pi}\left\{\frac{4}{\gamma_T{}^2+\omega^2}+\frac{3}{\left(\frac{3}{2}\gamma_T\right)^2+(\omega+\Omega_R)^2}+\frac{3}{\left(\frac{3}{2}\gamma_T\right)^2+(\omega-\Omega_R)^2}\right\}. \tag{14.2.34}$$

The three peaks arising from this spectrum are shown in Fig. 14.4, and correspond to side bands arising from motion of the spin vector at the Rabi frequency.

14.3 Master Equations for Harmonic Oscillator Systems

Harmonic oscillator systems are the foundation of the formalism for many-body Bose systems, so an understanding of the kinds of damping and gain which can arise is fundamental. This section covers the most basic forms, which are further elaborated in Part V.

14.3.1 Damping and Gain with a Harmonic Oscillator

We take the generic master equation form (13.1.21) for only one variable A_i, with the substitutions

$$A_i \to a, \qquad A_i^\dagger \to a^\dagger, \tag{14.3.1}$$

$$\mathcal{M}_i \to \mathcal{M}(\Omega), \qquad \mathcal{N}_i \to \mathcal{N}(\Omega), \tag{14.3.2}$$

with the usual harmonic oscillator operators satisfying $[a, a^\dagger] = 1$, and

$$H_{\mathrm{sys}} = \hbar\Omega a^\dagger a. \tag{14.3.3}$$

Thus, we can write down the master equation

$$\frac{\partial\rho}{\partial t} = -i[\Omega a^\dagger a, \rho] + \tfrac{1}{2}\gamma\mathcal{M}(\Omega)\left(2a\rho a^\dagger - \rho a^\dagger a - a^\dagger a\rho\right) + \tfrac{1}{2}\gamma\mathcal{N}(\Omega)\left(2a^\dagger\rho a - \rho a a^\dagger - a a^\dagger\rho\right). \tag{14.3.4}$$

a) Equations of Motion for Mean Values: Using the commutation relations and methodology of Ex. 13.1, we can deduce the equations of motion for the mean field and the mean population, defined by

$$\bar{a} = \langle a\rangle \quad = \mathrm{Tr}\{\rho a\}, \tag{14.3.5}$$

$$\bar{n} = \langle a^\dagger a\rangle = \mathrm{Tr}\left\{\rho a^\dagger a\right\}. \tag{14.3.6}$$

The equations of motion for these mean values take the form

$$\frac{d\bar{a}}{dt} = -\mathrm{i}\Omega\bar{a} + \tfrac{1}{2}\gamma\Big(\mathcal{N}(\Omega) - \mathcal{M}(\Omega)\Big)\,\bar{a}, \tag{14.3.7}$$

$$\frac{d\bar{n}}{dt} = \gamma\Big(\mathcal{N}(\Omega) - \mathcal{M}(\Omega)\Big)\,\bar{n} + \gamma\mathcal{N}(\Omega). \tag{14.3.8}$$

These equations of motion correspond to damping if $\mathcal{N}(\Omega) < \mathcal{M}(\Omega)$, but if $\mathcal{N}(\Omega) > \mathcal{M}(\Omega)$ then $\bar{a}$ and $\bar{n}$ increase exponentially, and we can see that gain can arise. The energy obviously comes out of the heat bath, which needs to be maintained in this state, which is known as an inverted state, since the population of the lower energy level is less than that of the upper energy level. This kind of heat bath can be used to model a laser, as is done in Sect. 17.3.

b) Stationary Solutions: If the situation corresponds to damping, there are stationary solutions of these equations with

$$\bar{a}_\mathrm{s} = 0, \tag{14.3.9}$$

$$\bar{n}_\mathrm{s} = \frac{\mathcal{N}(\Omega)}{\mathcal{M}(\Omega) - \mathcal{N}(\Omega)}. \tag{14.3.10}$$

c) Equations of Motion For Higher-Order Moments: The equations of motion for higher-order normally ordered moments are

$$\frac{d\left\langle (a^\dagger)^n a^m\right\rangle}{dt} = \{\mathrm{i}(n-m)\Omega + \tfrac{1}{2}(n+m)\gamma(\mathcal{N}(\Omega) - \mathcal{M}(\Omega))\}\left\langle (a^\dagger)^n a^m\right\rangle$$
$$+nm\gamma\mathcal{N}(\Omega)\left\langle (a^\dagger)^{n-1} a^{m-1}\right\rangle, \tag{14.3.11}$$

and these are consistent with the equations (14.3.7, 14.3.8) for the mean amplitude and occupation.

The stationary solutions for these equations are

$$\left\langle (a^\dagger)^n a^n\right\rangle = n!\,\bar{n}^n, \tag{14.3.12}$$

$$\left\langle (a^\dagger)^n a^m\right\rangle = 0, \qquad \text{for } n \neq m. \tag{14.3.13}$$

Exercise 14.3 Equations of Motion for Moments: Show that

i) The commutators with powers are

$$[a, (a^\dagger)^n] = n(a^\dagger)^{n-1}, \tag{14.3.14}$$

$$[a^\dagger, a^m] = m a^{m-1}. \tag{14.3.15}$$

ii) If X is an arbitrary operator and $\langle X\rangle \equiv \mathrm{Tr}\{X\rho\}$, then

$$\frac{d\langle X\rangle}{dt} = \mathrm{i}\Omega\big\langle[a^\dagger a, X]\big\rangle + \mathcal{M}(\Omega)\big\langle a^\dagger[X,a] + [a^\dagger, X]a\big\rangle + \mathcal{N}(\Omega)\big\langle a[X,a^\dagger] + [a,X]a^\dagger\big\rangle. \tag{14.3.16}$$

iii) By setting $X = (a^\dagger)^n a^m$, the equations of motion (14.3.11) follow from these results.

Exercise 14.4 Decay of Off-Diagonal Amplitudes: Assuming that $\mathcal{M}(\Omega) > \mathcal{N}(\Omega)$, show that the off-diagonal amplitudes decay at a rate which increases with $|n-m|$, by using the following procedure.

i) First show that

$$\langle a^m\rangle(t) = \exp\left(-\mathrm{i}\Omega t - \frac{\gamma m\big(\mathcal{M}(\Omega)-\mathcal{N}(\Omega)\big)}{2}t\right)\langle a^m\rangle(0). \tag{14.3.17}$$

ii) Use the fact that the equations of motion (14.3.11) couple the terms $\langle (a^\dagger)^r a^{m+r}\rangle$ to $\langle (a^\dagger)^{r-1} a^{m+r-1}\rangle$ to generate an iterative procedure which demonstrates that the slowest decay rate of $\langle (a^\dagger)^r a^{m+r}\rangle(t)$ is given by

$$\left|\left\langle (a^\dagger)^r a^{m+r}\right\rangle(t)\right| \sim \exp\left(-\frac{\gamma m\big(\mathcal{M}(\Omega)-\mathcal{N}(\Omega)\big)}{2}t\right). \tag{14.3.18}$$

Thus, the off-diagonal coherences are damped at a rate directly proportional to m, which is the measure of their distance from the diagonal.

d) Time Correlation Function Using the Quantum Regression Theorem: We would like to compute the correlation function $\langle a^\dagger(t)a(t')\rangle$ for times $t \geqslant t'$. We have derived the linear equation (14.3.7) for the mean value, and its solution is

$$\bar{a}(t) = \mathrm{e}^{(-\mathrm{i}\Omega-\Gamma/2)(t-t')}\,\bar{a}(t'), \tag{14.3.19}$$

$$\Gamma \equiv \gamma\big(\mathcal{M}(\Omega)-\mathcal{N}(\Omega)\big), \tag{14.3.20}$$

and thus

$$\langle a^\dagger(t)\rangle = \mathrm{e}^{(\mathrm{i}\Omega-\Gamma/2)(t-t')}\langle a^\dagger(t')\rangle. \tag{14.3.21}$$

The quantum regression theorem then says that the *stationary* time correlation function is obtained by using the stationary averages as initial conditions thus

$$\langle a^\dagger(t)a(t')\rangle = \langle a^\dagger(t')a(t')\rangle \mathrm{e}^{(\mathrm{i}\Omega-\Gamma/2)(t-t')}, \tag{14.3.22}$$

$$= \bar{n}_{\mathrm{s}}\mathrm{e}^{(\mathrm{i}\Omega-\Gamma/2)(t-t')}. \tag{14.3.23}$$

e) Spectrum: Taking into account the issues raised in Sect. 13.4.4, for this case we define the spectrum as

$$S(\omega) = \frac{1}{2\pi}\int_{-\infty}^{\infty} d\tau\,\mathrm{e}^{\mathrm{i}\omega\tau}\langle a^\dagger(t+\tau)a(t)\rangle_{\mathrm{s}}. \tag{14.3.24}$$

Notice that, since the stationary correlation function depends only on the time differences,

$$\langle a^\dagger(t+\tau)a(t)\rangle_s^* = \langle a^\dagger(t)a(t+\tau)\rangle_s, \tag{14.3.25}$$

$$= \langle a^\dagger(t-\tau)a(t)\rangle_s. \tag{14.3.26}$$

Using this, the part of the integrand with $\tau < 0$ is the complex conjugate of the part with $\tau > 0$, so we can write in general

$$S(\omega) = \frac{1}{\pi}\mathrm{Re}\int_0^\infty d\tau\, \mathrm{e}^{\mathrm{i}\omega\tau}\langle a^\dagger(t+\tau)a(t)\rangle_s. \tag{14.3.27}$$

The spectrum we want to calculate is then

$$S(\omega) = 2\mathrm{Re}\int_0^\infty d\tau\, \mathrm{e}^{-\mathrm{i}\omega\tau} n_s \mathrm{e}^{(\mathrm{i}\Omega-\Gamma/2)\tau},$$

$$= \frac{n_s\Gamma}{2\pi}\frac{1}{(\omega-\Omega)^2+\frac{1}{4}\Gamma^2}, \tag{14.3.28}$$

which is a Lorentzian.

14.3.2 Formulation in Terms of Density Operator Matrix Elements

We use the number states $|n\rangle$, which satisfy

$$a|n\rangle = \sqrt{n}|n-1\rangle, \qquad a^\dagger|n\rangle = \sqrt{n+1}|n+1\rangle, \tag{14.3.29}$$

to derive equations for matrix elements of the density operator, $\rho_{nn'} = \langle n|\rho|n'\rangle$. The equations which arise from (14.3.4) are

$$\dot\rho_{nn'} = -\mathrm{i}\Omega(n-n')\rho_{nn'} + \tfrac{1}{2}\gamma\mathcal{M}(\Omega)\left\{2\sqrt{(n+1)(n'+1)}\rho_{n+1,n'+1} - (n+n')\rho_{nn'}\right\} + \tfrac{1}{2}\gamma\mathcal{N}(\Omega)\left\{2\sqrt{nn'}\rho_{n-1,n'-1} - (n+n'+2)\rho_{nn'}\right\}. \tag{14.3.30}$$

Notice that these equations decouple—that is, the equations relate $\rho_{nn'}$ to $\rho_{mm'}$ when $n-n' = m-m'$. The equation of most interest is that for $p(n) = \rho_{nn}$, the probability of occupying the n photon state—this is

$$\dot p(n) = \gamma\mathcal{M}(\Omega)\left\{(n+1)p(n+1) - np(n)\right\} + \gamma\mathcal{N}(\Omega)\left\{np(n-1) - (n+1)p(n)\right\}. \tag{14.3.31}$$

a) Stationary Solution: This is an ordinary stochastic master equation, with transition probability rates

$$t(n\to n+1) = \gamma\mathcal{N}(\Omega)(n+1), \qquad t(n\to n-1) = \gamma\mathcal{M}(\Omega)n, \tag{14.3.32}$$

as illustrated in Fig. 14.5. A stationary solution is obtained when the rates of gain and loss in any direction balance;

$$p(n)t(n\to n+1) = p(n+1)t(n+1\to n), \tag{14.3.33}$$

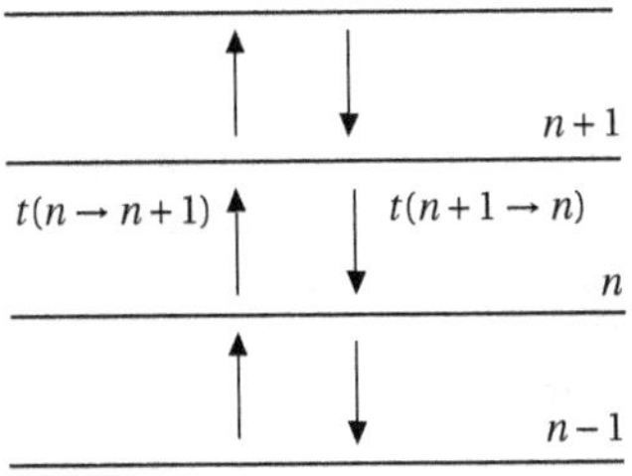

Fig. 14.5. Transitions between occupation levels as given by the master equation.

leading to

$$\frac{p(n+1)}{p(n)} = \frac{t(n \to n+1)}{t(n+1 \to n)} = \frac{\mathcal{N}(\Omega)(n+1)}{\mathcal{M}(\Omega)(n+1)} = \frac{\mathcal{N}(\Omega)}{\mathcal{M}(\Omega)}. \tag{14.3.34}$$

This is the detailed balance solution, and clearly direct substitution into (14.3.34) shows it is indeed a stationary solution.

b) Thermal Heat Bath: In this case, we can write the forward and backward coefficients in terms of the temperature

$$\mathcal{N}(\Omega) \to \bar{N}(\Omega) = \frac{1}{e^{\hbar\Omega/kT} - 1}, \qquad \mathcal{M}(\Omega) \to \bar{N}(\Omega) + 1, \tag{14.3.35}$$

so that $\mathcal{N}(\Omega)/\mathcal{M}(\Omega) = \exp(-\hbar\Omega/kT)$.

The solution (14.3.34) then becomes the correct canonical distribution.

$$p(n) = e^{-n\hbar\Omega/kT} \Big/ \left\{ \sum_{m=0}^{\infty} e^{-m\hbar\Omega/kT} \right\}. \tag{14.3.36}$$

Exercise 14.5 Off-Diagonal Matrix Elements: Show that the stationary off-diagonal matrix elements in this case are zero. This is consistent with the results of Ex. 14.3 and Ex. 14.4, which also show that the exponent of the rate of approach to zero is the same as (14.3.18).

c) Time-Dependent Solutions: The $p(n)$ equations, (14.3.31) provide a classical master equation, and can be solved by simulation methods similar to those used for the two-level atom. Because they are an infinite set of equations, no easy method of solution exists, though it is clear that we have the same picture of the various levels being occupied according to a random process algorithm.

14.3.3 The Phase-Damped Oscillator

The kind of damping we considered in Sect. 14.3.1 envisages the transformation of the oscillator quantum into a bath quantum; this corresponds to a system-bath Hamiltonian of the form

$$H_{\text{Int}} = \sum_i \left(\kappa_i \Gamma_i a^\dagger + \kappa_i^* \Gamma_i^\dagger a \right). \tag{14.3.37}$$

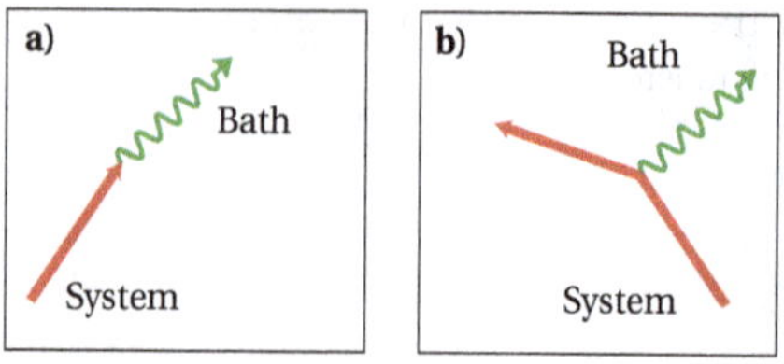

Fig. 14.6. Comparison of two ways in which a harmonic oscillator may experience damping. **a)** A system quantum is transformed into a bath quantum (or conversely). **b)** In phase damping, a system quantum is not destroyed when it emits a bath quantum.

This kind of damping destroys or creates the oscillator quanta, transferring energy between the system and the bath.

A different process is also possible, in which the oscillator quanta are conserved, but those of the bath are not; the corresponding interaction Hamiltonian is

$$H_{\text{Int}} = \sum_i \left(\bar{\kappa}_i \Gamma_i + \bar{\kappa}_i^* \Gamma_i^\dagger\right) a^\dagger a. \tag{14.3.38}$$

This interaction Hamiltonian commutes with the oscillator number operator $a^\dagger a$, but not with either of a or $a^\dagger$. Furthermore, it commutes with the oscillator Hamiltonian, and therefore does not transfer energy from the oscillator to the bath. The different kinds of damping are depicted diagrammatically in Fig. 14.6.

The corresponding master equation is of the form

$$\dot{\rho} = -\mathrm{i}\Omega[a^\dagger a, \rho] + \tfrac{1}{2}\gamma\big(\mathcal{M}_P(\Omega) + \mathcal{N}_P(\Omega)\big)\left\{2a^\dagger a\, \rho\, a^\dagger a - (a^\dagger a)^2 \rho - \rho (a^\dagger a)^2\right\}. \tag{14.3.39}$$

Because the master equation involves only the number operator $a^\dagger a$, the equations of motion for the matrix elements are are not coupled to each other, and in fact take the form

$$\frac{d\langle n|\rho|m\rangle}{dt} = \left\{-\mathrm{i}\Omega(n-m) - \tfrac{1}{2}\gamma\big(\mathcal{M}_P(\Omega) + \mathcal{N}_P(\Omega)\big)(n-m)^2\right\}\langle n|\rho|m\rangle, \tag{14.3.40}$$

with the solution

$$\langle n|\rho(t)\,|\,m\rangle = \mathrm{e}^{-\mathrm{i}\Omega(n-m)t}\mathrm{e}^{-\gamma(\mathcal{M}_P(\Omega)+\bar{\mathcal{N}}_P(\Omega))(n-m)^2 t/2}\langle n|\rho(0)|m\rangle. \tag{14.3.41}$$

i) The interaction always produces damping, not gain, since it depends only on the sum, $\mathcal{M}_P(\Omega) + \mathcal{N}_P(\Omega)$.

ii) The solutions are functions of $m - n$, and the off-diagonal matrix elements decay with an exponent proportional to $(n-m)^2$. When the system is macroscopic $n - m$ is very large, leading to a very fast decay.

iii) The diagonal matrix elements do not decay, and in fact are time-independent. The stationary solution of the master equation is therefore

$$\langle n|\rho_{\mathrm{s}}|m\rangle = \delta_{nm}\langle n|\rho(0)|m\rangle. \tag{14.3.42}$$

iv) Thus, this kind of interaction does not thermalize the occupation probabilities; rather, it *preserves* the occupation probabilities, while *erasing* all coherences.

v) This example is treated using a Wigner function method in Sect. 16.2.2, where the connection to classical phase damping is made.

14.3.4 Decoherence

The two examples of damping we have given for the harmonic oscillator, that is linear damping as in Sect. 14.3.1, and phase damping as in Sect. 14.3.3, both exhibit the approach to equilibrium in which off-diagonal matrix elements $\rho_{m+r,m}$ are damped exponentially, with a damping exponent which grows with the distance from the diagonal. For the linearly damped harmonic oscillator the exponent is proportional to $|r|$, while for the phase-damped oscillator the exponent is proportional to r^2. In both cases, if the numbers m and r are macroscopic this will represent an extremely rapid decay. Therefore a superposition state of the form $\alpha|m\rangle+\beta|m+r\rangle$, where m and r are macroscopic will almost instantaneously turn into a statistical mixture. In the case of the phase-damped oscillator, the diagonal matrix elements are not changed, whereas this is not the case for linear damping. However, because the off-diagonal elements are mostly damped much more rapidly than the rate at which the diagonal elements approach equilibrium, the initial effect is similar.

This effect, whereby macroscopic quantum superpositions are very rapidly degraded to statistical mixtures, is known as *decoherence*, and is an almost universal effect of irreversible processes.

14.4 A Simple Model of Laser Cooling

When an atom absorbs a photon, it experiences both a jump from the initial state to an excited state, and as well a recoil because the photon absorbed carries momentum as well as energy. This is a mechanical effect of absorption and emission of light, and it forms the basis of *laser cooling*, which is central to the preparation and manipulation of ultra-cold atoms. The momentum transferred to an atom by a single photon is in practice very small compared to the momentum of the atom, but the number of photons available using a laser is enormous, making laser cooling quite practicable.

In this section, we will present the simplest model of laser cooling, in which the mechanical effects combine with the spectral selectivity associated with an atomic transition to enable the atom to be slowed down to microKelvin temperatures. More detailed modelling is presented in *Book II.*

14.4.1 Formulation of the Model

Consider a two-level atom initially prepared in its ground state interacting with a near-resonant travelling light wave, as in Fig. 14.7. The atom can interact with the electromagnetic field in three ways:

i) *Absorption of a Laser Photon*: An atom in the ground state, moving with momentum $\boldsymbol{p}$, experiences the transition

$$|g,\boldsymbol{p}\rangle \xrightarrow{\hbar\omega_L} |e,\boldsymbol{p}+\hbar\boldsymbol{k}_L\rangle. \tag{14.4.1}$$

The transition is thus associated with a momentum transfer of $\hbar\boldsymbol{k}_L$ (the "recoil kick") to the atom.

ii) *Induced Emission into the Laser Beam*: This process returns the atom to its ground state;

$$|e,\boldsymbol{p}+\hbar\boldsymbol{k}_L\rangle \xrightarrow{\hbar\omega_L} |g,\boldsymbol{p}\rangle. \tag{14.4.2}$$

Thus, this process undoes the "recoil kick" experienced during the absorption.

iii) *Return to the Ground State by Spontaneous Emission*: This process takes place with rate Γ,

$$|e,\boldsymbol{p}+\hbar\boldsymbol{k}_L\rangle \xrightarrow{\hbar\omega_s} |g,\boldsymbol{p}+\hbar\boldsymbol{k}_L+\hbar\boldsymbol{k}_s\rangle. \tag{14.4.3}$$

Here $\boldsymbol{k}_s = k_{eg}\boldsymbol{n}$ (with $k_{eg} = \omega_{eg}/c$) is the wave vector associated with the spontaneously emitted photon. Note that the spontaneous photon is emitted in a *random direction*, which we can describe in terms of the angular distribution $d\Gamma/d\Omega_{\boldsymbol{n}}$.

In practice the processes take place using laser light tuned close to the transition frequency, so that $|\boldsymbol{k}_L|$ and k_{eg}, while not equal to each other, are of the same order of magnitude.

a) The Laser Cooling Cycle: We can therefore expect that the atom will undergo repeated absorption and emission processes. The processes which return the photon to the laser field have no net mechanical effect. Therefore, let us consider the momentum transfer cycle which takes place in laser cooling:

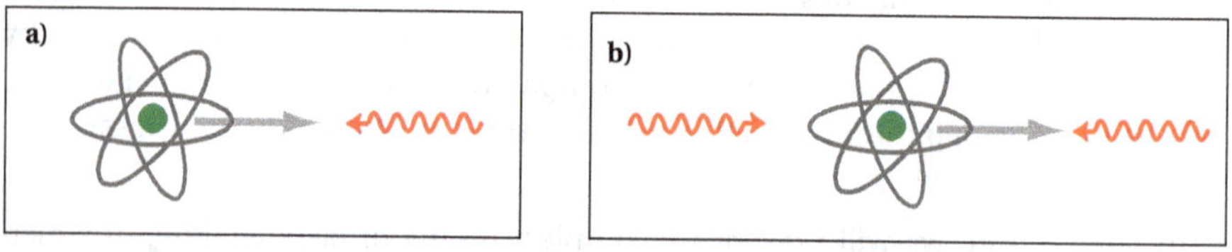

Fig. 14.7. a) A moving atom interacting with a travelling wave formed from a single laser beam; **b)** A moving atom interacting with a standing wave formed from a two counter-propagating laser beams.

i) Absorption of a laser photon, leading to excitation from the ground state to excited state.

ii) The spontaneous emission of a photon, returning the atom to the ground state.

For each cycle this changes the momentum of the atom according to

$$\boldsymbol{p} \to \boldsymbol{p} + \hbar \boldsymbol{k}_L + \hbar \boldsymbol{k}_S. \tag{14.4.4}$$

b) Momentum Transfer to the Atom: Consider a time interval Δt containing a number $n \gg 1$ absorption-emission cycles. At the end of this time interval, the momentum transfer along the laser direction $\boldsymbol{k}_L = k_L \boldsymbol{e}_z$ is

$$p = p_0 + n\hbar k_L + \sum_{l=1}^{n} \hbar k_L \cos\theta_l. \tag{14.4.5}$$

Here both the number n and the projection $\hbar k_L \cos\theta_l$ of the spontaneous emission momentum along z are random variables, which we can reasonably take to be independent of each other.

c) Mean and Variance of the Momentum: The mean value and variance of the momentum are therefore,

$$\langle p \rangle = p_0 + \langle n \rangle \hbar k_L, \tag{14.4.6}$$

$$\langle \Delta p^2 \rangle = \langle \Delta p_0^2 \rangle + \hbar^2 k_L^2 \left(\mathrm{var}\,[n] + \sum_{l,m} \langle \cos\theta_l \cos\theta_m \rangle \right). \tag{14.4.7}$$

Because each photon emission is independent of the others, we can write

$$\sum_{l,m} \langle \cos\theta_l \cos\theta_m \rangle = \sum_{l} \langle \cos^2\theta_l \rangle = \alpha \langle n \rangle, \tag{14.4.8}$$

where α is a purely geometrical factor of order of magnitude 1, which reflects the angular dependence of probability of emission. We therefore obtain

$$\langle \Delta p^2 \rangle = \langle \Delta p_0^2 \rangle + \hbar^2 k_L^2 \left(\mathrm{var}\,[n] + \alpha \langle n \rangle \right). \tag{14.4.9}$$

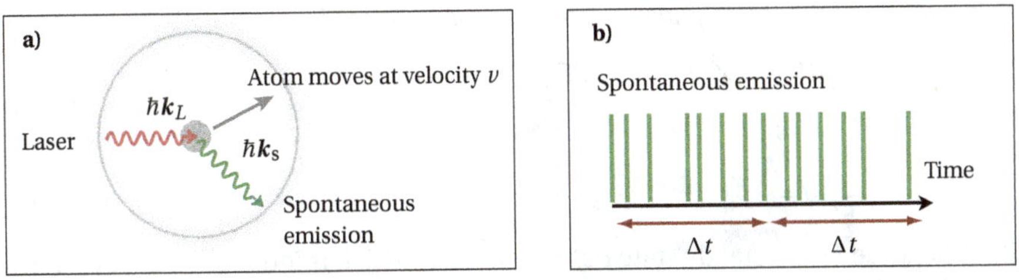

Fig. 14.8. Modelling Doppler cooling: **a)** Laser light detuned from resonance is absorbed, leading to spontaneous emission at the resonant frequency; **b)** Many spontaneous emissions happen in the averaging time interval Δt.

14.4.2 Doppler Cooling

Doppler cooling depends on the absorption of detuned laser light, which is on resonance with the Doppler shifted absorption frequency of a moving atom, as illustrated in Fig. 14.8.

a) Radiation Pressure Force Exerted by a Travelling Wave: The average radiation pressure force F on the atom is thus given by Newton's law to be

$$F = \frac{\langle p\rangle - p_0}{\Delta t} = \hbar k_L \frac{\langle n\rangle}{\Delta t}. \tag{14.4.10}$$

If ρ_{ee} represents the occupation fraction of the excited state, the number of spontaneous emissions in the time Δt will be

$$\langle n\rangle = \Gamma \rho_{ee}\,\Delta t = \frac{\frac{1}{4}\Gamma\Omega_R^2}{(\omega_L - \omega_{eg} - k_L v)^2 + \frac{1}{4}\Gamma^2}, \tag{14.4.11}$$

where we have used the result (14.2.15) for the excited state population, and have assumed that $\Omega_R \ll \Gamma$. We have also included a *Doppler shift* term $k_L v$, where

$$v \equiv \frac{\langle p\rangle}{m}, \tag{14.4.12}$$

because the atom moves with respect to the laser field. (There is no similar first-order correction to the transition frequency; the only correction would be a relativistic time-dilation effect, which is second order in v/c.) Putting these together, the force becomes

$$F = \hbar k_L \times \Gamma \rho_{ee} \equiv \frac{\frac{1}{4}\hbar k_L \Gamma \Omega_R^2}{(\omega_L - \omega_{eg} - k_L v)^2 + \frac{1}{4}\Gamma^2}. \tag{14.4.13}$$

b) Velocity Capture Range: The force is only appreciable near the maximum of the resonance, that is when $k_L v \sim \Gamma$, and this requirement sets the range of velocities within which the atomic motion is most effectively damped by the Doppler effect.

c) Small Velocity Expansion: When $k_L v \ll \Gamma$ (for sodium atoms this gives $v \ll 6$ m/s) we can make the expansion (for compactness setting $\langle p\rangle \to p$)

$$F(p) = \frac{\frac{1}{4}\hbar k_L \Gamma {\Omega_R}^2}{(\omega_L - \omega_{eg})^2 + \frac{1}{4}\Gamma^2} + \frac{\frac{1}{2}\Gamma {\Omega_R}^2(\omega_L - \omega_{eg})}{\left((\omega_L - \omega_{eg})^2 + \frac{1}{4}\Gamma^2\right)^2}\left(\frac{\hbar k_L^2}{2m}\right) p + \dots, \tag{14.4.14}$$

$$= F_0 - \beta v + \dots. \tag{14.4.15}$$

i) The force F_0 is a steady state *deflection force*, which is independent of velocity, while βv is a damping force, which vanishes at zero velocity.

ii) There will be *damping* when $\beta > 0$ for *red* detuning, that is $\omega_L < \omega_{eg}$. The maximum friction force occurs when at $\omega_L - \omega_{eg} \sim -\frac{1}{2}\Gamma/\sqrt{3}$.

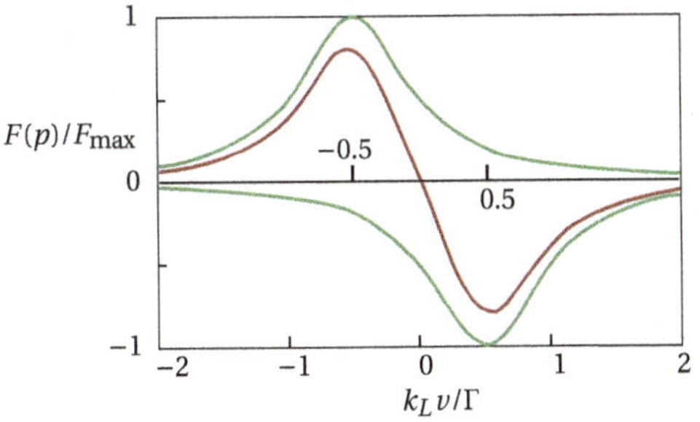

Fig. 14.9. The forces involved in Doppler cooling, as given by the formula (14.4.19). The green curves represent the two terms in (14.4.19), and the brown curve is the net cooling frictional force given by their difference $F(p)$. In this figure, we have set $\delta \equiv \omega_{eg} - \omega_L = \frac{1}{2}\Gamma$, and have plotted the force normalized by dividing by the maximum $F_{\max} = \hbar k_L \Omega_R^2/\Gamma$.

d) The Recoil Frequency ω_R: This is a quantity which is significant in all aspects of the mechanical action of light. The recoil frequency ω_R is defined by

$$\omega_R = \frac{\hbar k_L^2}{2m}, \tag{14.4.16}$$

and is the frequency corresponding the kinetic energy of an atom with the same momentum $\hbar k_L$ as a laser photon. For laser cooling of sodium, $\omega_R \approx 2\pi \times 23\,\text{kHz}$.

i) Note that for $\Omega_R \approx \frac{1}{2}\Gamma$ we have

$$F_0 \approx \tfrac{1}{2}\Gamma\hbar k_L, \qquad \beta \approx m\omega_R. \tag{14.4.17}$$

ii) The damping $m\dot{v} = -\beta v$ gives rise to a loss of kinetic energy at the rate

$$\dot{E} = -\beta v^2 = -2\omega_R E, \tag{14.4.18}$$

so that the time scale for cooling is $1/\omega_R$, the inverse recoil frequency.

14.4.3 Radiation Pressure Force Exerted by a Weak Standing Wave

For two counterpropagating laser beams we can in lowest order in the laser intensity just add up the forces from each laser, as in Fig. 14.9. We obtain

$$F(p) \equiv \hbar k_L \Gamma \left(\frac{\frac{1}{4}\Omega_R{}^2}{(\omega_L - \omega_{eg} - k_L v)^2 + \frac{1}{4}\Gamma^2} - \frac{\frac{1}{4}\Omega_R{}^2}{(\omega_L - \omega_{eg} + k_L v)^2 + \frac{1}{4}\Gamma^2} \right), \tag{14.4.19}$$

$$\approx -2\beta v. \tag{14.4.20}$$

The deflection forces cancel, while the damping forces add to each other.

14.4.4 Fluctuations

We can reasonably model the fluctuations which arise from spontaneous emission using a stochastic differential equation of Fokker–Planck formalism, since the jumps in momentum are very small. This could be done in the same kind of way as was used in Sect. 6.1.1a, but we can proceed in a slightly simpler and more phenomenological way here.

a) Stochastic Differential Equation Model: Let us model the momentum p using the stochastic differential equation

$$dp = \left(F_0 - \beta p/m\right) dt + \sqrt{D}\, dW(t), \tag{14.4.21}$$

where we have to determine D. Using Ito rules, we find that

$$\frac{d}{dt}\langle \Delta p^2 \rangle \equiv \frac{d}{dt}\left(\langle p^2 \rangle - \langle p \rangle^2\right) = 2D. \tag{14.4.22}$$

However, from (14.4.9), we know that

$$\frac{\langle \Delta p^2 \rangle - \langle \Delta p_0^2 \rangle}{\Delta t} = \hbar^2 k_L^2 \frac{\langle n \rangle}{\Delta t}\left(\frac{\mathrm{var}\,[n]}{\langle n \rangle} + \alpha\right). \tag{14.4.23}$$

Finally, $\langle n \rangle = \Gamma \rho_{ee} \Delta t$, since $\Gamma \rho_{ee}$ represents the rate of spontaneous emissions, so that we can conclude

$$D = \tfrac{1}{2}\hbar^2 k_L^2 \Gamma \rho_{ee}\,(1 + Q + \alpha). \tag{14.4.24}$$

The quantity $Q = \mathrm{var}\,[n]\,/\langle n \rangle - 1$ is called the *Mandel Q-parameter*. If the emissions are Poissonian $Q = 0$, but we will see in *Book II* that the light emitted from a two-level system is only Poissonian for weak driving and very strong driving fields. Even so, this represents only a small correction.

b) Contributions to Fluctuations: The diffusion term reflects fluctuations in the emitted light. There are two contributions.

i) The first term, $1+Q$, between the parentheses reflects *fluctuations in the number of photons n* emitted during the time interval Δt.

ii) The second term reflects fluctuations due to spontaneous emission in *random spatial directions.* It involves a geometric factor α, which can be derived by the more detailed methods presented in *Book II.*

c) Fokker–Planck Equation for Doppler Cooling: This takes the form

$$\frac{\partial}{\partial t} P(p,t) = \left\{ -\frac{\partial}{\partial p}\left(F_0 - \beta p/m\right) + D\frac{\partial^2}{\partial p^2} \right\} P(p,t). \tag{14.4.25}$$

The argument presented above assumes that the momentum jumps are small and this assumption can be made quantitative by requiring

$$\hbar k_L / \Delta p \ll 1, \tag{14.4.26}$$

where in this context Δp is to be interpreted as the breadth of the momentum distribution described by the Fokker–Planck equation, as illustrated in Fig. 14.10. In *Book II*, we make the smallness of this parameter the starting point of a systematic expansion in a semiclassical theory of laser cooling.

d) Solutions of the Equations: The mean momentum decays according to an exponential law

$$\langle p \rangle = p_0 e^{-\beta t/m} = p_0 e^{-\omega_R t}. \tag{14.4.27}$$

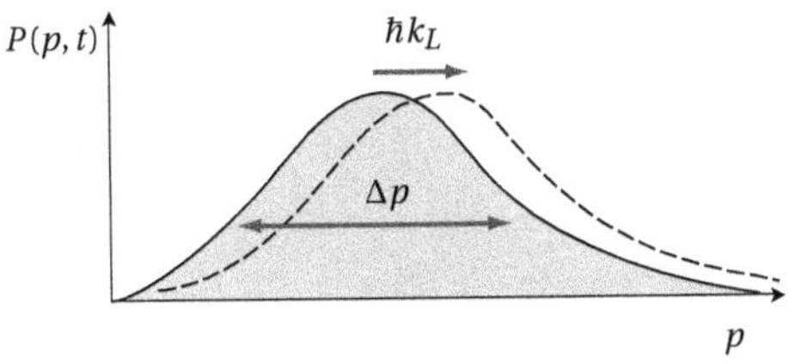

Fig. 14.10. Relative sizes of the momentum kick $\hbar k_L$ and the breadth Δp of the probability distribution.

The stationary distribution is

$$P(p) \propto \exp\left(-\frac{\beta p^2}{2mD}\right), \tag{14.4.28}$$

with mean kinetic energy

$$\left\langle \frac{p^2}{2m} \right\rangle \approx \frac{D}{\beta} \approx \frac{\hbar^2 k_L^2 \Gamma \rho_{ee}}{\frac{1}{2}\hbar k_L^2} \approx \hbar\Gamma, \tag{14.4.29}$$

which gives the ultimate limit of Doppler cooling.

V PHASE SPACE METHODS

15. Phase Space Representations for Bosons

Highly degenerate quantum states, that is, states in which the occupations of a number of modes of the field become very much larger than one, are of very great significance both for quantum optics and for ultra-cold Bose gases. For the electromagnetic field, these highly degenerate states correspond to the classical limit, and are described accurately by the classical Maxwell equations. In contrast, for ultra-cold Bosonic gases, the most relevant such state corresponds to a Bose–Einstein condensate, which is usually seen as state with extreme quantum properties. The perception is misleading, since a description in terms of a c-number matter wave field obeying the Gross–Pitaevskii equation is then valid. This field has a classical c-number equation of motion, but unlike the electromagnetic field, its equation of motion contains Planck's constant, so that the physics described is definitely of a quantum nature. Consequently, because the occupation of the condensate mode can be made as high as 10,000,000, the quantum behaviour is visible at an almost macroscopic scale. Furthermore, in an ultra-cold gas the quantum states which are of importance are very often highly quantum degenerate even when the system is not Bose-condensed, and in this case as well, the quantum description is often very closely related to a classical description.

The techniques for describing the dynamics of the highly degenerate fields found in lasers and in other situations involving laser fields were developed in the 1960s and continue to develop to this day. The basic technique used was based on the inclusion of quantum and thermal noise into the description by means of the techniques of quantum Markov processes, even though in the earlier period of development, this was often not made explicit. Central to the application of these techniques was the use of formalisms involving *coherent states*, whose significance in the field was realized by *Glauber* [55].

The two techniques[1] of most use are based on the use of the *P-representation*, which is usually most appropriate for quantum optics, and the *Wigner function*, which is most practical for the study of ultra-cold Bose systems. There is a common formalism behind these methods, strongly connected with the concept of

[1]A more comprehensive description of other techniques, including in particular the *positive P-representation*, can be found in *Quantum Noise*.

a coherent state—both can be regarded as a way of generating a c-number stochastic description which is equivalent to the quantum description, either exactly so, or approximately, depending on the system under investigation. In this chapter we will introduce the common background, and then in the subsequent two chapters, develop the two formalisms separately.

15.1 The Quantum Characteristic Function

The Fourier transform of a classical probability distribution is known as its characteristic function, and moments of the distribution are proportional to derivatives of the characteristic function. In quantum systems, we deal with a density operator, not a probability, and need a modification of the definition of the characteristic function to account for the fact that the Fourier transform must be done with respect to operator quantities. There are different possible operator orderings for computing moments, and correspondingly there are different characteristic functions. There are three principal characteristic functions for Bose operators, as follows.

15.1.1 Normally Ordered Quantum Characteristic Function

This is defined by

$$\chi_N(\lambda,\lambda^*) = \mathrm{Tr}\left\{\rho \exp(\lambda a^\dagger)\exp(-\lambda^* a)\right\}. \tag{15.1.1}$$

The *normally ordered* moments are given by its derivatives at $\lambda = 0$; precisely

$$\langle a^{\dagger r} a^s\rangle = (-)^s \frac{\partial^{r+s}}{\partial^r\lambda\partial^s\lambda^*}\chi_S(\lambda,\lambda^*)\bigg|_{\lambda=0}. \tag{15.1.2}$$

15.1.2 Antinormally Ordered Quantum Characteristic Function

This is defined by

$$\chi_A(\lambda,\lambda^*) = \mathrm{Tr}\left\{\rho \exp(-\lambda^* a)\exp(\lambda a^\dagger)\right\}. \tag{15.1.3}$$

The *antinormally ordered* moments are given by its derivatives at $\lambda = 0$;

$$\langle a^s a^{\dagger r}\rangle = (-)^s \frac{\partial^{r+s}}{\partial^r\lambda\partial^s\lambda^*}\chi_A(\lambda,\lambda^*)\bigg|_{\lambda=0}. \tag{15.1.4}$$

15.1.3 Symmetrically Ordered Quantum Characteristic Function

This is defined by

$$\chi_S(\lambda,\lambda^*) = \mathrm{Tr}\left\{\rho \exp(-\lambda^* a + \lambda a^\dagger)\right\}. \tag{15.1.5}$$

The *symmetrically ordered* moments are given by its derivatives at $\lambda = 0$;

$$\left\langle \left\{ a^s a^{\dagger r} \right\}_{\text{sym}} \right\rangle = (-)^s \frac{\partial^{r+s}}{\partial^r \lambda \partial^s \lambda^*} \chi_S(\lambda, \lambda^*) \bigg|_{\lambda=0} . \tag{15.1.6}$$

The notation $\{a^r (a^\dagger)^s\}_{\text{sym}}$ means the symmetrical product, which is the average of all ways of ordering the operators, for example

$$\left\{ a^2 a^{\dagger 2} \right\}_{\text{sym}} = \frac{a^{\dagger 2} a^2 + a^\dagger a a^\dagger a + a^\dagger a^2 a^\dagger + a a^{\dagger 2} a + a a^\dagger a a^\dagger + a^2 a^{\dagger 2}}{6} . \tag{15.1.7}$$

15.1.4 Relationship between the Different Quantum Characteristic Functions

Using the Baker–Hausdorff theorem (10.5.5, 10.5.6) the three characteristic functions are related by

$$\chi_S(\lambda, \lambda^*) = \chi_N(\lambda, \lambda^*) e^{-|\lambda|^2/2} = \chi_A(\lambda, \lambda^*) e^{|\lambda|^2/2} . \tag{15.1.8}$$

Exercise 15.1 Number Statistics: The variance and mean of the number operator $N = a^\dagger a$ are often of importance. Show that these are related to variance and mean of $|\alpha|^2$ for the P-function and the Wigner functions as follows.

$$\langle N \rangle = \langle |\alpha|^2 \rangle_P , \tag{15.1.9}$$

$$\text{var}[N] = \text{var}\left[|\alpha|^2 \right]_P + \langle N \rangle , \tag{15.1.10}$$

$$\langle N \rangle = \langle |\alpha|^2 \rangle_W + \tfrac{1}{2} , \tag{15.1.11}$$

$$\text{var}[N] = \text{var}\left[|\alpha|^2 \right]_W - \langle N \rangle - \tfrac{1}{4} . \tag{15.1.12}$$

15.1.5 Properties of the Quantum Characteristic Functions

The three different functions are so directly related to each other, that it is not necessary to discuss them separately. All of the properties listed below are proved in *Quantum Noise*.

a) Existence of the Quantum Characteristic Functions: For any density operator, the definition (15.1.5) can be shown to converge, thus proving the existence of the symmetrically ordered quantum characteristic function, and hence, using (15.1.8), proving the existence of all three.

b) The Quantum Characteristic Functions Determine the Density Operator: Thus, if we know a quantum characteristic function, we know the density operator.

c) Examples of Quantum Characteristic Functions: These will all be given for the symmetrically ordered quantum characteristic function; other orderings can be deduced using (15.1.8).

i) *Coherent State*

$$\rho = |\alpha_0\rangle\langle\alpha_0|; \qquad \chi_S(\lambda,\lambda^*) = \exp\left(\lambda\alpha_0^* - \alpha_0\lambda^* - \tfrac{1}{2}|\lambda|^2\right). \tag{15.1.13}$$

ii) *Number State*

$$\rho = |n\rangle\langle n|; \qquad \chi_S(\lambda,\lambda^*) = \exp(-|\lambda|^2/2)\sum_{m=0}^{n}\frac{n!(-|\lambda|^2)^m}{(m!)^2(n-m)!}. \tag{15.1.14}$$

iii) *Thermal State*

If we consider a harmonic oscillator with Hamiltonian $\hbar\omega a^\dagger a$, the density operator for the corresponding thermal state at temperature T is then

$$\rho_{\mathrm{T}} = (1-z)\sum_n z^n |n\rangle\langle n|, \tag{15.1.15}$$

where

$$z = \mathrm{e}^{-\hbar\omega/k_{\mathrm{B}}T}. \tag{15.1.16}$$

The corresponding quantum characteristic function is

$$\chi_S(\lambda,\lambda^*) = \exp\left\{-|\lambda|^2\tanh\frac{\hbar\omega}{2k_{\mathrm{B}}T}\right\}. \tag{15.1.17}$$

15.2 Phase Space Representations of the Density Operator

The concept of a phase space representation comes about when we construct the classical Fourier inverse transform of a quantum characteristic function. This produces a c-number function which behaves like a classical probability distribution for the variables α, α^*. Although the three different kinds are all trivially connected by (15.1.8), the simple proportionality of that formula becomes transformed into a much less transparent relationship when Fourier transformed.

15.2.1 Wigner Function

We define the Wigner function by

$$W(\alpha,\alpha^*) = \frac{1}{\pi^2}\int d^2\lambda \exp(-\lambda\alpha^* + \lambda^*\alpha)\chi_S(\lambda,\lambda^*). \tag{15.2.1}$$

The Wigner function so defined always exists as an ordinary real function (that is, not a generalized function such as a delta function or its derivatives), but is not necessarily positive. Thus, it is called a *quasiprobability*. It was introduced by *Wigner* [56] as long ago as 1932.

a) Moments: If we use (15.1.6) and integrate by parts, we can see that

$$\left\langle \left\{ a^s a^{\dagger r} \right\}_{\text{sym}} \right\rangle = \int d^2\alpha\, \alpha^s \alpha^{*r} W(\alpha,\alpha^*) \equiv \langle \alpha^s \alpha^{*r} \rangle. \tag{15.2.2}$$

This means that the classically calculated moments of the Wigner function give the symmetrically ordered quantum operator moments. Thus the Wigner function does truly behave as a probability distribution, apart from the fact that it may be negative.

b) Coherent State Wigner Function: If the density operator corresponds to a coherent state, then Fourier transforming (15.1.13) gives

$$W_{\text{coh}}(\alpha,\alpha^*) = \frac{2}{\pi}\exp(-2|\alpha-\alpha_0|^2). \tag{15.2.3}$$

This is a Gaussian distribution with mean and second moments given by

$$\langle a \rangle_{\text{sym}} = \langle \alpha \rangle = \int d^2\alpha\, \alpha W_{\text{coh}}(\alpha,\alpha^*) = \alpha_0, \tag{15.2.4}$$

$$\langle a^2 \rangle_{\text{sym}} = \langle \alpha^2 \rangle = \int d^2\alpha\, \alpha^2 W_{\text{coh}}(\alpha,\alpha^*) = \alpha_0^2, \tag{15.2.5}$$

$$\langle a^\dagger a \rangle_{\text{sym}} = \langle \alpha^* \alpha \rangle = \int d^2\alpha\, |\alpha|^2 W_{\text{coh}}(\alpha,\alpha^*) = |\alpha_0|^2 + \tfrac{1}{2}. \tag{15.2.6}$$

This distribution is positive, and therefore can be viewed as a genuine probability distribution; we see from equation (15.2.6), that the Wigner function has a characteristic uncertainty of 1/2, essentially it behaves as if there is an extra half quantum added to the intensity $|\alpha_0|^2$.

In fact we can say that the Wigner function makes a mapping

$$a \to \alpha_0 + w_{\text{vac}}, \tag{15.2.7}$$

where w_{vac} is a complex "vacuum noise", satisfying

$$\langle w_{\text{vac}} \rangle = \langle w_{\text{vac}}^2 \rangle = \langle w_{\text{vac}}^{*\,2} \rangle = 0, \tag{15.2.8}$$

$$\langle |w_{\text{vac}}|^2 \rangle = \tfrac{1}{2}. \tag{15.2.9}$$

This simple-minded idea can be elaborated to a field theory level, where each mode can be regarded as a classical variable plus a vacuum noise term. This is the foundation of the c-field method for Bose gases which is treated in *Book III.*

c) Thermal State: For a thermal state at temperature T

$$W_T(\alpha,\alpha^*) = \frac{2}{\pi}\tanh(\hbar\omega/2k_B T)\exp\left[-2|\alpha|^2 \tanh(\hbar\omega/2k_B T)\right]. \tag{15.2.10}$$

This is a positive function which behaves like a genuine probability density.

d) A Number State: We show in *Quantum Noise* that for a number state density operator, $|n\rangle\langle n|$, the Wigner function is given by

$$W_n(\alpha,\alpha^*) = \frac{2(-)^n}{\pi}\exp(-2|\alpha|^2)\mathrm{L}_n(4|\alpha|^2). \tag{15.2.11}$$

In this case the Wigner function is negative in certain regions, and care is needed in its use.

15.2.2 The P-Function

We define the P-function by the Fourier transform of the *normally ordered* quantum characteristic function

$$P(\alpha,\alpha^*) = \frac{1}{\pi^2}\int d^2\lambda \exp(-\lambda\alpha^* + \lambda^*\alpha)\chi_N(\lambda,\lambda^*). \tag{15.2.12}$$

The P-function does *not* always exist as an ordinary function, but can be defined in terms of a *distribution* or *generalized function.*

a) Moments: If we use (15.1.2) and integrate by parts, we can see that *normally ordered* moments are given by

$$\langle a^{\dagger r} a^s\rangle = \int d^2\alpha\, \alpha^s \alpha^{*r} P(\alpha,\alpha^*) \equiv \langle \alpha^s \alpha^{*r}\rangle. \tag{15.2.13}$$

b) Coherent State P-Function: The normally ordered quantum characteristic function for a coherent state follows from (15.1.8) and (15.1.13), or alternatively, it is a trivial consequence of the definition (15.1.1). Fourier transforming gives a delta function

$$P_{\mathrm{coh}}(\alpha,\alpha^*) = \delta^2(\alpha-\alpha_0). \tag{15.2.14}$$

From this it follows that if the P-function exists, the density operator can be written as

$$\rho = \int d^2\alpha\, P(\alpha,\alpha^*)|\alpha\rangle\langle\alpha|. \tag{15.2.15}$$

The proof involves calculating the quantum characteristic function for this density operator, and noting that the P-function derived from that is the function $P(\alpha,\alpha^*)$ as used in this equation.

c) Thermal P-Function: This can be computed to be

$$P_T(\alpha,\alpha^*) = \frac{\mathrm{e}^{\hbar\omega/k_BT}-1}{\pi}\exp\left(-|\alpha|^2\left(\mathrm{e}^{\hbar\omega/k_BT}-1\right)\right). \tag{15.2.16}$$

d) Number State P-Function: This can only be defined in terms of derivatives of delta functions, and is not of much practical use. For details see *Quantum Noise.*

15.2.3 The Q-Function

We define the Q-function by the Fourier transform of the *antinormally ordered* quantum characteristic function

$$Q(\alpha,\alpha^*) = \frac{1}{\pi^2}\int d^2\lambda \exp(-\lambda\alpha^* + \lambda^*\alpha)\chi_A(\lambda,\lambda^*). \tag{15.2.17}$$

If we use the trace formula (10.5.14) to evaluate the trace used to define the characteristic function, it is straightforward to show that

$$Q(\alpha,\alpha^*) = \frac{1}{\pi}\langle\alpha|\rho|\alpha\rangle. \tag{15.2.18}$$

Clearly, therefore, the Q-function always exists and is positive. However, not every positive normalizable function of α and α^* is a valid Q-function.

a) Moments: If we use (15.1.4) and integrate by parts, we can see that *antinormally ordered* moments are given by

$$\langle a^s a^{\dagger r}\rangle = \int d^2\alpha\, \alpha^s \alpha^{*r} Q(\alpha,\alpha^*) \equiv \langle \alpha^s \alpha^{*r}\rangle. \tag{15.2.19}$$

b) Coherent State Q-Function: Using (15.2.18) and (10.5.11) we get

$$Q_{\text{coh}}(\alpha,\alpha^*) = \frac{1}{\pi}\exp\left(-|\alpha-\alpha_0|^2\right). \tag{15.2.20}$$

c) Thermal Q-Function: This can be computed to be

$$Q_T(\alpha,\alpha^*) = \frac{1-\mathrm{e}^{-\hbar\omega/k_B T}}{\pi}\exp\left(-|\alpha|^2\left(1-\mathrm{e}^{-\hbar\omega/k_B T}\right)\right). \tag{15.2.21}$$

d) Number State Q-Function: From (15.2.18) and (10.5.20) we get

$$Q_n(\alpha,\alpha^*) = \frac{1}{\pi}\frac{\mathrm{e}^{-|\alpha|^2}|\alpha|^{2n}}{n!}. \tag{15.2.22}$$

15.2.4 Multitime Correlation Functions

The formulae derived in Sect. 13.4.2 can be mapped into corresponding formulae for the various phase space representations. Let us consider the correlation function

$$\langle a^\dagger(t+\tau)a(t)\rangle = \text{Tr}_{\text{sys}}\left\{a^\dagger V(t+\tau)\big(a\rho(t)\big)\right\}, \tag{15.2.23}$$

$$= \text{Tr}_{\text{sys}}\left\{V(t+\tau)\big(a\rho(t)\big)a^\dagger\right\}. \tag{15.2.24}$$

We can evaluate this in stages:

i) The quantity $V(t+\tau)\big(a\rho(t)\big)$ behaves like a density operator, and we can consider the P-function corresponding to this; let us call it $P_V(\alpha,\alpha^*,t+\tau)$. In terms of this P-function, we can use the moment formula (15.2.13) to write

$$\text{Tr}_{\text{sys}}\left\{V(t+\tau)\big(a\rho(t)\big)a^\dagger\right\} = \int d^2\alpha\, \alpha^* P_V(\alpha,\alpha^*,t+\tau). \tag{15.2.25}$$

The value of $P_V(\alpha,\alpha^*,t+\tau)$ at $\tau=0$ is the P-function corresponding to the density operator $a\rho(t)$. If $P(\alpha,\alpha^*,t)$ is the P-function corresponding to $\rho(t)$, the moment formula (15.2.13) then requires that

$$P_V(\alpha,\alpha^*,t) = \alpha P(\alpha,\alpha^*,t). \tag{15.2.26}$$

ii) Define $P(\alpha,\alpha^*,t+\tau|\alpha_0,\alpha_0^*,t)$ as the value of $P_V(\alpha,\alpha^*,t+\tau)$ with the initial condition

$$P(\alpha,\alpha^*,t|\alpha_0,\alpha_0^*,t) = \delta^2(\alpha-\alpha_0). \tag{15.2.27}$$

Then the P-function corresponding to the initial condition (15.2.26) is given by setting

$$P_V(\alpha,\alpha^*,t+\tau) = \int d^2\alpha_0\, P(\alpha,\alpha^*,t+\tau|\alpha_0,\alpha_0^*,t)\alpha_0 P(\alpha_0,\alpha_0^*,t). \tag{15.2.28}$$

iii) Now substitute this into (15.2.25) to get

$$\langle a^\dagger(t+\tau)a(t)\rangle = \int d^2\alpha\, d^2\alpha_0\, \alpha P(\alpha,\alpha^*,t+\tau|\alpha_0,\alpha_0^*,t)\alpha_0 P(\alpha_0,\alpha_0^*,t), \tag{15.2.29}$$

$$\equiv \langle\alpha^*(t+\tau)\alpha(t)\rangle_P. \tag{15.2.30}$$

Thus, the *normally ordered* time correlation function (15.2.23) is equal to the correlation function of the c-number variables in the P-representation.

For both the Wigner representation and the Q-representation there are similar relationships. In summary these are:

$$\langle\alpha^*(t+\tau)\alpha(t)\rangle_P = \langle a^\dagger(t+\tau)a(t)\rangle \equiv \mathrm{Tr}_{\mathrm{sys}}\left\{a^\dagger V(t+\tau)\big(a\rho(t)\big)\right\} \tag{15.2.31}$$

$$\langle\alpha^*(t+\tau)\alpha(t)\rangle_Q = \langle a(t)a^\dagger(t+\tau)\rangle \equiv \mathrm{Tr}_{\mathrm{sys}}\left\{a^\dagger V(t+\tau)\big(\rho(t)a\big)\right\} \tag{15.2.32}$$

$$\langle\alpha^*(t+\tau)\alpha(t)\rangle_W = \tfrac{1}{4}\mathrm{Tr}_{\mathrm{sys}}\left\{\Big[a^\dagger, V(t+\tau)\big([a,\rho(t)]_+\big)\Big]_+\right\}. \tag{15.2.33}$$

Exercise 15.2 Multitime Correlation Functions in Other Representations: Prove the the Wigner and Q-function relationships (15.2.32) and (15.2.33).

16. Wigner Function Methods

The Wigner function provides a transformation of quantum-mechanical equations into phase space which is adapted particularly well to the treatment of Bose–Einstein condensates and other systems in which the mode occupations are large. Because the equations of motion often yield derivatives of the third or higher orders, an exact mapping to a stochastic equation is often impossible. However, approximations based on the idea of large mode occupations often yield very simple equations of motion, which have a kind of semiclassical interpretation.

The Wigner function averages corresponds to *symmetrically* ordered quantum averages, and this means that even the vacuum state corresponds to the non-zero occupation $\langle|\alpha|^2\rangle = \frac{1}{2}$. Consequently, the initial conditions for any set of either exact or approximate equations of motion for the Wigner function are intrinsically random, and thus noisy. In fact, it is quite common to find that the equations of motion used (which may be approximate) contain no quantum noise terms—thus all noise effects are represented by the initial conditions.

16.1 Operator Correspondences and Equations of Motion

We consider the simplest possible example: If the Wigner function corresponding to a density operator ρ is $W(\alpha,\alpha^*)$, what is the Wigner function corresponding to $a\rho$? By looking at (15.1.3), it is easy to see that the *normally ordered* characteristic function corresponding to $a\rho$ is

$$\mathrm{Tr}\left\{a\rho\exp(\lambda a^\dagger)\exp(-\lambda^* a)\right\} = \mathrm{Tr}\left\{\rho\exp(\lambda a^\dagger)\exp(-\lambda^* a)a\right\}, \tag{16.1.1}$$

$$= -\frac{\partial}{\partial\lambda^*}\chi_N(\lambda,\lambda^*). \tag{16.1.2}$$

Next using (15.1.8), we see that the *symmetrically ordered* quantum characteristic function corresponding to $a\rho$ is

$$\left(\tfrac{1}{2}\lambda - \frac{\partial}{\partial\lambda^*}\right)\chi_N(\lambda,\lambda^*). \tag{16.1.3}$$

Now use the definition (15.2.1) of the Wigner function and integrate by parts, and we find that the Wigner function corresponding to $a\rho$ is

$$\left(\alpha + \frac{1}{2}\frac{\partial}{\partial\alpha^*}\right)W(\alpha,\alpha^*). \tag{16.1.4}$$

It is straightforward to generalize to all left and right products to get the table of mappings

$$a\rho \longleftrightarrow \left(\alpha + \frac{1}{2}\frac{\partial}{\partial\alpha^*}\right) W(\alpha,\alpha^*), \tag{16.1.5a}$$

$$a^\dagger\rho \longleftrightarrow \left(\alpha^* - \frac{1}{2}\frac{\partial}{\partial\alpha}\right) W(\alpha,\alpha^*), \tag{16.1.5b}$$

$$\rho a \longleftrightarrow \left(\alpha - \frac{1}{2}\frac{\partial}{\partial\alpha^*}\right) W(\alpha,\alpha^*), \tag{16.1.5c}$$

$$\rho a^\dagger \longleftrightarrow \left(\alpha^* + \frac{1}{2}\frac{\partial}{\partial\alpha}\right) W(\alpha,\alpha^*). \tag{16.1.5d}$$

Exercise 16.1 Useful Mappings: From these mappings derive the following mappings, which are useful in applications:

$$[a,\rho] \longleftrightarrow \frac{\partial}{\partial\alpha^*} W(\alpha,\alpha^*), \tag{16.1.6a}$$

$$[a^\dagger,\rho] \longleftrightarrow -\frac{\partial}{\partial\alpha} W(\alpha,\alpha^*), \tag{16.1.6b}$$

$$[a^\dagger a,\rho] \longleftrightarrow \left(-\frac{\partial}{\partial\alpha}\alpha + \frac{\partial}{\partial\alpha^*}\alpha^*\right) W(\alpha,\alpha^*), \tag{16.1.6c}$$

$$[a,[a^\dagger,\rho]] \longleftrightarrow -\frac{\partial^2}{\partial\alpha^*\partial\alpha} W(\alpha,\alpha^*), \tag{16.1.6d}$$

$$[a^\dagger,[a,\rho]] \longleftrightarrow -\frac{\partial^2}{\partial\alpha^*\partial\alpha} W(\alpha,\alpha^*). \tag{16.1.6e}$$

The above five mappings have the same form as the corresponding mappings for the P-function (17.1.6a–17.1.7a). In contrast, the following mappings differ from the corresponding P-function mappings.

$$2a\rho a^\dagger - \rho a^\dagger a - a^\dagger a\rho \longleftrightarrow \left(\frac{\partial}{\partial\alpha}\alpha + \frac{\partial}{\partial\alpha^*}\alpha^* + \frac{\partial^2}{\partial\alpha\partial\alpha^*}\right) W, \tag{16.1.7a}$$

$$2a^\dagger\rho a - \rho a a^\dagger - a a^\dagger\rho \longleftrightarrow \left(-\frac{\partial}{\partial\alpha}\alpha - \frac{\partial}{\partial\alpha^*}\alpha^* + \frac{\partial^2}{\partial\alpha\partial\alpha^*}\right) W, \tag{16.1.7b}$$

$$[a^\dagger a,\rho]_+ \longleftrightarrow \left(2|\alpha^2| - 1 - \tfrac{1}{2}\frac{\partial^2}{\partial\alpha^*\partial\alpha}\right) W. \tag{16.1.7c}$$

16.1.1 The Driven Harmonic Oscillator

We consider the Hamiltonian

$$H = \hbar\omega(a^\dagger a + \tfrac{1}{2}) + g a^\dagger + g^* a, \tag{16.1.8}$$

for which the equation of motion of the density operator is

$$i\hbar\frac{\partial\rho(t)}{\partial t} = \hbar\omega[a^\dagger a,\rho(t)] + \lambda[a^\dagger,\rho(t)] + \lambda^*[a,\rho(t)]. \tag{16.1.9}$$

We now turn this into an equation for $W(\alpha,\alpha^*,t)$ by using the mappings (16.1.6a–16.1.6c). Thus we find the equation of motion

$$\frac{\partial W}{\partial t} = i\left(\omega\frac{\partial}{\partial\alpha}\alpha - \omega\frac{\partial}{\partial\alpha^*}\alpha^* + \frac{g}{\hbar}\frac{\partial}{\partial\alpha} - \frac{g^*}{\hbar}\frac{\partial}{\partial\alpha^*}\right)W. \tag{16.1.10}$$

This corresponds to a Liouville equation for the Wigner function, as described in Sect. 5.1.2. The solutions can be represented in terms of the complex differential equations

$$\frac{d\alpha}{dt} = -i(\omega\alpha + g/\hbar), \tag{16.1.11}$$

$$\frac{d\alpha^*}{dt} = i(\omega\alpha^* + g^*/\hbar), \tag{16.1.12}$$

which have the solutions

$$\alpha(t) = -g/\hbar\omega + \alpha_0 e^{-i\omega t}, \tag{16.1.13}$$

$$\alpha^*(t) = -g^*/\hbar\omega + \alpha_0^* e^{i\omega t}. \tag{16.1.14}$$

If the initial value of the Wigner function is $W_0(\alpha,\alpha^*)$, then the solution of the Liouville equation (16.1.10) is

$$W(\alpha,\alpha^*,t) = \int d^2\alpha_0\, W_0(\alpha_0,\alpha_0^*)\,\delta\left(\alpha + g/\hbar\omega - \alpha_0 e^{-i\omega t}\right)\delta\left(\alpha^* + g^*/\hbar\omega - \alpha_0^* e^{i\omega t}\right). \tag{16.1.15}$$

Exercise 16.2 Coherent State Solution: Show that if the initial condition is a coherent state, the Wigner function is at all times that of a coherent state.

Exercise 16.3 Time-Dependent Driving: Consider the case of a time-dependent driving term, so that (16.1.8) is modified by setting $g \to g(t)$. Show how to solve the equations of motion, and that in this case also, an initial coherent state leads to a solution that is always a coherent state.

Exercise 16.4 Random Time-Dependent Driving: Suppose that $g(t)$ is a random process. Find the equation of motion for the Wigner function. In the case that $g(t)$ is an Ornstein–Uhlenbeck process satisfying (**I**) $dg(t) = -\kappa g(t)\,dt + \sqrt{n}\,dW(t)$, find:

i) The joint Fokker–Planck equation for the Wigner function $W(\alpha,\alpha^*,g,t)$.

ii) The limiting Fokker–Planck equation in the case that the relaxation constant κ is very large.

16.1.2 The Anharmonic Oscillator

The relevant Hamiltonian is

$$H = \hbar\omega a^\dagger a + \tfrac{1}{2}u a^\dagger a^\dagger aa, \tag{16.1.16}$$

and this is the prototype of the Hamiltonian for Bose–Einstein condensation, corresponding to the highest-order term in the Bogoliubov Hamiltonian, as is explained in detail in *Book III.*

a) Derivation of Equation of Motion: The derivations of equations of motion can become quite complicated—in this example we indicate the kind of procedure that can simplify the calculations.

i) We write

$$N = a^\dagger a, \tag{16.1.17}$$

$$a^\dagger a^\dagger aa = N^2 - N, \tag{16.1.18}$$

$$[N^2, \rho] = \left[N, [N,\rho]_+\right]. \tag{16.1.19}$$

ii) We now use the mappings (16.1.6c, 16.1.7c) and get the Wigner function equation of motion corresponding to (16.1.16)

$$\frac{\partial W}{\partial t} = \mathrm{i}\left\{\left(\frac{\partial}{\partial\alpha}\alpha - \frac{\partial}{\partial\alpha^*}\alpha^*\right)\left(\omega + u(|\alpha|^2 - \tfrac{1}{2})\right) + \frac{u}{4}\left(\frac{\partial^3}{\partial\alpha^2\partial\alpha^*}\alpha - \frac{\partial^3}{\partial\alpha\partial\alpha^{*2}}\alpha^*\right)\right\}W. \tag{16.1.20}$$

b) System Size Scaling: The third-order derivatives in the equation of motion (16.1.20) cannot be easily applied, but for large systems it can be reasonable to drop them, leaving a simple Liouville equation. Let us use the Bogoliubov scaling as in *Book III,* which introduces a system volume V, and assumes that the number of particles is proportional to V. In the phase space variable α, this corresponds to making the substitutions

$$\alpha = \sqrt{V}\xi, \tag{16.1.21}$$

$$u = \frac{\tilde{u}}{V}. \tag{16.1.22}$$

The second substitution means that the non-linear energy $u|\alpha|^4/2$ is also proportional to the volume V.

The equation of motion (16.1.20) in terms of these variables becomes

$$\begin{aligned}\frac{\partial W}{\partial t} = \mathrm{i}\Bigg\{&\left(\frac{\partial}{\partial\xi}\xi - \frac{\partial}{\partial\xi^*}\xi^*\right)\left(\omega + \tilde{u}|\xi|^2\right) \\ &- \frac{\tilde{u}}{2V}\left(\frac{\partial}{\partial\xi}\xi - \frac{\partial}{\partial\xi^*}\xi^*\right) \\ &+ \frac{\tilde{u}}{4V^2}\left(\frac{\partial^3}{\partial\xi^2\partial\xi^*}\xi - \frac{\partial^3}{\partial\xi\partial\xi^{*2}}\xi^*\right)\Bigg\}W.\end{aligned} \tag{16.1.23}$$

i) We now take the limit of large V at fixed $\tilde{u}$ and ξ. In this limit, the typical number of particles becomes large, and the three terms in this expansion become successively smaller. The minimum quantum fluctuations in $|\alpha^2|$ are of order of magnitude 1/2, and are expressed in the initial condition. They correspond to fluctuations of order of magnitude $1/V$ in $|\xi|$. The term with third-order derivative is of order $1/V^2$, and therefore is negligible compared to the quantum fluctuations.

ii) The second line is of order of magnitude $1/V$, and is thus comparable with quantum fluctuations. However, it is really a small correction to the first line, and thus its main effect is to alter the trajectory of any solution slightly, but not the noise. It is often omitted even when the quantum noise is being evaluated.

iii) Therefore, if V is sufficiently large, we can neglect all but the first line, leaving a Liouville equation, equivalent to the differential equations

$$\left.\begin{aligned} \frac{d\xi}{dt} &= -\,\mathrm{i}\big(\omega\xi + \tilde{u}|\xi|^2\xi\big), \\ \frac{d\xi^*}{dt} &= \mathrm{i}\big(\omega\xi^* + \tilde{u}|\xi|^2\xi^*\big). \end{aligned}\right\} \tag{16.1.24}$$

These equations correspond to the zero-dimensional equivalent of the Gross–Pitaevskii equation, which we have introduced in Sect. 2.4.1c. The quantum fluctuations do not appear in the equation of motion, but do appear in the initial condition. When the full Gross–Pitaevskii equation is implemented in an analogous way, this provides a very powerful technique for describing partially thermalized Bose–Einstein condensates. This technique, known as the *c-field* method, has been reviewed in [57], and is also described in *Book III*.

16.1.3 The Bogoliubov Hamiltonian

The excitation spectrum of the Bogoliubov theory of a Bose condensed gas is given by the Bogoliubov Hamiltonian

$$H_{\mathrm{Bog}} = \sum_{\boldsymbol{k}} \left\{ \left(\frac{\hbar^2 k^2}{2m} + \tfrac{1}{2}\mu \right) \left(a^\dagger_{\boldsymbol{k}} a_{\boldsymbol{k}} + a^\dagger_{-\boldsymbol{k}} a_{-\boldsymbol{k}} \right) + \tfrac{1}{2}\mu \left(a^\dagger_{\boldsymbol{k}} a^\dagger_{-\boldsymbol{k}} + a_{\boldsymbol{k}} a_{-\boldsymbol{k}} \right) \right\}. \tag{16.1.25}$$

The Hamiltonian is the sum of independent terms, one for each pair of equal and opposite momenta $\boldsymbol{k}, -\boldsymbol{k}$. Making the substitutions $a_{\boldsymbol{k}} \to a$, $a_{-\boldsymbol{k}} \to b$, this Hamiltonian can be written as a sum of terms of the form

$$H_{a,b} = \hbar v(a^\dagger a + b^\dagger b) + \hbar u(a^\dagger b^\dagger + ba). \tag{16.1.26}$$

The Wigner function equation of motion for the coupled modes a and b becomes

$$\frac{\partial W}{\partial t} =$$
$$\mathrm{i}\left(\frac{\partial}{\partial\alpha}(v\alpha+u\beta^*)-\frac{\partial}{\partial\alpha^*}(v\alpha^*+u\beta)+\frac{\partial}{\partial\beta}(v\beta+u\alpha^*)-\frac{\partial}{\partial\beta^*}(v\beta^*+u\alpha)\right)W, \tag{16.1.27}$$

which is a Liouville equation corresponding to

$$\left.\begin{aligned}\frac{d\alpha}{dt} &= -\,\mathrm{i}(v\alpha+u\beta^*), & \frac{d\beta^*}{dt} &= \mathrm{i}(v\beta^*+u\alpha),\\ \frac{d\beta}{dt} &= -\,\mathrm{i}(v\beta+u\alpha^*), & \frac{d\alpha^*}{dt} &= \mathrm{i}(v\alpha^*+u\beta).\end{aligned}\right\} \tag{16.1.28}$$

As in the case of the anharmonic oscillator, quantum fluctuations appear only in the initial conditions.

16.2 Damped and Driven Systems

In this section we will include simple damping terms arising from appropriate master equations. For these cases it is possible to show that the Wigner function Fokker–Planck equations are very similar to those of analogous classical, namely complex variable systems as introduced in Sect. 7.3. Thus, the kinds of damping represented by quantum Markov processes are the natural generalizations of classical damped systems.

16.2.1 The Harmonic Oscillator Including Loss or Gain

Here we take a master equation, as in (14.3.4), for a damped harmonic oscillator, in the form

$$\begin{aligned}\frac{\partial\rho}{\partial t} &= -\mathrm{i}[\Omega a^\dagger a,\rho] + \frac{\gamma}{2}\mathcal{M}\{2a\rho a^\dagger - \rho a^\dagger a - a^\dagger a\rho\}\\ &\quad + \frac{\gamma}{2}\mathcal{N}\{2a^\dagger\rho a - \rho a a^\dagger - a a^\dagger\rho\}.\end{aligned} \tag{16.2.1}$$

i) Using the mappings (16.1.7a, 16.1.7b) we derive the Wigner function equation of motion

$$\begin{aligned}\frac{\partial W}{\partial t} &= \left\{\mathrm{i}\Omega\left(\frac{\partial}{\partial\alpha}\alpha-\frac{\partial}{\partial\alpha^*}\alpha^*\right)\right.\\ &\quad\left.+\tfrac{1}{2}\gamma(\mathcal{M}-\mathcal{N})\left(\frac{\partial}{\partial\alpha^*}\alpha^*+\frac{\partial}{\partial\alpha}\alpha\right)+\tfrac{1}{2}\gamma(\mathcal{M}+\mathcal{N})\frac{\partial^2}{\partial\alpha\partial\alpha^*}\right\}W.\end{aligned} \tag{16.2.2}$$

ii) This is a Fokker–Planck equation, and is of exactly the same form as that of the thermalized oscillator already treated in Sect. 7.3.2. It is equivalent to the stochastic differential equations

$$
\begin{aligned}
&(\mathrm{I})\, d\alpha &&= \left(-\mathrm{i}\Omega - \tfrac{1}{2}\gamma(\mathcal{M}-\mathcal{N})\right)\alpha\, dt + \sqrt{\tfrac{1}{2}\gamma(\mathcal{M}+\mathcal{N})}\, dw(t), && (16.2.3)\\
&(\mathrm{I})\, d\alpha^* &&= \left(\mathrm{i}\Omega - \tfrac{1}{2}\gamma(\mathcal{M}-\mathcal{N})\right)\alpha^*\, dt + \sqrt{\tfrac{1}{2}\gamma(\mathcal{M}+\mathcal{N})}\, dw^*(t), && (16.2.4)\\
&dw(t) &&\equiv \frac{dW_1(t) + \mathrm{i}\, dW_2(t)}{\sqrt{2}}, && (16.2.5)\\
&dw(t)\, dw^*(t) = dt, && dw(t)^2 = dw(t)^{*2} = 0. && (16.2.6)
\end{aligned}
$$

a) Parameters for the Thermal Case: When the oscillator is interacting with a thermal bath at temperature T, the parameters are as in (14.3.35)

$$
\mathcal{N} \to \bar{N}(T,\Omega) = \frac{1}{e^{\hbar\Omega/k_B T} - 1}, \qquad \mathcal{M} \to \bar{N}(T,\Omega) + 1. \tag{16.2.7}
$$

b) Classical Equivalent: This equation is then of the same form as the thermalized oscillator equation treated in Sect. 7.3.2, with the parameter correspondences

$$
\left.\begin{aligned}
\omega &\to \Omega,\\
\gamma &\to \gamma(\mathcal{M}-\mathcal{N}) \to \gamma,\\
\bar{N} &\to \tfrac{1}{2}(\mathcal{M}+\mathcal{N}) \to \bar{N}(T,\Omega) + \tfrac{1}{2}.
\end{aligned}\right\} \tag{16.2.8}
$$

Exercise 16.5 Stationary Solution: Show that the thermal stationary solution of (16.2.2), using the notation of (16.2.7), has the form

$$
W_s(\alpha,\alpha^*) = \frac{1}{\pi\left(\bar{N}(T,\Omega) + 1/2\right)} \exp\left(-\frac{|\alpha|^2}{\bar{N}(T,\Omega) + 1/2}\right), \tag{16.2.9}
$$

$$
\langle |\alpha^2| \rangle_s = \bar{N}(T,\Omega) + 1/2. \tag{16.2.10}
$$

Exercise 16.6 Quantum Characteristic Function for Harmonic Motion with Loss or Gain: Show that one can use the Wigner function form (16.2.3–16.2.6) of the equations of motion to compute the quantum characteristic function as follows:

i) Transfer to a rotating frame in which the terms in Ω are eliminated.

ii) Using the techniques for complex Ito stochastic differential equations developed in Sect. 7.3, show that the solution of (16.2.3) becomes

$$
\alpha(t) = e^{-\gamma(\mathcal{M}-\mathcal{N})t/2}\alpha(0) + \sqrt{\frac{\gamma(\mathcal{M}+\mathcal{N})}{2}} \int_0^t e^{-\gamma(\mathcal{M}-\mathcal{N})(t-t')/2}\, dw(t'). \tag{16.2.11}
$$

iii) Using the result of Ex. 7.3 on the mean of the exponential of a Gaussian variable, and the inverse of the definition (15.2.1) of the Wigner function in terms of the quantum characteristic function $\chi_S(\lambda,\lambda^*)$, show that the time-dependent quantum characteristic function in this case is

$$\chi_S(\lambda,\lambda^*,t) = \chi_S\left(\lambda e^{-\gamma(\mathcal{M}-\mathcal{N})t/2}, \lambda^* e^{-\gamma(\mathcal{M}-\mathcal{N})t/2}, 0\right) \times \exp\left\{-\tfrac{1}{2}|\lambda|^2 \frac{\mathcal{M}+\mathcal{N}}{\mathcal{M}-\mathcal{N}}\left(1-e^{-\gamma(\mathcal{M}-\mathcal{N})t}\right)\right\}. \tag{16.2.12}$$

iv) Using the result (15.1.13), show that if the initial state of the system is the coherent state $|\alpha_0\rangle$, then the time-dependent quantum characteristic function is

$$\chi_S(\lambda,\lambda^*,t) = \exp\left(\lambda\alpha_0^*(t) - \lambda^*\alpha_0(t) - \tfrac{1}{2}|\lambda|^2\right) \times \exp\left\{-|\lambda|^2 \frac{\mathcal{N}}{\mathcal{M}-\mathcal{N}}\left(1-e^{-\gamma(\mathcal{M}-\mathcal{N})t}\right)\right\}, \tag{16.2.13}$$

where $\alpha_0(t) \equiv \alpha(0)e^{-\gamma(\mathcal{M}-\mathcal{N})t/2}$. (16.2.14)

v) In the case that $\mathcal{N}=0$, that is, when there is only loss with no gain, show that this corresponds to a coherent state $|\alpha_0(t)\rangle$. Thus, in the case of a purely damped oscillator, a coherent state evolves into a coherent state with damped amplitude $\alpha_0(t)$.

c) Amplification: Let us use the notation

$$r(t) \equiv \frac{2\mathcal{N}}{\mathcal{M}-\mathcal{N}}\left(1-e^{-\gamma(\mathcal{M}-\mathcal{N})t}\right), \tag{16.2.15}$$

in the formula (16.2.13) for the characteristic function, which then takes the form

$$\chi_S(\lambda,\lambda^*,t) = \exp\left(\lambda\alpha_0^*(t) - \lambda^*\alpha_0(t) - \tfrac{1}{2}|\lambda|^2\left(1+r(t)\right)\right). \tag{16.2.16}$$

Using the defining formula (15.2.1), the Wigner function for the amplified coherent state becomes

$$W_{\mathrm{Amp}}(\alpha,\alpha^*) = \frac{2}{\pi\left(1+r(t)\right)}\exp\left(-\frac{2|\alpha-\alpha_0(t)|^2}{1+r(t)}\right). \tag{16.2.17}$$

This differs from the Wigner function (15.2.1), only by the inclusion of the term $r(t)$, so that instead of the coherent state results (15.2.4–15.2.6), we have

$$\langle a\rangle_{\mathrm{sym}} = \langle\alpha\rangle = \alpha_0(t), \tag{16.2.18}$$

$$\langle a^2\rangle_{\mathrm{sym}} = \langle\alpha^2\rangle = \alpha_0(t)^2, \tag{16.2.19}$$

$$\langle a^\dagger a\rangle_{\mathrm{sym}} = \langle|\alpha|^2\rangle = |\alpha_0(t)|^2 + \tfrac{1}{2}\left(1+r(t)\right). \tag{16.2.20}$$

Substituting the definitions of $\alpha_0(t)$ and $r(t)$, the last of these becomes

$$\langle a^\dagger a\rangle_{\mathrm{sym}} = \langle|\alpha|^2\rangle = \left(|\alpha_0|^2 + \frac{\mathcal{N}}{\mathcal{N}-\mathcal{M}}\right)e^{-\gamma(\mathcal{M}-\mathcal{N})t} + \frac{1}{2}\frac{\mathcal{M}+\mathcal{N}}{\mathcal{M}-\mathcal{N}} \tag{16.2.21}$$

for some real $s \ge 0$.

d) Amplification and Attenuation of a Gaussian State: We will now consider as an initial state a Gaussian density operator whose quantum characteristic function is of the form

$$\chi_{\text{Gauss}}(\lambda,\lambda^*) = \exp\left(\lambda\alpha_0^* - \lambda^*\alpha_0 - \tfrac{1}{2}|\lambda|^2(1+s)\right). \tag{16.2.22}$$

The corresponding Wigner function is of the same form as (16.2.22), but with $r(t) \to s$, and this represents coherent state with some added thermal fluctuations. This form of course excludes states in which $\langle a^2\rangle \neq \langle a\rangle^2$.

After amplification or attenuation, the quantum characteristic function is

$$\chi_{\text{Gauss}}(\lambda,\lambda^*,t) = \exp\left(\lambda\alpha_0^*(t) - \lambda^*\alpha_0(t) - \tfrac{1}{2}|\lambda|^2\right) \times \exp\left[-\tfrac{1}{2}|\lambda|^2\left(\frac{2\mathcal{N}}{\mathcal{M}-\mathcal{N}}\left(1-e^{-\gamma(\mathcal{M}-\mathcal{N})t}\right) + s e^{-\gamma(\mathcal{M}-\mathcal{N})t}\right)\right]. \tag{16.2.23}$$

We now consider the principal situations for a distribution appropriate to the case of a large initial coherent component α_0 with a finite amount of added fluctuations, that is, we will require

$$|\alpha_0| \gg \sqrt{s}. \tag{16.2.24}$$

i) *Pure Attenuation*: $\mathcal{N} = 0$. Since $|\alpha_0|$ is large, there will be a time T at which $|\alpha_0(T)| = |\alpha_0|\exp\left(-\tfrac{1}{2}\gamma\mathcal{M}T\right)$ is quite large, but $s\exp\left(-\gamma\mathcal{M}T\right) \ll 1$. The resulting quantum characteristic function is then very close to that of a coherent state of amplitude $\alpha_0(T)$.

   This means that a pure attenuator can reduce the relative noise in a Gaussian quantum state to the minimum allowed by quantum mechanics.

ii) *Net Attenuation*: $\mathcal{M} > \mathcal{N}$. In this case the net result is a quantum characteristic function of the Gaussian form (16.2.22), but with

$$1+s \longrightarrow \frac{\mathcal{M}+\mathcal{N}}{\mathcal{M}-\mathcal{N}}. \tag{16.2.25}$$

   The fluctuations in addition to those of a coherent state are replaced by fluctuations derived from the medium supplying gain and loss. For $\mathcal{N}$ not too large, this can be a state of reduced relative fluctuations.

iii) *Net Gain*: $\mathcal{N} > \mathcal{M}$. In this case the exponentials grow indefinitely, so that we get an asymptotic result, that the quantum characteristic function is of the Gaussian form (16.2.22), but with

$$1+s \longrightarrow \left(\frac{2\mathcal{N}}{\mathcal{N}-\mathcal{M}} + s\right)e^{\gamma(\mathcal{N}-\mathcal{M})t}, \tag{16.2.26}$$

$$\alpha_0(t) \longrightarrow \alpha_0 e^{\gamma(\mathcal{N}-\mathcal{M})t/2}. \tag{16.2.27}$$

   Here the fluctuations in excess of the coherent value (the term involving s) are amplified, and as well there is an added noise component, which grows at the same rate. However, the *relative* fluctuations are only modestly increased:

$$\frac{\langle a^\dagger a\rangle_{\text{sym}} - |\langle a\rangle|^2}{|\langle a\rangle|^2} \longrightarrow \frac{1}{|\alpha_0|^2}\left(\frac{\mathcal{N}}{\mathcal{N}-\mathcal{M}} + \tfrac{1}{2}s\right). \tag{16.2.28}$$

The best performance is achieved by the *ideal amplifier*, that is when $\mathcal{M} = 0$ and the amplifier supplies only gain. If we take $s = 0$, that is an initial coherent state, the noise ratio becomes

$$\frac{\langle a^\dagger a\rangle_{\text{sym}} - |\langle a\rangle|^2}{|\langle a\rangle|^2} \longrightarrow \frac{1}{|\alpha_0|^2}, \tag{16.2.29}$$

which is exactly twice the value for a coherent state.

Exercise 16.7 Almost Perfect Amplification: Take a coherent state $|\alpha_0\rangle$ and amplify it using an ideal amplifier, with a gain $\exp(\gamma\mathcal{N}t) = 10^{12}$. Now take the result, and attenuate it using a pure attenuator, with attenuation $\exp(-\gamma\mathcal{M}t) = 10^{-6}$. Show the result is very close to the amplified coherent state $|10^3\alpha_0\rangle$, with almost no added noise.

Perfect amplification is only possible for an initial coherent state, or for states close to one. For example, we will see in Sect. 19.4 that if the initial state is a superposition of two coherent states $|\alpha_0\rangle + |\beta_0\rangle$, both the process of attenuation and that of amplification turn the superposition into an incoherent mixture of the two states with density operator $|\alpha_0(t)\rangle\langle\alpha_0(t)| + |\beta_0(t)\rangle\langle\beta_0(t)|$.

16.2.2 The Phase-Damped Harmonic Oscillator

Here we consider the master equation which was introduced in Sect. 14.3.3, with $N \equiv a^\dagger a$,

$$\frac{\partial\rho}{\partial t} = \frac{\kappa}{2}(2\bar{N}+1)\left\{2N\rho N - \rho N^2 - N^2\rho\right\}. \tag{16.2.30}$$

i) Note that

$$2N\rho N - \rho N^2 - N^2\rho = -\left[N,[N,\rho]\right]. \tag{16.2.31}$$

ii) Use the mapping (16.1.6c) recursively to deduce that the equation of motion for the Wigner function is

$$\frac{\partial W}{\partial t} = \frac{\kappa}{2}(2\bar{N}+1)\left(\frac{\partial}{\partial\alpha}\alpha + \frac{\partial}{\partial\alpha^*}\alpha^* + 2\frac{\partial^2}{\partial\alpha\partial\alpha^*}|\alpha|^2 - \frac{\partial^2}{\partial\alpha^2}\alpha^2 - \frac{\partial^2}{\partial\alpha^{*2}}\alpha^{*2}\right)W. \tag{16.2.32}$$

iii) This is equivalent to the stochastic differential equations

$$(\mathrm{I})\, d\alpha = -\tfrac{1}{2}\kappa(2\bar{N}+1)\alpha\, dt + \mathrm{i}\alpha\sqrt{\kappa(2\bar{N}+1)}\, dW(t), \tag{16.2.33}$$

$$(\mathrm{I})\, d\alpha^* = -\tfrac{1}{2}\kappa(2\bar{N}+1)\alpha^*\, dt - \mathrm{i}\alpha^*\sqrt{\kappa(2\bar{N}+1)}\, dW(t), \tag{16.2.34}$$

which are exactly those of the classical random frequency (or phase-damped) oscillator of Sect. 7.3.1 with $\omega_0 = 0$. A non-zero value of ω_0 can be included by adding a Hamiltonian term $-\mathrm{i}[\omega_0 a^\dagger a, \rho]$ to the equation of motion.

16.3 The Wigner Distribution Function $f(x,p)$

Historically, the Wigner function was first defined in terms of the position and momentum operators rather than the creation and destruction operators, with the aim of providing a distribution function in classical phase space. To achieve this, instead of the definition of the Wigner function (15.2.1) in terms of the complex variable α, we introduce an equivalent definition in terms of the real phase space co-ordinates $\boldsymbol{x}$ and $\boldsymbol{p}$ by

$$f(\boldsymbol{x},\boldsymbol{p}) \equiv \frac{1}{(2\pi\hbar)^3}\int d^3\boldsymbol{u}\,\langle \boldsymbol{x}+\tfrac{1}{2}\boldsymbol{u}|\,\rho\,|\boldsymbol{x}-\tfrac{1}{2}\boldsymbol{u}\rangle\, e^{-i\boldsymbol{p}\cdot\boldsymbol{u}/\hbar} \tag{16.3.1}$$

$$= \frac{1}{(2\pi\hbar)^3}\int d^3\boldsymbol{q}\,\langle \boldsymbol{p}+\tfrac{1}{2}\boldsymbol{q}|\,\rho\,|\boldsymbol{p}-\tfrac{1}{2}\boldsymbol{q}\rangle\, e^{i\boldsymbol{q}\cdot\boldsymbol{x}/\hbar}. \tag{16.3.2}$$

We will call the function $f(\boldsymbol{x},\boldsymbol{p})$ so defined the *Wigner distribution function*, in contrast to the terminology *Wigner function* for $W(\alpha,\alpha^*)$ defined in (15.2.1).

We have written the above explicitly in three dimensions, since this is usually what is required. However, the formulation of relationships to the Wigner function as defined in (15.2.1), and of operator mappings, is best done in one dimension, which we shall use in most of what follows. The only difference in appearance is the change $(2\pi\hbar)^3 \to 2\pi\hbar$ in (16.3.1).

The relationship to the definition of (15.2.1) depends on the relationship between the creation and destruction operators a, $a^\dagger$ and the operators P and X, which is normally given in terms of a harmonic oscillator Hamiltonian; if the particle mass is m and the oscillator frequency is ω, then

$$a = \sqrt{\frac{\omega m}{2\hbar}}X + \frac{1}{\sqrt{2\hbar\omega m}}iP. \tag{16.3.3}$$

If we let $\kappa \equiv \sqrt{\omega m}$, then the set of possible relationships between the two kinds of variables is

$$a = \frac{1}{\sqrt{2\hbar}}\left(\kappa X + \frac{iP}{\kappa}\right). \tag{16.3.4}$$

Exercise 16.8 Equivalence of Formulations of the Wigner Function: Show that (16.3.1) and (16.3.2) are equivalent by writing

$$|\boldsymbol{p}\rangle = \frac{1}{(2\pi\hbar)^{3/2}}\int d^3\boldsymbol{x}\, e^{i\boldsymbol{p}\cdot\boldsymbol{x}}. \tag{16.3.5}$$

Exercise 16.9 Equivalence of $W(\alpha,\alpha^*)$ and $f(x,p)$: Show that the Wigner function as defined by (15.2.1) and as defined by the one-dimensional version of (16.3.1) are related by

$$f(x,p) = \frac{1}{2\hbar}W\left(\alpha(x,p),\alpha(x,p)^*\right), \tag{16.3.6}$$

$$\alpha(x,p) \equiv \frac{1}{\sqrt{2\hbar}}\left(\kappa x + \frac{ip}{\kappa}\right), \tag{16.3.7}$$

and where a, P, X are related by (16.3.4).

Exercise 16.10 Wigner Distribution Function for a Coherent State: Using (15.2.3) and the result of the previous exercise, show that for a coherent state with mean position $\bar{x}$ and mean momentum $\bar{p}$

$$f_{\text{coherent}}(x,p) = \frac{1}{\pi\hbar}\exp\left(-\frac{\kappa^2(x-\bar{x})^2}{\hbar} - \frac{(p-\bar{p})^2}{\hbar\kappa^2}\right). \tag{16.3.8}$$

16.3.1 The Wigner Distribution Function as a Quasiprobability

Ex. 16.9 shows that the Wigner distribution function $f(\boldsymbol{x},\boldsymbol{p})$ is fully equivalent to $W(\alpha,\alpha^*)$, and therefore shares all of its properties. However, because its behaviour is very like that of a joint probability distribution in phase space, it provides a particularly convenient technique for the study of the semiclassical limit of quantum problems involving the position and momentum of a particle. Proceeding directly from the definitions (16.3.2, 16.3.5), we can derive some properties which are particularly relevant to this kind of application.

i) *Position Distribution Function*: By integrating over the momentum variable

$$\int d^3\boldsymbol{p}\, f(\boldsymbol{x},\boldsymbol{p}) = \langle \boldsymbol{x}|\rho|\boldsymbol{x}\rangle \longrightarrow |\psi(\boldsymbol{x})|^2 \quad\text{when}\quad \rho = |\psi\rangle\langle\psi|. \tag{16.3.9}$$

ii) *Momentum Distribution Function*: Similarly, by integrating over the position variable

$$\int d^3\boldsymbol{x}\, f(\boldsymbol{x},\boldsymbol{p}) = \langle \boldsymbol{p}|\rho|\boldsymbol{p}\rangle \longrightarrow |\tilde{\psi}(\boldsymbol{p})|^2 \quad\text{when}\quad \rho = |\psi\rangle\langle\psi|, \tag{16.3.10}$$

where we use the notation for the momentum wavefunction

$$\psi(\boldsymbol{x}) = \frac{1}{(2\pi\hbar)^3}\int d^3\boldsymbol{p}\,\tilde{\psi}(\boldsymbol{p})e^{i\boldsymbol{p}\cdot\boldsymbol{x}/\hbar}. \tag{16.3.11}$$

iii) *Normalization*:

$$\int d^3\boldsymbol{p}\int d^3\boldsymbol{x}\, f(\boldsymbol{x},\boldsymbol{p}) = \mathrm{Tr}\{\rho\} = 1. \tag{16.3.12}$$

iv) *Mean Values*: For arbitrary functions $A(\boldsymbol{X})$ of $\boldsymbol{X}$ only, or $B(\boldsymbol{P})$ of $\boldsymbol{P}$ only, it follows from (16.3.9, 16.3.10) that

$$\langle\psi|A(\boldsymbol{X})|\psi\rangle = \int d^3\boldsymbol{p}\int d^3\boldsymbol{x}\, A(\boldsymbol{x})f(\boldsymbol{x},\boldsymbol{p}), \tag{16.3.13}$$

$$\langle\psi|B(\boldsymbol{P})|\psi,t\rangle = \int d^3\boldsymbol{p}\int d^3\boldsymbol{x}\, B(\boldsymbol{p})f(\boldsymbol{x},\boldsymbol{p}). \tag{16.3.14}$$

In particular, the potential $V(\boldsymbol{X})$ and the kinetic energy $\boldsymbol{P}^2/2m$ will be examples of such functions relevant in laser cooling.

It is important to point out that such simple results are *not* in general true for arbitrary functions $F(X,P)$ of both X and P.

v) *Heisenberg Uncertainty Principle*: We can write the quantum-mechanical variances of X and P in terms of those for the Wigner distribution function, using (16.3.13, 16.3.14), so that the Heisenberg uncertainty principle requires $f(\boldsymbol{x},\boldsymbol{p})$ to be a function such that

$$\operatorname{var}[x_i]\operatorname{var}[p_i] \geq \tfrac{1}{4}\hbar^2. \tag{16.3.15}$$

The specification of functions $f(\boldsymbol{x},\boldsymbol{p})$ which satisfy all quantum-mechanical requirements is by no means straightforward. However, since the evolution equation for the Wigner function must transform a valid Wigner function at one time into a valid Wigner function at any future time, the problem reduces to a choice of the correct initial condition for the Wigner function. We will address this problem further in Sect. 16.4.

16.3.2 Operator Mappings for the Wigner Distribution Function

The application of the Wigner distribution function depends directly on appropriate mappings, similar to those given in (16.1.5a–16.1.5d), and in (16.1.6a–16.1.7c). A large number of useful relations follow from the following fundamental maps:

$$X\rho \longleftrightarrow \left(x + \frac{i\hbar}{2}\frac{\partial}{\partial p}\right) f(x,p), \tag{16.3.16}$$

$$\rho X \longleftrightarrow \left(x - \frac{i\hbar}{2}\frac{\partial}{\partial p}\right) f(x,p), \tag{16.3.17}$$

$$P\rho \longleftrightarrow \left(p - \frac{i\hbar}{2}\frac{\partial}{\partial x}\right) f(x,p), \tag{16.3.18}$$

$$\rho P \longleftrightarrow \left(p + \frac{i\hbar}{2}\frac{\partial}{\partial x}\right) f(x,p). \tag{16.3.19}$$

The derivations of all of these are similar to each other. For example, the first relationship follows from (16.3.1) thus:

$$\int du \langle x + \tfrac{1}{2}u | X\rho | x - \tfrac{1}{2}u\rangle \, e^{-ipu/\hbar}$$

$$= (x + \tfrac{1}{2}u) \int du \langle x + \tfrac{1}{2}u | \rho | x - \tfrac{1}{2}u\rangle \, e^{-ipu/\hbar}, \tag{16.3.20}$$

$$= \left(x + \frac{i\hbar}{2}\frac{\partial}{\partial p}\right) \int du \langle x + \tfrac{1}{2}u | \rho | x + \tfrac{1}{2}u\rangle \, e^{-ipu/\hbar}. \tag{16.3.21}$$

The others are similarly derived from either (16.3.1) or (16.3.2).

a) Simple Commutator and Anticommutator Mappings: Commutators and anticommutators inevitably appear in applications, and the appropriate mappings are straightforward consequences of those above:

$$[X,\rho] \longleftrightarrow i\hbar\frac{\partial}{\partial p}f(x,p), \tag{16.3.22}$$

$$[P,\rho] \longleftrightarrow -i\hbar\frac{\partial}{\partial x}f(x,p), \tag{16.3.23}$$

$$[X,\rho]_+ \longleftrightarrow 2xf(x,p), \tag{16.3.24}$$

$$[P,\rho]_+ \longleftrightarrow 2pf(x,p). \tag{16.3.25}$$

b) Quadratic Commutators: Quadratic forms in X and P relate to kinetic energy and harmonic potentials, and the relevant mappings are as follows:

$$\left[\tfrac{1}{2}X^2,\rho\right] \longleftrightarrow i\hbar\frac{\partial}{\partial p}xf(x,p), \tag{16.3.26}$$

$$\left[\tfrac{1}{2}P^2,\rho\right] \longleftrightarrow -i\hbar\frac{\partial}{\partial x}pf(x,p), \tag{16.3.27}$$

$$[XP,\rho]=[PX,\rho] \longleftrightarrow i\hbar\left(p\frac{\partial}{\partial p}-x\frac{\partial}{\partial x}\right)f(x,p). \tag{16.3.28}$$

c) Double Commutators: When these involve the same operator, they correspond to diffusion terms such as occur in a classical Fokker–Planck equation:

$$2X\rho X-\rho X^2-X^2\rho=-[X,[X,\rho]] \longleftrightarrow \hbar^2\frac{\partial^2}{\partial p^2}f(x,p), \tag{16.3.29}$$

$$2P\rho P-\rho P^2-P^2\rho=-[P,[P,\rho]] \longleftrightarrow \hbar^2\frac{\partial^2}{\partial x^2}f(x,p). \tag{16.3.30}$$

d) Commutator-Anticommutator Forms: These are usually related to damped motion, and the appropriate mappings take the form

$$\left[X,[P,\rho]_+\right] \longleftrightarrow 2i\hbar\frac{\partial}{\partial p}pf, \tag{16.3.31}$$

$$\left[P,[X,\rho]_+\right] \longleftrightarrow -2i\hbar\frac{\partial}{\partial x}xf. \tag{16.3.32}$$

e) Fourier Transform Forms: Directly from the definitions (16.3.1, 16.3.2), one finds

$$e^{ikX}\rho \longleftrightarrow e^{ikx} f(x, p - \tfrac{1}{2}\hbar k), \tag{16.3.33}$$

$$\rho e^{ikX} \longleftrightarrow e^{ikx} f(x, p + \tfrac{1}{2}\hbar k), \tag{16.3.34}$$

$$e^{iPz/\hbar}\rho \longleftrightarrow e^{ipz/\hbar} f(x + \tfrac{1}{2}z, p), \tag{16.3.35}$$

$$\rho e^{iPz/\hbar} \longleftrightarrow e^{ipz/\hbar} f(x - \tfrac{1}{2}z, p). \tag{16.3.36}$$

16.3.3 Motion in a Potential

We will consider first the motion of a particle in a potential, so that the Hamiltonian is

$$H = \frac{\boldsymbol{p}^2}{2m} + V(\boldsymbol{X}). \tag{16.3.37}$$

We suppose that there is a Fourier transform representation for the potential

$$V(\boldsymbol{X}) = \int d^3\boldsymbol{k}\, e^{i\boldsymbol{k}\cdot\boldsymbol{X}} \tilde{V}(\boldsymbol{k}), \tag{16.3.38}$$

and then:

i) For the kinetic term we use the mapping (16.3.27).

ii) For the potential, we use the Fourier representation and use (16.3.33, 16.3.34).

The density operator evolution equation $\dot{\rho}(t) = -\frac{i}{\hbar}[H, \rho(t)]$ then maps to

$$\frac{\partial}{\partial t} f(\boldsymbol{x}, \boldsymbol{p}, t) = -\frac{\boldsymbol{p}}{m} \cdot \nabla_{\boldsymbol{x}} f(\boldsymbol{x}, \boldsymbol{p}, t) + \frac{i}{\hbar} \int d^3\boldsymbol{k}\, \tilde{V}(\boldsymbol{k}) e^{i\boldsymbol{k}\cdot\boldsymbol{x}} \left[f(\boldsymbol{x}, \boldsymbol{p} + \tfrac{1}{2}\hbar\boldsymbol{k}, t) - f(\boldsymbol{x}, \boldsymbol{p} - \tfrac{1}{2}\hbar\boldsymbol{k}, t) \right]. \tag{16.3.39}$$

a) The Classical Limit: If the typical value of $\hbar k$ is much smaller than the typical scale of variation of $f(\boldsymbol{x}, \boldsymbol{p}, t)$ in momentum, then we can make the expansion

$$f(\boldsymbol{x}, \boldsymbol{p} + \tfrac{1}{2}\hbar\boldsymbol{k}, t) = f(\boldsymbol{x}, \boldsymbol{p}, t) + \tfrac{1}{2}\hbar\boldsymbol{k} \cdot \nabla_{\boldsymbol{p}} f(\boldsymbol{x}, \boldsymbol{p}, t) + \ldots, \tag{16.3.40}$$

and in lowest order we get the *classical Liouville equation*

$$\frac{\partial}{\partial t} f(\boldsymbol{x}, \boldsymbol{p}, t) = -\frac{\boldsymbol{p}}{m} \cdot \nabla_{\boldsymbol{x}} f(\boldsymbol{x}, \boldsymbol{p}, t) - \boldsymbol{F}(\boldsymbol{x}) \cdot \nabla_{\boldsymbol{p}} f(\boldsymbol{x}, \boldsymbol{p}, t), \tag{16.3.41}$$

with force $\quad \boldsymbol{F}(\boldsymbol{x}) = -\nabla_{\boldsymbol{x}} V(\boldsymbol{x}).$ (16.3.42)

b) Harmonic Potential: For a harmonic potential $V(\boldsymbol{X})$ is quadratic in $\boldsymbol{X}$, and the semiclassical mapping (16.3.41) is *exact*. To show this we use the mappings (16.3.16, 16.3.17, 16.3.24) to generate the mapping

$$[X_i X_j, \rho] = X_i \left[X_j, \rho\right]_+ - \left[X_i, \rho\right]_+ X_j \longleftrightarrow \mathrm{i}\hbar\left(\frac{\partial}{\partial p_i} x_j + \frac{\partial}{\partial p_j} x_i\right) f(\boldsymbol{x}, \boldsymbol{p}). \quad (16.3.43)$$

From this, if we write the most general three-dimensional harmonic potential as $V(\boldsymbol{X}) = \sum_{i,j} w_{ij} X_i X_j$, we can generate the mapping for this potential

$$-\frac{\mathrm{i}}{\hbar}\left[V(\boldsymbol{X}), \rho\right] \longleftrightarrow \left(\nabla_{\boldsymbol{x}} V(\boldsymbol{x})\right) \cdot \nabla_{\boldsymbol{p}} f(\boldsymbol{x}, \boldsymbol{p}), \quad (16.3.44)$$

the same as for the semiclassical approximation.

16.3.4 A Free Particle in One Dimension

In the case of a free particle, with the momentum space wave function

$$\tilde{\psi}(p,t) = \tilde{\psi}_0(p)\mathrm{e}^{-\mathrm{i}p^2 t/2m\hbar}, \quad (16.3.45)$$

we can compute the Wigner function explicitly thus

$$f(x,p,t) = \int dq\, \tilde{\psi}(p+\tfrac{1}{2}q, t)\tilde{\psi}^*(p-\tfrac{1}{2}q, t)\mathrm{e}^{\mathrm{i}qx/\hbar}, \quad (16.3.46)$$

$$= \int dq\, \mathrm{e}^{\mathrm{i}q(x-pt/m)}\, \tilde{\psi}_0^*(p-\tfrac{1}{2}q)\tilde{\psi}_0(p+\tfrac{1}{2}q), \quad (16.3.47)$$

$$= f_0\left(x - \frac{pt}{m}, p\right). \quad (16.3.48)$$

As we would expect, this is the solution of the free particle evolution equation which follows from (16.3.41)

$$\left(\frac{\partial}{\partial t} + \frac{p}{m}\frac{\partial}{\partial x}\right) f(x,p,t) = 0. \quad (16.3.49)$$

a) The Gaussian Wavepacket: A Gaussian wavepacket, peaked in momentum around p_0, and in position around x_0 can be written as

$$\tilde{\psi}_0(p) = \frac{1}{(2\pi\sigma)^{1/4}}\mathrm{e}^{-(p-p_0)^2/4\sigma}\mathrm{e}^{-\mathrm{i}px_0}. \quad (16.3.50)$$

This is of course a pure quantum state, and its Wigner distribution function, computed using (16.3.2), has a simple factorized form

$$f_0(x,p) = \frac{1}{(2\pi\sigma)^{1/2}}\mathrm{e}^{-(p-p_0)^2/2\sigma} \times \frac{1}{(2\pi\Lambda)^{1/2}}\mathrm{e}^{-(x-x_0)^2/2\Lambda}, \quad (16.3.51)$$

where $\Lambda \equiv 1/4\sigma$.

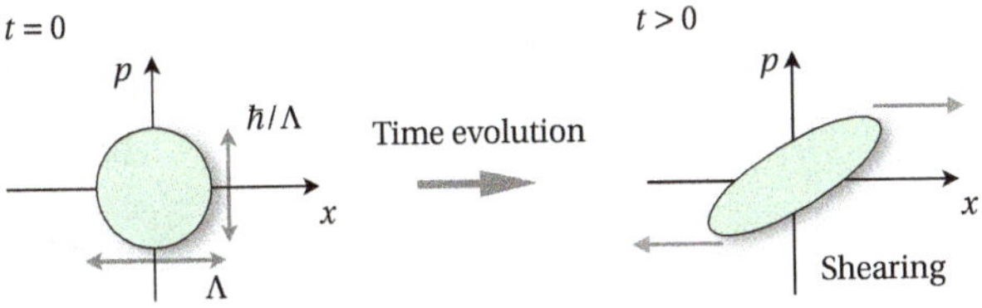

Fig. 16.1. Evolution of the shape of the Gaussian Wigner distribution (16.3.51) function during time evolution.

This Wigner distribution function is thus Gaussian in p and x, and since the variances are

$$\mathrm{var}\left[p\right] \equiv (\Delta p)^2 = \sigma, \tag{16.3.52}$$

$$\mathrm{var}\left[x\right] \equiv (\Delta x)^2 = \Lambda \equiv \frac{\hbar^2}{4\sigma}, \tag{16.3.53}$$

it follows that $\Delta x \Delta p = \frac{1}{2}\hbar$. Thus, a Gaussian wavepacket is necessarily a minimum uncertainty state, and thus corresponds to a coherent state wavefunction, as in Ex. 16.10

b) Positivity of the Wigner Function of a Gaussian Wavepacket: The distribution function (16.3.51) is positive, and this makes the analogy to a probability distribution quite appealing. However, this result is really quite special, since it has been proved [58] that the Wigner function for a pure state wavefunction is positive only if it is Gaussian.

c) Time Evolution: For this distribution function we can use (16.3.48), and explicitly this yields

$$f(x, p, t) = f_0\left(x - \frac{pt}{m}, p\right), \tag{16.3.54}$$

$$= \frac{1}{2\pi(\sigma\Lambda)^{1/2}} \exp\left(-\frac{(p-p_0)^2}{2\sigma} - \frac{(x - x_0 - pt/m)^2}{2\Lambda}\right). \tag{16.3.55}$$

The solution represents a kind of shearing motion in phase space, and is illustrated in Fig. 16.1. This kind of motion corresponds exactly to what would be expected for a classical phase space distribution.

16.3.5 Quantum Brownian Motion

The quantum version of Brownian motion can be described by a *non-Lindblad* master equation, which is derivable by a number of techniques, and is considered in some detail in *Quantum Noise*. It has the form

$$\frac{\partial \rho}{\partial t} = -\frac{\mathrm{i}}{\hbar}\left[H_{\mathrm{sys}}, \rho\right] - \frac{\mathrm{i}\gamma}{2\hbar}\left[X, [\dot{X}, \rho]_+\right] - \frac{\gamma k_{\mathrm{B}} T}{\hbar^2}\left[X, [X, \rho]\right]. \tag{16.3.56}$$

In this equation, we define

$$H_{\text{sys}} \equiv \frac{P^2}{2m} + V(X), \tag{16.3.57}$$

$$\dot{X} \equiv \frac{\mathrm{i}}{\hbar}\left[H_{\text{sys}}, X\right] = \frac{P}{m}. \tag{16.3.58}$$

a) Fokker–Planck Equation for the Wigner Distribution Function: We will consider the case where there is no potential $V(X)$, and use the mappings (16.3.27, 16.3.29, 16.3.31), which lead directly to the Fokker–Planck equation

$$\frac{\partial f}{\partial t} = \left\{-\frac{\partial}{\partial x}\frac{p}{m} - \gamma\frac{\partial}{\partial p}p + \gamma k_{\mathrm{B}}T\frac{\partial^2}{\partial p^2}\right\}f. \tag{16.3.59}$$

This is exactly the classical Brownian motion Fokker–Planck equation of Ex. 5.6.

b) Inclusion of a Potential Term: This can be done in the same way as was done in Sect. 16.3.3, and when the smooth potential approximation is made, we are led to the corresponding classical Fokker–Planck equation (5.3.15).

16.4 Quantum Fluctuations in Equations of Motion

The Wigner function provides the phase space description of quantum mechanics which most closely resembles the classical description. This happens because the symmetrical operator ordering implicit in the Wigner function often makes the second-order derivative terms in the equations of motion generated by the use of the mappings such as (16.1.5a–16.1.5d) cancel with each other. For example, the mappings (16.3.26–16.3.28) all have this property; the operators X^2, P^2, XP and PX would all produce second-order derivatives, but in the commutator the second-order terms cancel.

In such instances the equation of motion of a Wigner function can be a Liouville equation, which adds no noise. This happens, for example, in the following following examples which we have already met:

i) In (16.1.10), where the equation of motion is exact, and represents Hamiltonian motion.

ii) In the large V limit of (16.1.23), which leads to the approximate description by the equations (16.1.24).

iii) In the Bogoliubov equations of motion (16.1.28).

iv) In the Liouville equation for the Wigner distribution function (16.3.41), which is exact when the potential $V(x)$ vanishes.

Since a Liouville equation corresponds to noise-free motion, the only place that the fluctuations occur in this kind of situation is in the initial conditions. However, the choice of possible initial conditions is not arbitrary, since not every function

$h(\alpha, \alpha^*)$ is an allowable Wigner function. The most obvious restriction on a possible Wigner function is given by the Heisenberg uncertainty principle, which we have already addressed in Sect. 16.3.1 v). Fortunately, there are strategies for specifying the initial conditions.

i) We do know Wigner functions for a number of particular kinds of state, such as number states and coherent states.

ii) It is in principle possible to construct a Wigner function for any pure quantum state.

iii) However, of most utility is the fact that we do know that Gaussian wavefunctions, as in Sect. 16.3.4, and coherent states (which are essentially the same thing) as in Ex. 16.10 and (15.2.3) provide valid Wigner functions, and therefore provide valid initial conditions for the evolution equation. These states are sufficiently close to realistic initial conditions that they have been used to form the basis of the *c-field method* [57] for simulating Bose–Einstein condensates, which we treat in detail in *Book III*.

17. P-Function Methods

The P-function provides a phase space representation best adapted to quantum optics. This arises because of the close connection between normally ordered averages and the process of photon counting.

17.1 Operator Correspondences and Equations of Motion

We start from the representation (15.2.15), and compute the action of the creation and destruction operators on the operator $|\alpha\rangle\langle\alpha|$ from which the P-function is formed. To do this, we use the actions on the coherent states given in Sect. 10.5.1b. Thus, using the expression (15.2.15) that $\rho = \int d^2\alpha\, P(\alpha,\alpha^*,t)|\alpha\rangle\langle\alpha|$, we can write, for example

$$a\rho = \int d^2\alpha\, \alpha P(\alpha,\alpha^*,t)|\alpha\rangle\langle\alpha|, \tag{17.1.1}$$

$$a^\dagger\rho = \int d^2\alpha\, P(\alpha,\alpha^*,t)\left(\alpha^*|\alpha\rangle + \tfrac{1}{2}\frac{\partial}{\partial\alpha}|\alpha\rangle\right)\langle\alpha|, \tag{17.1.2}$$

$$= \int d^2\alpha\, P(\alpha,\alpha^*,t)\left(\alpha^* + \frac{\partial}{\partial\alpha}\right)\Big(|\alpha\rangle\langle\alpha|\Big), \tag{17.1.3}$$

$$= \int d^2\alpha\, |\alpha\rangle\langle\alpha|\left(\alpha^* - \frac{\partial}{\partial\alpha}\right)P(\alpha,\alpha^*,t). \tag{17.1.4}$$

The final step is an integration by parts, which assumes that there are no boundary terms. It is straightforward to generalize these results to all left and right products to get the table of mappings

$$a\rho \longleftrightarrow \alpha P(\alpha,\alpha*), \tag{17.1.5a}$$

$$a^\dagger\rho \longleftrightarrow \left(\alpha^* - \frac{\partial}{\partial\alpha}\right)P(\alpha,\alpha^*), \tag{17.1.5b}$$

$$\rho a \longleftrightarrow \left(\alpha - \frac{\partial}{\partial\alpha^*}\right)P(\alpha,\alpha^*), \tag{17.1.5c}$$

$$\rho a^\dagger \longleftrightarrow \alpha^* P(\alpha,\alpha^*). \tag{17.1.5d}$$

Exercise 17.1 Useful Mappings: From these mappings, derive the following mappings which are useful in applications:

$$[a,\rho] \longleftrightarrow \frac{\partial}{\partial\alpha^*}P(\alpha,\alpha^*), \tag{17.1.6a}$$

$$[a^\dagger,\rho] \longleftrightarrow -\frac{\partial}{\partial\alpha}P(\alpha,\alpha^*), \tag{17.1.6b}$$

$$[a^\dagger a,\rho] \longleftrightarrow \left(-\frac{\partial}{\partial\alpha}\alpha+\frac{\partial}{\partial\alpha^*}\alpha^*\right)P(\alpha,\alpha^*), \tag{17.1.6c}$$

$$[a,[a^\dagger,\rho]] \longleftrightarrow -\frac{\partial^2}{\partial\alpha^*\partial\alpha}P(\alpha,\alpha^*), \tag{17.1.6d}$$

$$[a^\dagger,[a,\rho]] \longleftrightarrow -\frac{\partial^2}{\partial\alpha^*\partial\alpha}P(\alpha,\alpha^*). \tag{17.1.6e}$$

The above five mappings have the same form as the corresponding mappings for the Wigner function (16.1.6a–16.1.7a). In contrast, the following mappings differ from the corresponding Wigner function mappings.

$$2a\rho a^\dagger-\rho a^\dagger a-a^\dagger a\rho \longleftrightarrow \left(\frac{\partial}{\partial\alpha}\alpha+\frac{\partial}{\partial\alpha^*}\alpha^*\right)P, \tag{17.1.7a}$$

$$2a^\dagger\rho a-\rho a a^\dagger-a a^\dagger\rho \longleftrightarrow \left(-\frac{\partial}{\partial\alpha}\alpha-\frac{\partial}{\partial\alpha^*}\alpha^*+2\frac{\partial^2}{\partial\alpha\partial\alpha^*}\right)P, \tag{17.1.7b}$$

$$[a^\dagger a,\rho]_+ \longleftrightarrow \left(2|\alpha^2|-\frac{\partial}{\partial\alpha}\alpha-\frac{\partial}{\partial\alpha^*}\alpha^*\right)P. \tag{17.1.7c}$$

17.1.1 The Driven Harmonic Oscillator

We consider as in Sect. 16.1.1 the Hamiltonian

$$H=\hbar\omega(a^\dagger a+\tfrac{1}{2})+ga^\dagger+g^*a. \tag{17.1.8}$$

Using the mappings (17.1.6a–17.1.6c), which are identical with the corresponding Wigner function mappings (16.1.6a–16.1.6c), the equation of motion for the P-function is exactly the same as that for the Wigner function. The difference between the two is exhibited solely in the relationship of the initial conditions to the initial quantum state. For example, if one has an initial coherent state $|\alpha_0\rangle$, the initial condition for the P-function is the delta function (15.2.15), whereas the initial condition for the Wigner function is the Gaussian (15.2.3).

Exercise 17.2 Random Time-Dependent Driving: Suppose that $g(t)$ is a random process. Find the equation of motion for the P-function. If $g(t)$ satisfies the Ornstein–Uhlenbeck stochastic differential equation (**I**) $dg(t)=-\kappa g(t)\,dt+\sqrt{n}\,dW(t)$, find:

i) The joint Fokker–Planck equation for the P-function $P(\alpha,\alpha^*,g,t)$.

ii) The limiting Fokker–Planck equation in the case that the relaxation constant κ is very large.

17.1.2 The Anharmonic Oscillator

As in Sect. 16.1.2 the relevant Hamiltonian is

$$H = \hbar\omega a^\dagger a + \tfrac{1}{2}u a^\dagger a^\dagger a a. \tag{17.1.9}$$

We follow the same procedure as for the Wigner function, but using now the mappings (17.1.6c, 17.1.7c), and get the equations of motion

$$\frac{\partial P}{\partial t} = \mathrm{i}\left\{\left(\frac{\partial}{\partial\alpha}\alpha - \frac{\partial}{\partial\alpha^*}\alpha^*\right)(\omega + u|\alpha|^2) + \tfrac{1}{2}u\left(\frac{\partial^2}{\partial\alpha^2}\alpha^2 - \frac{\partial^2}{\partial\alpha^{*2}}\alpha^{*2}\right)\right\} P. \tag{17.1.10}$$

In this equation of motion

i) The deterministic part (that is, the terms with first-order derivatives) is almost the same as that for the Wigner function, apart from the change $|\alpha|^2 - 1/2 \to |\alpha|^2$.

ii) However the remainder is completely different. There are no *third-order* derivatives, but there are *second-order* derivatives.

iii) The scaling argument used for the Wigner function to give the deterministic equation (16.1.24) cannot be used here, because the second-order terms are of order $1/V$, the same order of magnitude as the expected fluctuations in the scaled variable.

iv) The equation looks superficially like a Fokker–Planck equation, but the noise is *complex*. An interpretation as a genuine Fokker–Planck equation can be made using the *positive P-representation*.

17.2 Damped and Driven Systems

Methods based on the P-function also have strong resemblance similar to those based on the Wigner function, in particular to the methods introduced for complex variable oscillators in Sect. 7.3. However, it is not merely a matter of taste which method is used—there are advantages to both methods, and which method to use is determined by the nature of the problem being considered.

17.2.1 The Harmonic Oscillator Including Loss or Gain

Here we take a master equation, as in (14.3.4), for a damped harmonic oscillator, in the form

$$\frac{\partial\rho}{\partial t} = -\mathrm{i}\omega[a^\dagger a, \rho] + \tfrac{1}{2}\gamma\mathcal{M}\{2a\rho a^\dagger - \rho a^\dagger a - a^\dagger a\rho\}$$
$$+ \tfrac{1}{2}\gamma\mathcal{N}\{2a^\dagger\rho a - \rho a a^\dagger - a a^\dagger\rho\}. \tag{17.2.1}$$

i) Using the mappings (17.1.7a, 17.1.7b) we derive the P-function equation of motion

$$\frac{\partial P}{\partial t} = \Big\{ \mathrm{i}\omega \left(\frac{\partial}{\partial \alpha}\alpha - \frac{\partial}{\partial \alpha^*}\alpha^* \right) + \tfrac{1}{2}\gamma(\mathcal{M}-\mathcal{N})\left(\frac{\partial}{\partial \alpha}\alpha + \frac{\partial}{\partial \alpha^*}\alpha^* \right) + \gamma\mathcal{N}\frac{\partial^2}{\partial\alpha\partial\alpha^*} \Big\} P . \tag{17.2.2}$$

ii) This is a Fokker–Planck equation, and is of the same form as that of the thermalized oscillator already treated in Sect. 7.3.2. It is equivalent to the stochastic differential equations

$$(\mathrm{I})\ d\alpha = \left(-\mathrm{i}\omega - \tfrac{1}{2}\gamma(\mathcal{M}-\mathcal{N})\right)\alpha\, dt + \sqrt{\gamma\mathcal{N}}\, dw(t), \tag{17.2.3a}$$

$$(\mathrm{I})\ d\alpha^* = \left(\mathrm{i}\omega - \tfrac{1}{2}\gamma(\mathcal{M}-\mathcal{N})\right)\alpha^*\, dt + \sqrt{\gamma\mathcal{N}}\, dw^*(t), \tag{17.2.3b}$$

$$dw(t) \equiv \frac{dW_1(t) + \mathrm{i}\, dW_2(t)}{\sqrt{2}}, \tag{17.2.3c}$$

$$dw(t)\, dw^*(t) = dt, \qquad dw(t)^2 = dw(t)^{*2} = 0. \tag{17.2.3d}$$

a) Parameters for the Thermal Case: When the oscillator is interacting with a thermal bath at temperature T, the parameters are as in (14.3.35)

$$\mathcal{N} \to \bar{N}(T,\omega) = \frac{1}{e^{\hbar\omega/k_B T} - 1}, \qquad \mathcal{M} \to \bar{N}(T,\omega) + 1. \tag{17.2.4}$$

b) Classical Equivalent: This equation is then of exactly the same form as the thermalized oscillator equation treated in Sect. 7.3.2.

Exercise 17.3 Stationary Solution: Derive the stationary solutions

$$P_s(\alpha,\alpha^*) = \frac{\mathcal{M}-\mathcal{N}}{\pi\mathcal{N}} \exp\left(-\frac{|\alpha|^2(\mathcal{M}-\mathcal{N})}{\mathcal{N}} \right), \tag{17.2.5}$$

$$\langle|\alpha^2|\rangle_s = \frac{\mathcal{N}}{\mathcal{M}-\mathcal{N}}. \tag{17.2.6}$$

The stationary solutions in the thermal case are then

$$P_s(\alpha,\alpha^*) = \frac{1}{\pi\bar{N}(T,\omega)} \exp\left(-\frac{|\alpha|^2}{\bar{N}(T,\omega)} \right), \tag{17.2.7}$$

$$\langle|\alpha^2|\rangle_s = \bar{N}(T,\omega). \tag{17.2.8}$$

17.2.2 The Phase-Damped Harmonic Oscillator

The master equation (16.2.30), originally introduced in Sect. 14.3.3, can be written

$$\frac{\partial \rho}{\partial t} = -\frac{\kappa}{2}(2\bar{N}+1) - [N,[N,\rho]], \tag{17.2.9}$$

and involves only the mapping (17.1.6c), which is the same for the Wigner function and for the P-function. Hence the Fokker–Planck equation and stochastic differential equations are the same as given in Sect. 16.2.2 for the Wigner function.

17.3 The Laser

The laser is an engineered device for producing almost coherent light, and one whose theoretical description was found to *require* a formulation of quantum noise. Because the light produced can be almost perfectly coherent, a P-function description is very useful. In this section we will firstly set up a laser model, and then show how to use the P-function to give a description in terms of the classical stochastic differential equations which can be derived by this method.

17.3.1 A Simple Laser Model

As mentioned previously in Sect. 14.3.1, it is quite possible to have a master equation in which the "gain terms" are bigger than the "loss terms", so that it takes the form

$$\begin{aligned}\frac{\partial \rho}{\partial t} = -\frac{i}{\hbar}[H_{\text{sys}},\rho] &+ \tfrac{1}{2}\mathcal{L}\{2a\rho a^\dagger - \rho a^\dagger a - a^\dagger a\rho\} \\ &+ \tfrac{1}{2}\mathcal{G}\{2a^\dagger \rho a - \rho a a^\dagger - a a^\dagger \rho\}.\end{aligned} \tag{17.3.1}$$

In that case when the gain $\mathcal{G}$ is greater than the loss $\mathcal{L}$, the solutions of this equation would give a continually growing population of photons, and this provides the prototypical laser model. In any realistic laser the gain term must *saturate*, that is, as the population of the photons increases, the effective value of $\mathcal{G}$ must decrease. We therefore have two problems to address:

i) What is the physical mechanism by which saturation takes place?

ii) How do we incorporate this into a laser model?

a) Physical Mechanism: Let us consider a laser model as in Fig. 17.1. Here the gain medium is composed of two-level atoms in which the population of the upper levels is held by a pumping mechanism of some kind to be greater than that of the lower levels. In practice these atoms can be in a gas, as dopants in a solid, or even in a liquid, and the population inversion can be maintained by a wide variety of methods.

Thus we suppose that there are N_{At} two-level atoms, each of which is in contact with a *non-equilibrium* heat bath. This can be modelled by taking the density

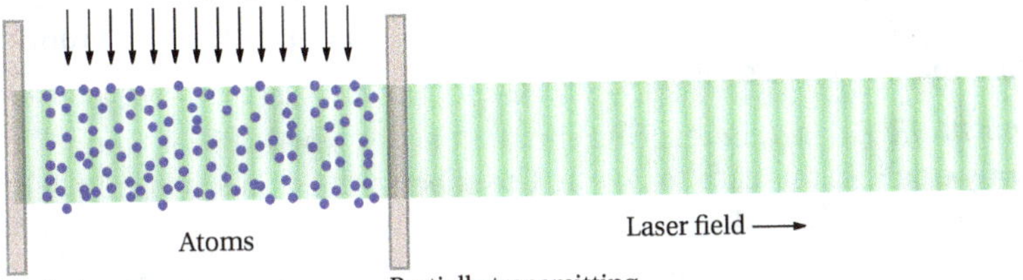

Fig. 17.1. Configuration of the essential elements of a laser whose gain medium is implemented by atoms in an excited state.

operator for each atom to obey the master equation for a two-level atom driven by an optical field, as in (14.1.3). For simplicity, we assume the driving field is at the same frequency as the transition between the atomic levels, so that each atom's density operator would satisfy the master equation

$$\begin{aligned}\frac{\partial \rho_{\text{sys}}^{k}}{\partial t} &= g\left[\sigma_k^{+}\mathcal{E} - \sigma_k^{-}\mathcal{E}^{*}, \rho_{\text{sys}}^{k}\right] \\ &\quad + \frac{\gamma}{2}\mathcal{A}\left(2\sigma_k^{-}\rho_{\text{sys}}^{k}\sigma_k^{+} - \rho_{\text{sys}}^{k}\sigma_k^{+}\sigma_k^{-} - \sigma_k^{+}\sigma_k^{-}\rho_{\text{sys}}^{k}\right) \\ &\quad + \frac{\gamma}{2}\mathcal{B}\left(2\sigma_k^{+}\rho_{\text{sys}}^{k}\sigma_k^{-} - \rho_{\text{sys}}^{k}\sigma_k^{-}\sigma_k^{+} - \sigma_k^{-}\sigma_k^{+}\rho_{\text{sys}}^{k}\right).\end{aligned} \tag{17.3.2}$$

The relevant procedure is essentially the same as that given in Sect. 14.2, but modified to take account of the non-equilibrium heat bath. This means that the notation used there is modified by the replacements

$$\left.\begin{array}{llll}\gamma\bar{N} \longrightarrow & \gamma\mathcal{B}, & \gamma(\bar{N}+1) \longrightarrow & \gamma\mathcal{A}, \\ \gamma \longrightarrow & \gamma(\mathcal{A}-\mathcal{B}), & \Gamma \longrightarrow & \tfrac{1}{2}\gamma(\mathcal{A}+\mathcal{B}).\end{array}\right\} \tag{17.3.3}$$

b) Mean Values of Atom Operators: Putting these replacements into equation (14.2.11), we find that the mean values of atom operators are given by

$$\langle\sigma_z^k\rangle = \frac{\gamma^2(\mathcal{B}^2-\mathcal{A}^2)}{\gamma^2(\mathcal{A}+\mathcal{B})^2+8g^2|\mathcal{E}|^2}, \tag{17.3.4}$$

$$\langle\sigma_k^{+}\rangle = \frac{\gamma(\mathcal{A}-\mathcal{B})(2g\mathcal{E}^{*})}{\gamma^2(\mathcal{A}+\mathcal{B})^2+8g^2|\mathcal{E}|^2}, \tag{17.3.5}$$

$$\langle\sigma_k^{-}\rangle = \frac{\gamma(\mathcal{A}-\mathcal{B})(2g\mathcal{E})}{\gamma^2(\mathcal{A}+\mathcal{B})^2+8g^2|\mathcal{E}|^2}. \tag{17.3.6}$$

The electromagnetic field amplitude $|\mathcal{E}|$ inside the cavity modifies the value of $\langle\sigma_z^k\rangle$, which in turn determines the amount of gain or loss available from the intracavity atoms. For sufficiently small $|\mathcal{E}|$, one can see two extreme situations when the laser field is small, and the effect as the laser field increases:

i) *All Atoms in the Ground State*: This situation arises when $\mathcal{B} \to 0$ and $|\mathcal{E}|$ is sufficiently *small.* In this case $\langle \sigma_z^k \rangle \to -1$. Since the atoms are in the ground state, their main effect is to absorb light.

ii) *All Atoms in the Excited State*: This situation arises when $\mathcal{A} \to 0$, and $|\mathcal{E}|$ is sufficiently small. In this case $\langle \sigma_z^k \rangle \to 1$. The effect of the atoms is to emit light into the laser beam, providing gain.

c) The Effect of Increasing $|\mathcal{E}|$—Saturation: This happens if $|\mathcal{E}|$ is sufficiently *large,* so that $\langle \sigma_z^k \rangle \to 0$, independently of the values of $\mathcal{A}$ and $\mathcal{B}$. Thus, as $|\mathcal{E}|$ increases, the value of $|\langle \sigma_z^k \rangle|$ decreases, and both gain and loss decrease—this is known as saturation of the non-linearity. It plays a fundamental role in the operation of the laser, limiting the strength of the laser field which can be produced.

d) The Heat Bath: To implement the bath of two-level atoms we use the method described in Sect. 13.3.1 a, using the bath operators

$$\Gamma_k \to \sigma_k^-, \qquad \Gamma_k^\dagger \to \sigma_k^+. \tag{17.3.7}$$

The actual correlation functions we will use in the laser theory are given by the following reasoning:

i) We use the form (17.3.4) for $\langle \sigma_z^k \rangle$.

ii) The algebraic relationships for the atom operators mean that the correlation functions can be written in terms of only $\langle \sigma_z^k \rangle$ thus:

$$\langle \sigma_k^+ \sigma_{k'}^- \rangle = \tfrac{1}{2}\delta_{kk'} \left\langle 1 + \sigma_z^k \right\rangle, \tag{17.3.8}$$

$$\langle \sigma_k^- \sigma_{k'}^+ \rangle = \tfrac{1}{2}\delta_{kk'} \left\langle 1 - \sigma_z^k \right\rangle. \tag{17.3.9}$$

iii) We know there are a large number $N_{\rm At}$ of atoms, which will be uncorrelated with each other, so that the effective values of the gain and loss quantities are obtained by adding up the individual terms from each of these atoms. As well as this, we have to include a coupling constant g between each individual atom and the electromagnetic field. Putting these together means that we use coefficients of the form

$$\text{Gain:} \quad \mathcal{N} = g^2 N_{\rm At} \frac{\mathcal{B}(\mathcal{A}+\mathcal{B}) + |F|^2}{(\mathcal{A}+\mathcal{B})^2 + 2|F|^2}, \tag{17.3.10}$$

$$\text{Loss:} \quad \mathcal{M} = g^2 N_{\rm At} \frac{\mathcal{A}(\mathcal{A}+\mathcal{B}) + |F|^2}{(\mathcal{A}+\mathcal{B})^2 + 2|F|^2}, \tag{17.3.11}$$

$$\text{in which} \quad F \equiv \frac{2g|\mathcal{E}|}{\gamma}. \tag{17.3.12}$$

iv) We can now put in the correct forms for the gain and loss functions into the laser master equation. The quantity $\mathcal{G}$ in (17.3.1) is simply given by $\mathcal{N}(\Omega)$ in (17.3.11). However the loss coefficient $\mathcal{L}$ is given by the sum of the term for $\mathcal{M}(\Omega)$ and another term, often much larger, which represents the loss of photons through the partially transmitting mirror, and any other absorption which may occur. We will therefore write $\mathcal{L}$ and $\mathcal{G}$ in the form

$$\mathcal{G} = g^2 N_{\text{At}} \frac{\mathcal{B}(\mathcal{A}+\mathcal{B}) + |F|^2}{(\mathcal{A}+\mathcal{B})^2 + 2|F|^2}, \tag{17.3.13}$$

$$\mathcal{L} = \kappa + g^2 N_{\text{At}} \frac{\mathcal{A}(\mathcal{A}+\mathcal{B}) + |F|^2}{(\mathcal{A}+\mathcal{B})^2 + 2|F|^2}, \tag{17.3.14}$$

where κ represents all photon absorption other than that caused by the presence of the atoms.

17.3.2 Implementation of the Laser Equation

The quantity $|F|^2$ which occurs in these equations is proportional to the intensity of the laser field inside the laser cavity, which is supposed to be fully quantized. Somehow we have to include this in the quantum-mechanical master equation (17.3.1).

The simplest way is to modify the gain-loss equation for a harmonic oscillator in contact with a (possibly non-equilibrium) heat bath, that is, the stochastic differential equation (17.2.3a), by including the substitutions (17.3.13, 17.3.14), and by setting

$$|F|^2 \longrightarrow r\,|\alpha|^2, \tag{17.3.15}$$

which implements the proportionality between $|F|^2$ and the P-function representation of $|\alpha|^2$, the intensity of the laser mode inside the laser cavity.

Starting from the gain-loss equations (17.2.3a–17.2.3d), the laser equation then becomes (together with its complex conjugate)

$$d\alpha = \left\{ -i\Omega - \tfrac{1}{2}\left(\kappa + \frac{g^2 N_{\text{At}}(\mathcal{A}^2 - \mathcal{B}^2)}{(\mathcal{A}+\mathcal{B})^2 + 2r\,|\alpha|^2} \right) \right\} \alpha\, dt$$

$$+ \sqrt{g^2 N_{\text{At}} \frac{\mathcal{B}(\mathcal{A}+\mathcal{B}) + r|\alpha|^2}{(\mathcal{A}+\mathcal{B})^2 + 2r|\alpha|^2}}\; dw(t). \tag{17.3.16}$$

a) Laser Parameters: Conventionally the following quantities are defined:

i) *The Saturation Photon Number* n_0 is defined as

$$n_0 = \frac{(\mathcal{A}+\mathcal{B})^2}{2r}. \tag{17.3.17}$$

This is a measure of the value of $|\alpha|^2$ at which the non-linearity of the coefficient of dt becomes significant, leading to saturation of the parts of the gain and loss coefficients arising from the gain medium.

ii) *The Cooperativity Parameter* C is defined as

$$C = \frac{g^2 N_{\text{At}}(\mathcal{B}-\mathcal{A})}{\kappa(\mathcal{B}+\mathcal{A})}, \tag{17.3.18}$$

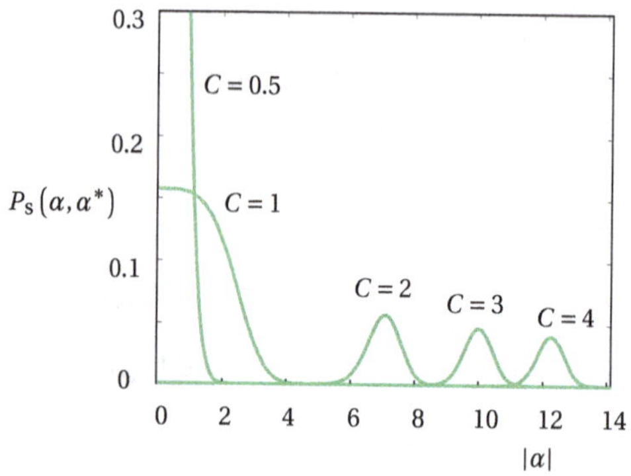

Fig. 17.2. Stationary laser distribution for $C = 0$ (well below threshold), $C = 1$ (at threshold), and $C = 2,3,4$ (above threshold).

and this is a measure of the *laser gain.* Since $\mathcal{A}$ and $\mathcal{B}$ are both positive, we can conclude that

$$|C| \le \frac{g^2 N_{\rm At}}{\kappa}\,. \tag{17.3.19}$$

b) Laser Stochastic Differential Equation: The laser equation (17.3.16) then becomes

$$d\alpha = \left\{-i\Omega - \tfrac{1}{2}\kappa\left(1 - \frac{C}{1+|\alpha^2|/n_0}\right)\right\}\alpha\, dt + \sqrt{\tfrac{1}{2}\left(g^2 N_{\rm At} + \frac{\kappa C}{1+|\alpha|^2/n_0}\right)}\, dw(t). \tag{17.3.20}$$

From this equation, it can be seen that there is gain that will lead the amplitude α to grow from zero only if $C > 1$. The gain becomes unity when $|\alpha|^2 = n_0(C-1)$, and this must represent approximately the stationary intensity, if the noise terms can be neglected. Thus, the saturation of the gain as $|\alpha|^2$ increases is clearly visible in the equation of motion.

c) Laser Fokker–Planck Equation: From the laser stochastic differential equation we can write the laser Fokker–Planck equation for the P-function as

$$\frac{\partial P}{\partial t} = \left\{ i\Omega\left(\frac{\partial}{\partial\alpha}\alpha - \frac{\partial}{\partial\alpha^*}\alpha^*\right) + \tfrac{1}{2}\kappa\left(\frac{\partial}{\partial\alpha^*}\alpha^* + \frac{\partial}{\partial\alpha}\alpha\right)\left(1 - \frac{C}{1+|\alpha|^2/n_0}\right)\right.$$
$$\left. + \tfrac{1}{2}\frac{\partial^2}{\partial\alpha\partial\alpha^*}\left(g^2 N_{\rm At} + \frac{\kappa C}{1+|\alpha|^2/n_0}\right)\right\} P. \tag{17.3.21}$$

17.3.3 Solutions of the Laser Equations

The laser equations can be solved in various degrees of approximation, the principal requirement being that the fluctuations are small relative to the laser field, which is the most interesting case. In principle, the effect of the fluctuations can be made very small.

a) Stationary Solution of the Laser Fokker–Planck Equation: As in the case of the van der Pol equation treated in Sect. 7.3.4, the stationary probability distribution is a function of $|\alpha|^2$; specifically if $P_s(\alpha,\alpha^*) \equiv F(|\alpha|^2)$ is the stationary solution, then, as substituting into (17.3.21) shows

$$\left(g^2 N_{At} + \frac{\kappa C}{1+z/n_0}\right)\frac{dF(z)}{dz} = -2\kappa\left(1 - \frac{C}{1+z/n_0}\right)F(z). \tag{17.3.22}$$

This leads to the solution

$$P_s(\alpha,\alpha^*) = \mathcal{N}\exp\left(\frac{2\kappa}{g^2 N_{At}}\left[-|\alpha|^2 + n_0 C\left(1+\frac{\kappa}{g^2 N_{At}}\right)\log\left(|\alpha|^2 + n_0\left(1+\frac{\kappa C}{g^2 N_{At}}\right)\right)\right]\right). \tag{17.3.23}$$

The distribution is plotted in Fig. 17.2, for various values of the parameter C, and shows that the behaviour is very different depending on whether the laser is below threshold ($C < 1$), where the most probable photon number is zero, or above threshold ($C > 1$), where there is a well-defined non-zero number of photons, with only a small spread about the peak value.

Exercise 17.4 Approximate Mean and Variance: Define a "photon number" in the P-representation $n \equiv |\alpha|^2$, and then set $n = n_0(C-1)+x$, so that x represents the fluctuation around the above threshold stationary mean $\bar{n}_s = n_0(C-1)$. By substituting in the stationary distribution (17.3.23) show that we can write approximately

$$P_s(n) \propto \exp\left(-\frac{\kappa x^2}{n_0 C(\kappa + g^2 N_{At})}\right). \tag{17.3.24}$$

This means that above threshold we can write the stationary mean and variance of the photon number as

$$\bar{n}_s = n_0(C-1), \tag{17.3.25}$$

$$\mathrm{var}[n]_s = \tfrac{1}{2} n_0 C\left(1 + \frac{g^2 N_{At}}{\kappa}\right). \tag{17.3.26}$$

Thus, the number fluctuations have a *Poissonian* nature in the sense that they are of the same order of magnitude as the mean. In this model, (17.3.18) shows that $C \le g^2 N_{At}/\kappa$, and the fluctuations are always greater than the exact Poissonian.

b) The Laser for Large Photon Number: The most precisely defined laser field arises when the ratio of the fluctuations to the mean number is small. The mean and variance (17.3.25, 17.3.26) are for $n_s \equiv |\alpha|^2$, the P-function variable. The mean and variance of the actual photon number operator N are then, using (15.1.9, 15.1.10),

$$\langle N\rangle_s = \bar{n}_s = n_0(C-1), \tag{17.3.27}$$

$$\mathrm{var}[N]_s = \mathrm{var}[n]_s + \bar{n}_s = n_0(C-1) + \tfrac{1}{2} n_0 C\left(1+\frac{g^2 N_{At}}{\kappa}\right). \tag{17.3.28}$$

The ratio of the variance to the mean is given by

$$\frac{\mathrm{var}[N]_s}{\langle N\rangle_s} = 1 + \tfrac{1}{2}\frac{C}{C-1}\left(1+\frac{g^2 N_{At}}{\kappa}\right) \geqslant 1 + \frac{C(1+C)}{2(C-1)}. \tag{17.3.29}$$

The final inequality results from the lower bound $g^2 N_{At}/\kappa \ge C$ given in (17.3.19). The fluctuations are thus always greater than expected from a Poisson distribution for which this ration would be exactly one. Nevertheless, for moderate values of C, the ratio is close to the Poissonian result. The closest approach to a Poissonian result happens when $g^2 N_{At}/\kappa = C$, and when (17.3.29) is minimized, which happens when $C = 1+\sqrt{2} \approx 2.414$. Thus in this case $\mathrm{var}[N]_s/\langle N\rangle_s \to 5/2+\sqrt{2} \approx 3.914$.

This result is for the light *inside* the laser. The light beam which emerges from the cavity is highly attenuated, since the exit mirror has a transmission coefficient t which is very small, and therefore for the transmitted photon number $N_t \equiv tN_s$, we find

$$\frac{\mathrm{var}[N_t]}{\langle N_t\rangle} = t\frac{\mathrm{var}[N_s]}{\langle N_s\rangle}. \tag{17.3.30}$$

This ratio can be made very small, but we must also include the randomness induced by the transmission process, and this adds extra fluctuations which make the transmitted light very accurately Poissonian.

Exercise 17.5 The van der Pol Approximation: If we use the van der Pol equation of Sect. 7.3.4 we get a simplified description of small fluctuations. Write $|\alpha|^2 = x + C(N_0-1)$, and show that to lowest order in x, we can write

$$1 - \frac{C}{1+|\alpha|^2/n_0} \approx \frac{1}{C} - 1 + \frac{|\alpha|^2}{n_0 C}. \tag{17.3.31}$$

Approximate the noise coefficient by its value at equilibrium, and derive the van der Pol equation in the form given in Sect. 7.3.4.

c) Approximate Equations for Phase and Amplitude: We consider a situation in which the fluctuations are very small relative to the laser field, and approximate the noise coefficient in the stochastic differential equation (17.3.20) by setting $|\alpha|^2 \to \bar{n}_s = n_0(C-1)$, so that it becomes

$$d\alpha = \left\{-i\Omega - \tfrac{1}{2}\kappa\left(1 - \frac{C}{1+|\alpha^2|/n_0}\right)\right\}\alpha\, dt + \sqrt{\tfrac{1}{2}\left(g^2 N_{At}+\kappa\right)}\, dw(t). \tag{17.3.32}$$

Let us change variables to phase and amplitude, so that

$$\alpha(t) = a(t)\exp\left(i\phi(t)\right). \tag{17.3.33}$$

Following the method of Sect. 7.3.3, the equations for these variables corresponding to the stochastic differential equation (17.3.20) are

$$d\phi(t) = -\Omega\, dt + \frac{\sqrt{g^2 N_{\mathrm{At}} + \kappa}}{2a(t)}\, dW_\phi(t), \tag{17.3.34}$$

$$\begin{aligned} da(t) &= \left\{ -\tfrac{1}{2}\kappa \left(1 - \frac{C}{1 + |a(t)^2|/n_0} \right) a(t) + \frac{g^2 N_{\mathrm{At}} + \kappa}{4a(t)} \right\} dt \\ &\qquad + \tfrac{1}{2}\sqrt{g^2 N_{\mathrm{At}} + \kappa}\, dW_a(t), \end{aligned} \tag{17.3.35}$$

$$dW_\phi(t)^2 = dW_a(t)^2 = dt, \qquad dW_\phi(t)\, dW_a(t) = 0. \tag{17.3.36}$$

d) Laser Linewidth: If the laser is well above threshold so that the number fluctuations are small, we can set $a(t) \approx \sqrt{\bar{n}_{\mathrm{s}}}$, and then solve the phase equation (17.3.34) exactly and write

$$\alpha(t) \approx \sqrt{\bar{n}_{\mathrm{s}}} \exp\left(-\mathrm{i}\Omega t + \mathrm{i}\frac{\sqrt{g^2 N_{\mathrm{At}} + \kappa}}{2\sqrt{\bar{n}_{\mathrm{s}}}} W(t) \right). \tag{17.3.37}$$

As shown in Sect. 15.2.4, the quantum-mechanical correlation function, whose Fourier transform is the measured spectrum, is equal to the P-function correlation function. We can use the method of Sect. 7.3.1 to evaluate the stationary P-function correlation function as

$$\langle \alpha^*(t)\alpha(t') \rangle = n_{\mathrm{s}} \mathrm{e}^{\mathrm{i}\Omega(t-t')} \exp\left(-\frac{g^2 N_{\mathrm{At}} + \kappa}{8\bar{n}_{\mathrm{s}}} |t - t'| \right), \tag{17.3.38}$$

and the spectrum of $\alpha(t)$ then takes the form

$$S(\omega) = \frac{\bar{n}_{\mathrm{s}}}{\pi} \frac{\gamma}{(\omega - \Omega)^2 + \gamma^2}, \tag{17.3.39}$$

$$\text{with } \gamma \equiv \frac{g^2 N_{\mathrm{At}} + \kappa}{8\bar{n}_{\mathrm{s}}}. \tag{17.3.40}$$

The quantity γ is the *laser linewidth*, and this can be very small, decreasing as the laser intensity increases.

VI Quantum Measurement Theory

18. Foundations and Formalism of Quantum Measurement

From its very beginning, the relationship between quantum mechanics, as a description of the microscopic world, and the measurements that can be made on such microscopic systems, yielding as they do macroscopic results, has been an issue which has been very hard to put in a convincing form. The uncertainty principle, as formulated by *Heisenberg*, was the first way in which it was demonstrated that any measurement must be a physical process which would disturb the system under observation. The wave-particle duality inherent in quantum mechanics is the fundamental basis for the quantitative form of uncertainty principle in the form $\Delta x \Delta p \geq \hbar/2$, which shows that the effect becomes negligible as the quantities measured become macroscopic.

18.1 Formulations of Quantum Mechanics

The original formulation of quantum mechanics in terms of state vectors (i.e., wavefunctions) in a Hilbert space and Hermitian operators is still the most convenient for everyday calculations, but it is not at all clear that this formulation is fundamental. The path-integral formulation of quantum mechanics [44] does not require the concept of a wavefunction, since it is formulated in terms of probability amplitudes for observable processes. However, because it is mandatory in this formulation to consider every possible path by which a system can proceed from an initial state to a final state, and add up the *amplitudes* for each path to get the amplitude for the process, the path-integral formulation is not practical for every kind of problem, even though it can be a very powerful tool for a wide range of problems.

In this chapter we want to formulate measurement theory using the conventional Hilbert space formulation, but we will take the view that this is merely a framework within which one can compute the *amplitudes* for processes of interest. Thus, we will not see the *wavefunction* as a fundamental object, but rather as a mathematical construct which provides a method for the computation of the amplitudes, whose squares give probabilities.

18.1.1 Fundamental Postulates

Within the Hilbert space formulation, the interpretation of quantum mechanics relies on the three postulates:

I. Probability Interpretation:

i) The quantum state of a system is described by a *wavefunction* or *vector in Hilbert space* $|\psi, t\rangle$.

ii) A physically measurable quantity is described by a *Hermitian operator* A.

iii) If the quantity A is measured, the only values obtainable are the eigenvalues a of A, and the probability of obtaining the eigenvalue a is $|\langle a|\psi, t\rangle|^2$.

II. Unitary Evolution: The wavefunction evolves according to the Schrödinger equation

$$i\hbar\partial_t|\psi, t\rangle = H(t)|\psi, t\rangle, \tag{18.1.1}$$

or equivalently

$$|\psi, t\rangle = U(t, t_0)|\psi, t_0\rangle, \tag{18.1.2}$$

where $U(t, t_0)$ is a unitary operator given by

$$i\hbar\partial_t U(t, t_0) = H(t)U(t, t_0), \qquad U(t_0, t_0) = 1. \tag{18.1.3}$$

III. Non-Unitary Evolution—Collapse of the Wavefunction: After a measurement of A which results in the value a, the wavefunction of the system is $|a\rangle$, an eigenstate of A with eigenvalue a.

18.1.2 Status of the Postulates

The first two postulates are probably all that is required to interpret quantum mechanics, provided we know enough about the physical world to use these requirements.

i) The first postulate is the most fundamental, and tells us how information is organized in the quantum world. It is a distillation of the concept of wave-particle duality which has been with quantum mechanics from the time of *Einstein's* [59] theory of the photoelectric effect. However, there is no intrinsic reality ascribed to the wavefunction $|\psi, t\rangle$ other than its role in calculating the probabilities of the kind $|\langle a|\psi, t\rangle|^2$.

ii) The second postulate is essentially the superposition principle, since the linear operator nature of the equation of motion is essential to obtain the physically observed fact that a superposition state $a_1|\psi_1, 0\rangle + a_2|\psi_2, 0\rangle$ evolves with time into the superposition state $a_1|\psi_1, t\rangle + a_2|\psi_2, t\rangle$. It is then not too difficult to show that the structure of the evolution must be that given by the second postulate. However, nothing is said about the nature of the Hamiltonian operator $H(t)$, or the way the states fit into the physical world. Relating the *details* of physics to the fundamental structure is what keeps theoretical physicists in business. At one extreme, we have the physics of fundamental particles, and the other, phenomenological model building based on the facts of electromagnetism and atomic physics.

iii) However, the third interpretative postulate is usually made, and is referred to as the *collapse of the wavefunction.* Nevertheless, the concept of a *non-unitary* "collapse of the wavefunction" sits uncomfortably with the unitary evolution of the second postulate, especially as in practice measurement is carried out by instruments which we *design using the laws of unitary evolution.* It is worth looking more carefully into how real physical measurements are made, because it is possible to see that the collapse of the wavefunction is really already implied by the first two postulates, as a compact *Markovian* description of a process of quantum-mechanical *entanglement.*

18.2 Modelling a Measurement—Tracks in a Cloud Chamber

A question posed by *Mott* [60] at the very beginning of quantum mechanics was:

> In the theory of radioactive disintegration, as presented by Gamow, the α-particle is represented by a spherical wave which slowly leaks out of the nucleus. On the other hand, the α-particle, once emerged, has particle-like properties, the most striking being the ray tracks that it forms in a Wilson cloud chamber. It is a little difficult to picture how it is that an outgoing spherical wave can produce a straight track; we think intuitively that it should ionize atoms at random throughout space.

Mott used quantum mechanics to show that ionizations arising from a single α particle must all lie along very close to a single straight line proceeding from the decaying nucleus, but that the orientation of this line would be random. This problem contains within it the essence of the problem of the collapse of the wavefunction. We can, following Mott's argument, show that the collapse of the wavefunction is a consequence of unitary evolution and the probability interpretation of quantum mechanics.

18.2.1 Excitation of an Atom

Consider an α-particle moving in a cloud chamber, ionizing an atom as it passes close by. We can model this by a wavefunction $\psi(\boldsymbol{x}, \boldsymbol{y}, t)$, where $\boldsymbol{x}$ is the position of the α-particle, and $\boldsymbol{y}$ is the position of the electron in an atom localized at some position $\boldsymbol{a}$—for simplicity we do not consider the motion of the atom as a whole.

The Hamiltonian which describes this system will be taken as

$$H = H_\alpha + H_e + v(\boldsymbol{x}-\boldsymbol{y}), \tag{18.2.1}$$

where $$H_\alpha = -\frac{\hbar^2}{2M}\nabla_x^2, \tag{18.2.2}$$

and $$H_e = -\frac{\hbar^2}{2m}\nabla_y^2 + V(\boldsymbol{y}). \tag{18.2.3}$$

Here:

i) $V(\boldsymbol{y})$ is the potential which binds the electron in the atom, and is centred at the origin.

ii) $v(\boldsymbol{x}-\boldsymbol{y})$ represents the interaction between the α-particle and the electron.

a) Expansion of the Wavefunction: If we expand the wavefunction in eigenfunctions of $\xi_n(\boldsymbol{y})$ of H_e, with eigenvalues ϵ_n

$$\psi(\boldsymbol{x},\boldsymbol{y},t) = \sum_n \psi_n(\boldsymbol{x},t)\xi_n(\boldsymbol{y}), \tag{18.2.4}$$

the Schrödinger equation becomes

$$\mathrm{i}\hbar\partial_t\psi_n(\boldsymbol{x},t) = (H_\alpha+\epsilon_n)\psi_n(\boldsymbol{x},t) + \sum_m u_{nm}(\boldsymbol{x})\psi_m(\boldsymbol{x},t), \tag{18.2.5}$$

in which

$$u_{nm}(\boldsymbol{x}) = \int d^3\boldsymbol{y}\,\xi_n^*(\boldsymbol{y})v(\boldsymbol{x}-\boldsymbol{y})\xi_m(\boldsymbol{y}). \tag{18.2.6}$$

b) Solution Using the Green's Function: We can solve this as a scattering problem, using Green's function techniques and the Born approximation. We want to find the amplitude to scatter an incident α-particle wave $\exp(\mathrm{i}\boldsymbol{k}\cdot\boldsymbol{x})$ in a certain asymptotic direction and leave the electron in the n-th energy level. If we define k_n by

$$\frac{\hbar^2 k_n^2}{2M} = \frac{\hbar^2\boldsymbol{k}^2}{2M} - \epsilon_n, \tag{18.2.7}$$

we will need the Green's function

$$G_n(\boldsymbol{x},\boldsymbol{x}') = -\frac{\mathrm{e}^{\mathrm{i}k_n|\boldsymbol{x}-\boldsymbol{x}'|}}{4\pi|\boldsymbol{x}-\boldsymbol{x}'|}, \tag{18.2.8}$$

which satisfies the equation

$$(\nabla^2+k_n^2)G_n(\boldsymbol{x},\boldsymbol{x}') = \delta(\boldsymbol{x}-\boldsymbol{x}'). \tag{18.2.9}$$

Adapting the viewpoint of scattering theory, we consider the solution of the stationary Schrödinger equation corresponding to (18.2.5)

$$(E-\epsilon_n)\psi_n(\boldsymbol{x}) = -\frac{\hbar^2}{2M}\nabla_x^2\psi_n(\boldsymbol{x}) + \sum_m u_{nm}(\boldsymbol{x})\psi_m(\boldsymbol{x}). \tag{18.2.10}$$

The scattering solution we want should correspond to a spherical wave incident on the atom, a wave which originates at the decaying nucleus which is the source

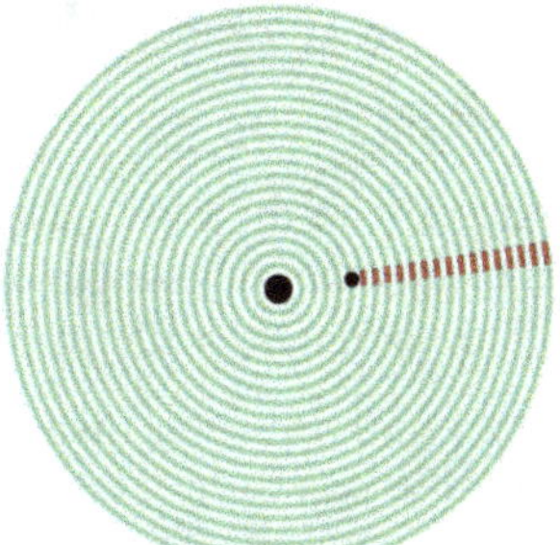

Fig. 18.1. Ionization of an atoms by an α-particle. The α-particle is emitted by the nucleus (large black dot at the centre,) and has a spherically symmetrical wavefunction (green), which overlaps an atom (small black dot). If an atom is ionized, the part of the α-particle wavefunction coloured brown is entangled with its ionized state, while the green part of the wavefunction is entangled with the atom's non-ionized state. Subsequent ionizations can only occur where the brown part of the wavefunction is non-zero, and will hence lie on a ray joining all ionizations to the nucleus.

of the alpha particle. However, on the scale of the atom being excited, this will look essentially like a plane wave—hence we take an incident plane wave for the α-particle, and take the electron in the atom to be in the ground state. The solution then satisfies the equation

$$\psi_n(\boldsymbol{x}) = \delta_{n0}\,\mathrm{e}^{\mathrm{i}\boldsymbol{k}\cdot\boldsymbol{x}} + \int d^3\boldsymbol{x}'\,G_n(\boldsymbol{x},\boldsymbol{x}')\sum_m u_{nm}(\boldsymbol{x}')\,\psi_m(\boldsymbol{x}'). \tag{18.2.11}$$

c) Approximations: Now we make some approximations often used in scattering theory:

i) We look at the long distance behaviour in which $|\boldsymbol{x}| \gg |\boldsymbol{x}'|$.

ii) We make the Born approximation, and consequently make the replacement $\psi_m(\boldsymbol{x}') \to \delta_{m0}\,\mathrm{e}^{\mathrm{i}\boldsymbol{k}\cdot\boldsymbol{x}'}$ in the integrand. These lead to the solution in the form

$$\psi_n(\boldsymbol{x}) \sim \delta_{n0}\,\mathrm{e}^{\mathrm{i}\boldsymbol{k}\cdot\boldsymbol{x}} - \frac{\mathrm{e}^{\mathrm{i}k_n|\boldsymbol{x}|}}{4\pi|\boldsymbol{x}|}\int d^3\boldsymbol{x}'\,\mathrm{e}^{\mathrm{i}(\boldsymbol{k}-\boldsymbol{k}_n)\cdot\boldsymbol{x}'}\,u_{n0}(\boldsymbol{x}'), \tag{18.2.12}$$

where $\boldsymbol{k}_n \equiv k_n\hat{\boldsymbol{x}}$, that is, a vector with magnitude k_n in the direction of $\boldsymbol{x}$.

iii) Finally, we take account of the definition of $u_{nm}(\boldsymbol{x}')$ in (18.2.6) to write

$$\int d^3\boldsymbol{x}'\,\mathrm{e}^{\mathrm{i}(\boldsymbol{k}-\boldsymbol{k}_n)\cdot\boldsymbol{x}'}\,u_{n0}(\boldsymbol{x}') = W_{n0}(\boldsymbol{k}-\boldsymbol{k}_n)\,\tilde{v}(\boldsymbol{k}-\boldsymbol{k}_n), \tag{18.2.13}$$

in which the two quantities in this expression are

$$W_{n0}(\boldsymbol{q}) \equiv \int d^3\boldsymbol{z}\,\xi_n^*(\boldsymbol{z})\xi_m(\boldsymbol{z})\mathrm{e}^{\mathrm{i}\boldsymbol{q}\cdot\boldsymbol{z}}, \tag{18.2.14}$$

which represents the amplitude for a transition $m \to n$ induced by a plane wave, and

$$\tilde{v}(\boldsymbol{q}) \equiv \int d^3\boldsymbol{z}\,v(\boldsymbol{z})\mathrm{e}^{\mathrm{i}\boldsymbol{q}\cdot\boldsymbol{z}}, \tag{18.2.15}$$

is the Fourier transform of the interaction potential.

iv) Put together, these mean that we can write

$$\psi_n(\boldsymbol{x}) \sim \delta_{n0}\,\mathrm{e}^{\mathrm{i}\boldsymbol{k}\cdot\boldsymbol{x}} - \frac{\mathrm{e}^{\mathrm{i}k_n|\boldsymbol{x}|}}{4\pi|\boldsymbol{x}|}W_{n0}(\boldsymbol{k}-\boldsymbol{k}_n)\,\tilde{v}(\boldsymbol{k}-\boldsymbol{k}_n). \tag{18.2.16}$$

18.2.2 Interpretation

We now take into account the physical aspects of the problem.

a) The α-Particle is Very Energetic: The typical energies are on the MeV scale, while the energies in the atom are on an eV scale. This means the wavelength of the α-particle is very short—of order of magnitude 10^{-28}m, so that $|\boldsymbol{k}| \approx 10^{28}\mathrm{m}^{-1}$.

b) The Atom is Rather Large: An atom is of dimensions $r_A \approx 10^{-9}$m—this means that the amplitude (18.2.14) is only significant for $|\boldsymbol{q}| < 1/r_A \ll |\boldsymbol{k}|$.

c) The Range of the Interaction is Large: The range r_{int} is in fact infinite, since this would typically be a Coulomb force, but in any consideration is very much larger than the α-particle wavelength. Thus, we can say that $\tilde{v}(\boldsymbol{q})$ as given in (18.2.15) is only significant when $|\boldsymbol{q}| < 1/r_{\text{int}} \ll |\boldsymbol{k}|$.

Putting these together, we conclude that the scattered part of $\psi_n(\boldsymbol{x})$ for $n \neq 0$ is an outgoing spherical wave modulated by a function of $\boldsymbol{k}'_n = k_n\hat{\boldsymbol{x}}$ which is only non-zero when $\hat{\boldsymbol{x}}$ is very close to the direction $\boldsymbol{k}$, as illustrated in Fig. 18.1.

d) Collapse of the Wavefunction: The part of the wavefunction corresponding to an excitation having taken place is the second term in (18.2.16)—an outgoing wave emanating from the position of the atom, modulated by a function which is only significant in the forward direction. The future evolution of the full wavefunction will mean that the part corresponding to an excitation of this atom behaves as if the outgoing spherical wave *collapsed* down to the volume containing the atom, while essentially retaining its momentum. This corresponds to measuring the position of the atom to an accuracy of the volume of the atom—and after measurement, the future wavefunction is given by propagating from the collapsed wavefunction.

18.2.3 Entanglement

The resultant wavefunction can be written in a formal sense to a good degree of approximation as

$$\psi_n(\boldsymbol{x}) = (\delta_{n0} + S_n(\boldsymbol{x}-\boldsymbol{a},\boldsymbol{k}))\,\mathrm{e}^{\mathrm{i}\boldsymbol{k}\cdot\boldsymbol{x}}. \tag{18.2.17}$$

Here $\boldsymbol{a}$ is the position of the atom being excited, and S_n is defined by

$$S_n(\boldsymbol{x}-\boldsymbol{a},\boldsymbol{k}) \equiv \frac{\mathrm{e}^{\mathrm{i}k_m|\boldsymbol{x}-\boldsymbol{a}|}}{4\pi|\boldsymbol{x}-\boldsymbol{a}|}W_{n0}(\boldsymbol{k}-\boldsymbol{k}'_n)\,\tilde{v}(\boldsymbol{k}-\boldsymbol{k}'_n). \tag{18.2.18}$$

This wavefunction represents an *entangled state.* By this it is meant that the multicomponent wavefunction

$$\Psi(\boldsymbol{x}) \equiv \{\psi_0(\boldsymbol{x}),\psi_1(\boldsymbol{x}),\psi_2(\boldsymbol{x}),\ldots\}, \tag{18.2.19}$$

is the sum of components in which *the state of the α-particle depends on the state of the atom.*

a) Many Atoms: Proceeding now to include a number of atoms each located at r different positions $\boldsymbol{a}_i$, we can write the wavefunction in the form

$$\psi_{n_1,n_2,n_3,\dots,n_r}(\boldsymbol{x}) = \mathrm{e}^{\mathrm{i}\boldsymbol{k}\cdot\boldsymbol{x}} \prod_i \left(\delta_{n_i 0} + S_{n_i}(\boldsymbol{x}-\boldsymbol{a}_i,\boldsymbol{k}_i)\right). \tag{18.2.20}$$

b) Probability of Ionizing Several Atoms: The probability of finding atoms in states $n_a, n_b, n_c, \dots$ at positions $\boldsymbol{a}_a, \boldsymbol{a}_b, \boldsymbol{a}_c, \dots$ is then

$$P(a,b,c,\dots) \propto \left|S_{n_a}(\boldsymbol{x}-\boldsymbol{a}_a,\boldsymbol{k}_a)S_{n_b}(\boldsymbol{x}-\boldsymbol{a}_b,\boldsymbol{k}_b)S_{n_c}(\boldsymbol{x}-\boldsymbol{a}_c,\boldsymbol{k}_c)\dots\right|^2. \tag{18.2.21}$$

This can only be non-zero when all of the functions $S_{n_i}(\boldsymbol{x}-\boldsymbol{a}_i,\boldsymbol{k}_i)$ are non-zero, and because of the "searchlight-like" nature of these functions this can only happen if all of the excited atoms are essentially in a straight line.

c) Nature of the state: The argument does not determine the position of the first excitation in a series—but that there is no probability of a multiparticle excitation unless these are in a straight line—thus we must expect straight line tracks in random orientations.

18.2.4 Interpretation as Collapse of the Wavefunction

The argument above views the state of the system after a certain time, and asserts that a measurement of all of the excitations is made at that time. However, it is clear that solution of the Schrödinger equation can be made continuously, and the entangled state can be built up continuously.

The concept of the *collapse of the wavefunction* is a kind of convenience which avoids having to think of joint probabilities. What we get out of the above analysis (given the initial state—that of a nucleus which has just decayed producing an outgoing α-particle wave) is an expression for the joint probability of obtaining a number of excitations of atoms at positions $\boldsymbol{a}_a$, $\boldsymbol{a}_d$, $\boldsymbol{a}_c \dots$, and these probabilities have two principal properties:

i) If we observe two or more excitations, they will lie on a straight line passing through the source of the original α-particle.

ii) If we want to compute the probability of future excitations, this involves only the part of the wavefunction which corresponds to the previous excitations—this is equivalent to the collapse of the wavefunction.

a) The Observation Process: The above argument gives the probabilities of excitations of atoms, which in practice lead to the formation of clouds of droplets in the supersaturated vapour of a cloud chamber. Essentially the same pattern of droplets is observed for every possible set of excitations of the atoms, independently of each atom's precise final state of excitation. This brings into consideration the concept of a *macrostate,* that is, a state which corresponds to a macroscopic observation, such as a set of vapour droplets at certain positions, the position of a pointer, or a set of numbers recorded in a computer memory.

b) Macroscopic Systems and Macrostates: Making a measurement requires a transfer of information from the microscopic domain—described by quantum mechanics—to the macroscopic world—described using classical mechanics. The measurements which we make in a laboratory are *always* macroscopic. The concept of a macroscopic state is not one of a separate kind of physics, but rather simply of size. A very clear formulation of this issue has been given by *van Kampen* [61] as follows:

> A macroscopic system, such as a certain amount of gas, a crystal, or a pointer on a meter, is composed of an inordinately large number of particles. As a consequence, its energy levels lie inordinately dense on any energy scale used in the laboratory. The typical distance δE between two successive energy levels is much smaller than the inaccuracy ΔE of the best energy measurement of the best experimenter. Hence such a system can never be prepared in a single eigenstate of the energy operator. Rather, the wavefunction ψ of a macroscopic system is always in a superposition of an enormous number (namely $\Delta E/\delta E$) of eigenstates. A macroscopic system does obey the laws of quantum mechanics, but *the familiar picture of individual eigenvalues and eigenstates is no longer adequate.* Other features become prominent; they constitute the subject matter of macroscopic physics. (A loose analogy within the realm of classical theory is formed by statistical mechanics: A many body system has features, such as pressure and temperature, which do not exist for a few particles; they are the subject matter of thermodynamics.)
>
> When an experimenter has prepared the system well defined state, he refers to the *macrostate.* He does not pretend to know its Hilbert space vector, he only knows that it lies within a certain subspace of Hilbert space with $\Delta E/\delta E$ dimensions.
>
> When a macroscopic pointer indicates a macroscopic point on a dial the number of microscopic eigenstates involved has been estimated to be 10^{50}. When the observer shines light in order to read the position of the pointer, the photons do perturb the ψ of the pointer, but the perturbation does not affect the macrostate. The vector is moved round a bit in these 10^{50} dimensions, but its components outside the subspace remain negligible. That is the reason why macroscopic observations can be recorded objectively, independently of the observations and the observer, and may therefore be the subject of scientific study.

c) Metastability and Gain: Certain systems, such as for example the supersaturated vapour we are presently considering, possess metastable states. Such a state is a macrostate in which the system can reside for a long time before its ultimate transition to an equilibrium state. In many cases this transition can be triggered by a minute perturbation, such as the excitation of a single atom particle by an α-particle in a cloud chamber, or a single photon hitting a photomultiplier.

The production of a macroscopic signal from a microscopic signal can also be described as *gain,* so that another way of looking at this concept is from the point of view of high sensitivity amplification. Here the issue is not so much that the system is metastable, but rather that it is maintained out of equilibrium by some non-equilibrium process, such as that envisaged in the laser model of Sect. 17.3.1.

18.2.5 The Essential Features

The process of measurement we have outlined for this example of tracks in a cloud chamber has the nature of a cascade of processes. The α-particle passes by a sequence of atoms, and we can describe the process by which each atom becomes excited in a way dependent strongly on previous excitations, leading to an entangled state in which the resultant probability of observing a number of excitations is zero unless they all lie very close to a straight line. For any given straight line there are very many possible microscopic quantum states, corresponding to the different possible final states of the ions and the electrons. We can compute the probabilities of each of these quantum-mechanically, and add them all up. The excitations then individually trigger the condensation process of the metastable macroscopic vapour, giving the observed result, a vapour trail, corresponding to a certain *macrostate.*

A very accurate description of this process can be given by considering each droplet in the vapour trail to correspond to an excitation, and each excitation to a *measurement* of the position of the α-particle at a time given kinematically in terms of its velocity and initial position. After each of these measurements, the particle's wavefunction can be considered to be a modulated plane wave as in Fig. 18.1, that is, the the particle's wavefunction can be considered to have *collapsed* into this wavefunction after each process of observation as given by an excitation leading to the condensation of a water droplet.

18.3 Formal Quantum Measurement Theory

The description of the measurement process as a sequence of entanglements yielding joint probabilities is so complex that an abstraction in terms of the collapse of the wavefunction after each measurement is obviously an essential tool, even if it is not a fundamental concept. This section will formulate quantum measurement theory along these lines. Such a formulation is essentially Markovian, since the measurement process is regarded as instantaneous. In fact this is never exactly true, and thus this is an approximate theory. However, it is, like the theory of quantum Markov processes, also a logically complete and self-consistent theory, and we will show there is an intimate relationship between these two extensions of quantum mechanics to non-unitary evolution.

In the form written in Sect. 18.1.1, if the state upon which the measurement is performed is already an eigenstate $|a_0\rangle$ of A, the value a_0 will certainly be obtained, and the state after measurement will be again $|a_0\rangle$. This is called a measurement of the *first kind,* or alternatively, a *projective measurement.* Such a measurement is rarer than one might imagine, for one of the most common measurements is to count photons using a photodetector, which absorbs the photons as it counts them. In this case if the system has exactly one photon in it, then after detecting this photon there will be none. This is called a measurement of the *second kind*, or a *non-projective measurement*, that is a measurement in which an eigenstate of the variable being measured becomes a different state after measurement. Obviously, the ubiquity of photodetectors shows that this is a useful kind of measurement. We therefore need a measurement formalism which encapsulates all possibilities.

18.3.1 Measurement Operators

Let us abstractly formulate a general concept of the measurement of a physical observable whose operator is A.

i) When we do a measurement, we get a value $\bar{a}$ which is one of the eigenvalues a_i of A, belonging to the quantum states $|i\rangle$, which form a complete set of states.

ii) It is possible that the measurement is inefficient, for example, a photodetector may absorb a photon, but give no detectable signal. Thus, when we get the value zero for the number of photons, this could mean there were no photons to detect, or that there was a photon which we failed to detect. This means that the same measured value $\bar{a}$ may be obtained by more than one measurement operator. We will use the symbol α as a label to distinguish different measurement operators which give the same measured result.

a) Definition of a Measurement Operator: We define a measurement operator $E(\bar{a},\alpha)$ for the measurement $(\bar{a},\alpha)$ as a linear operator defined in terms of some coefficients $e_{ij}(\bar{a},\alpha)$ thus:

$$E(\bar{a},\alpha)=\sum_{i,j} e_{ij}(\bar{a},\alpha)|i\rangle\langle j|. \tag{18.3.1}$$

b) Probability of the Result $\bar{a}$: If the state being measured is $|\psi\rangle$ then the probability of the result $\bar{a}$ being obtained from the measurement $(\bar{a},\alpha)$ is

$$P(\bar{a},\alpha)=\langle\psi|E^\dagger(\bar{a},\alpha)E(\bar{a},\alpha)|\psi\rangle, \tag{18.3.2}$$

and the probability of the result $\bar{a}$ from any method is given

$$P(\bar{a})=\sum_\alpha P(\bar{a},\alpha). \tag{18.3.3}$$

c) State after the Measurement: This is found by applying the operator $E(\bar{a},\alpha)$ to the state being measured, and then adjusting the normalization to unity:

$$|\psi_{\bar{a}}\rangle=\frac{E(\bar{a},\alpha)|\psi\rangle}{\sqrt{P(\bar{a},\alpha)}}. \tag{18.3.4}$$

d) Completeness: The sum of all the probabilities must add to one, so that

$$\sum_{\bar{a},\alpha} P(\bar{a},\alpha)=\langle\psi|E^\dagger(\bar{a},\alpha)E(\bar{a},\alpha)|\psi\rangle=1. \tag{18.3.5}$$

This must be true for *all* states $|\psi\rangle$, so that we must require the *completeness relationship*

$$\sum_{\bar{a},\alpha} E^\dagger(\bar{a},\alpha)E(\bar{a},\alpha)=1. \tag{18.3.6}$$

18.3.2 Density Operator Formulation of Measurement Theory

We can express the quantum theory of measurement quite naturally within the density operator formalism as follows. Supposing that the initial density operator is ρ, the probability of the outcome $(\bar{a}, \alpha)$ will be

$$P(\bar{a},\alpha) = \mathrm{Tr}\left\{E(\bar{a},\alpha)\rho E^{\dagger}(\bar{a},\alpha)\right\}. \tag{18.3.7}$$

The density operator after the measurement will then be

$$\rho(\bar{a},\alpha) = \frac{E(\bar{a},\alpha)\rho E^{\dagger}(\bar{a},\alpha)}{P(\bar{a},\alpha)}. \tag{18.3.8}$$

These results follow simply by applying the measurement theory in terms of state vectors to the individual state vectors in the density operator.

18.3.3 Example—The Two-Level Atom

Let us consider a two-level atom, described by the two states

$$u(+) = \begin{pmatrix}1\\0\end{pmatrix}, \quad u(-) = \begin{pmatrix}0\\1\end{pmatrix}, \tag{18.3.9}$$

and assume that we wish to test whether the atom is in the excited state, $u(+)$, by absorbing the energy in it, and thus leaving it in the lower state.

We would not expect the method of measurement to be perfect; even if the atom is in the upper state, we may absorb the energy and nevertheless not detect it, leaving the atom in the ground state. We will introduce the probability ϵ of detecting the energy that is released as the atom makes a transition to the ground state. This means that there are three possibilities, which will be characterized by three parameters; these are ϵ, and two other parameters λ and μ such that $|\lambda|^2 + |\mu|^2 = 1$. In terms of these, the operations are:

i) $E(1) = \begin{pmatrix}0 & 0\\ \lambda & 0\end{pmatrix}$ The atom goes into the ground state. With probability ϵ, the energy released is detected.

ii) $E(2,\alpha) = \begin{pmatrix}0 & 0\\ \lambda & 0\end{pmatrix}$ The atom goes into the ground state. But, with probability $1-\epsilon$, the energy is not detected.

iii) $E(2,\beta) = \begin{pmatrix}\mu & 0\\ 0 & 1\end{pmatrix}$ The atom makes no transition; no energy is released, and therefore none is detected.

The two operations $E(2,\alpha)$ and $E(2,\beta)$ give the same observable outcomes. The density operator after performing the detection, and observing outcome 2 (no energy detected) is the average of the two. This means that the probability of outcome 2 is

$$P(2) \equiv \mathrm{Tr}\left\{E(2,\beta)\,\rho\,E^\dagger(2,\beta) + (1-\epsilon)\,E(2,\alpha)\,\rho\,E^\dagger(2,\alpha)\right\}. \tag{18.3.10}$$

After measurement, the density operator is

$$\rho(2) = P(2)^{-1}\left\{E(2,\beta)\,\rho\,E^\dagger(2,\beta) + (1-\epsilon)\,E(2,\alpha)\,\rho\,E^\dagger(2,\alpha)\right\}. \tag{18.3.11}$$

On the other hand the probability of outcome 1 is

$$P(1) = \mathrm{Tr}\left\{\epsilon\,E(1)\,\rho\,E^\dagger(1)\right\}, \tag{18.3.12}$$

and

$$\rho(1) = P(1)^{-1}\,\epsilon\,E(1)\,\rho\,E^\dagger(1). \tag{18.3.13}$$

Exercise 18.1 Completeness: Check that the operators satisfy the completeness requirement, i.e.,

$$\epsilon\,E^\dagger(1)E(1) + (1-\epsilon)E^\dagger(2,\alpha)E(2,\alpha) + E^\dagger(2,\beta)E(2,\beta) = 1. \tag{18.3.14}$$

18.4 Multitime Measurements

We will now extend the measurement theory to handle sequences of measurements at subsequent times. This is relatively straightforward; after each measurement we have a new density operator, which then evolves according to Hamiltonian time development until the next measurement. This gives a new density operator which depends on the outcome of the measurement. The probability of each outcome is determined by the single-time measurement theory of the previous section. The final result of this procedure is the multitime joint probability for obtaining a sequence of measured values corresponding to the measurements carried out sequentially.

18.4.1 Sequences of Measurements

For simplicity, we will consider only the case where we start with a well-defined initial quantum state. We consider measurements described by measurement operators $E_{\bar{a}}$—for simplicity and compactness, we use a slightly different notation from that introduced in (18.3.1), using a subscript notation, and treating only the case where the measured value determines the measurement and the parameter α is not necessary.

a) Unitary Evolution: During the times between successive measurements the system will evolve unitarily; between the measurements at times t and $t+\tau$,

$$|\psi,t+\tau\rangle = U(t+\tau,t)|\psi,t\rangle. \tag{18.4.1}$$

b) Formulation of Sequential Measurements: The process of measurement and its analysis is an iterative process as follows.

i) We first define the sequences of times and results under consideration:

a) At the initial time t_0 the state of the system is $|\psi_0,t_0\rangle$.

b) The measurements occur at times t_r, where $r=1,2,3,\ldots$, and $t_r > t_{r-1}$.

c) For $r=1,2,3,\ldots$, the result of the r-th measurement is $\bar{a}_r$.

ii) The effect of the unitary evolution between measurements is now introduced.

a) The state of the system just after the $(q-1)$-th measurement depends on all of the previous measurements, and we therefore use a notation which makes this dependence explicit; namely $|\psi_{\bar{a}_1,\bar{a}_2,\ldots\bar{a}_{q-1}},t_{q-1}\rangle$.

b) The state just before the q-th measurement is given by the unitary time evolution from the state just after the $(q-1)$-th measurement, namely

$$|\psi_{\bar{a}_1\bar{a}_2,\ldots\bar{a}_{q-1}},t_q\rangle = U(t_q,t_{q-1})|\psi_{\bar{a}_1\bar{a}_2,\ldots\bar{a}_{q-1}},t_{q-1}\rangle. \tag{18.4.2}$$

c) We now make the measurement at time t_q. The probability of obtaining the result $\bar{a}_q$ at the q-th measurement, given the results of all previous samplings is, according to (18.3.2)

$$P\left(\bar{a}_q\middle|\bar{a}_1,\bar{a}_2,\ldots,\bar{a}_{q-1}\right) = \left\langle\psi_{\bar{a}_1,\ldots\bar{a}_{q-1}},t_q\middle|E^{\dagger}_{\bar{a}_q}E_{\bar{a}_q}\middle|\psi_{\bar{a}_1,\ldots\bar{a}_{q-1}},t_q\right\rangle, \tag{18.4.3}$$

and the wavefunction after the measurement is

$$|\psi_{\bar{a}_1,\ldots\bar{a}_q},t_q\rangle = \frac{E_{\bar{a}_q}|\psi_{\bar{a}_1,\ldots\bar{a}_{q-1}},t_q\rangle}{\sqrt{P(\bar{a}_q|\bar{a}_1,\bar{a}_2,\ldots,\bar{a}_{q-1})}}. \tag{18.4.4}$$

d) We can now iterate these equations to get

$$|\psi_{\bar{a}_1,\ldots\bar{a}_q},t_q\rangle = \frac{\left[\prod_{r=1}^{q}E_{\bar{a}_r}U(t_r,t_{r-1})\right]|\psi_0,t_0\rangle}{\sqrt{P(\bar{a}_1;\bar{a}_2;\ldots;\bar{a}_q)}}, \tag{18.4.5}$$

where the product in the numerator is ordered so that the times increase to the left, and

$$P(\bar{a}_1;\ldots;\bar{a}_q) = \prod_{r=1}^{q}P(\bar{a}_r|\bar{a}_1,\ldots,\bar{a}_{r-1}). \tag{18.4.6}$$

e) Since the state is normalized, we deduce that

$$P(\bar{a}_1;\ldots;\bar{a}_q) = \left\langle \psi_0, t_0 \left| \left[\prod_{r=1}^{q} E_{\bar{a}_r} U(t_r, t_{r-1}) \right]^\dagger \prod_{r=1}^{q} E_{\bar{a}_r} U(t_r, t_{r-1}) \right| \psi_0, t_0 \right\rangle . \tag{18.4.7}$$

18.4.2 Expression as a Correlation Function

We will write the Heisenberg picture operators corresponding to the $E_{\bar{a}}$ as

$$E_{\bar{a}}(t) = U^\dagger(t, t_0) E_{\bar{a}} U(t, t_0). \tag{18.4.8}$$

Using this notation, we can write the formula (18.4.7) as

$$P(\bar{a}_1;\bar{a}_2;\ldots\bar{a}_q) = \left\langle \psi_0, t_0 \left| E^\dagger_{\bar{a}_1}(t_1)\ldots E^\dagger_{\bar{a}_q}(t_q) E_{\bar{a}_q}(t_q)\ldots E_{\bar{a}_1}(t_1) \right| \psi_0, t_0 \right\rangle . \tag{18.4.9}$$

This is a particular kind of correlation function of the operators $E_{\bar{a}}(t)$ with each other, in which there are two sets of products; a sequence of increasing times $t_1 \leqslant t_2 \leqslant t_3 \cdots \leqslant t_q$, and the reverse sequence of decreasing times $t_q \geqslant \cdots t_3 \geqslant t_2 \geqslant t_1$.

18.4.3 General Correlation Functions

The polarization identity

$$A^\dagger M B = \tfrac{1}{4}\Big\{ (A+B)^\dagger M (A+B) - (A-B)^\dagger \, M(A-B) - \mathrm{i}(A+\mathrm{i}B)^\dagger M(A+\mathrm{i}B) + \mathrm{i}(A-\mathrm{i}B)^\dagger M(A-\mathrm{i}B) \Big\}, \tag{18.4.10}$$

enables us to generate more general correlations by linear combinations of probabilities. Provided we have a complete set of operations $E(\bar{a}, t)$ (in the sense that any operator, including the identity, can be expressed in terms of a linear combination of them) we can generate correlation functions of the kind

$$\langle \psi_0, t_0 | A_1(t_1) A_2(t_2) \ldots A_n(t_n) B_m(s_m) \ldots B_2(s_2) B_1(s_1) | \psi_0, t_0 \rangle . \tag{18.4.11}$$

Here the A_i and the B_i are arbitrary operators, and s_i and t_i are times such that

$$t_1 \leqslant t_2 \leqslant \ldots \leqslant t_n, \qquad s_1 \leqslant s_2 \leqslant \ldots \leqslant s_m. \tag{18.4.12}$$

Such correlation functions are generally called *time-ordered correlation functions*, and are the only kind of correlation functions susceptible to direct measurement in this form. The order of the operators in the fundamental result, (18.4.10), represents the order in which the measurements are performed—it is therefore logically mandatory that the time sequence increases in the same order. Other correlation functions, with time sequences not of this form, can be written down, but involve paradoxical sequences of measurements, in which the operator sequence and the time sequence do not correspond.

19. Continuous Measurements

It is intriguing to note that almost every measurement we do in practice involves the observation of photons, be it with the eye or a photodetector. The measurement of photons usually involves counting them, and any detector is a device which waits for a photon to arrive. Experimental physics therefore turns out to involve mainly waiting for something to happen—we cannot order a photon count to take place at a precise time. In practice light intensities can be so large that one does not have to wait for long on the time scale of everyday life, but even so, on the scale of times between successive measurements—the time to detect a given photon—is quite short, and a Markovian description is appropriate.

This chapter will give an outline of how such measurements can be described in terms of the measurement theory of the previous chapter. It turns out that the measurement process involved in photon counting can be described by a quantum Markov process and the corresponding master equation. However, in addition, another formulation of quantum Markov processes known as the *stochastic Schrödinger equation* automatically arises. The stochastic Schrödinger equation is a very powerful tool, and is formulated in detail in *Quantum Noise*, as well as in *Book II.*

19.1 Photon Counting

A photodetector is a device which continuously checks whether or not a photon is present, a process which we can formulate by considering repeated measurements taking place at times t and separated by time intervals dt. For simplicity, let us consider a detector with 100% efficiency, so that there are only two outcomes, a photon is absorbed and detected, or no photon is absorbed.

There are two corresponding measurement operators given by

$$\left.\begin{aligned} E(1) &= a\sqrt{\gamma\,dt}, && \text{(Photon detected)} \\ E(2) &= 1-\tfrac{1}{2}\gamma a^{\dagger} a\,dt, && \text{(No photon detected)} \end{aligned}\right\} \qquad (19.1.1)$$

where a is the photon destruction operator. Notice that

$$E^\dagger(1)E(1) + E^\dagger(2)E(2) = 1 + \tfrac{1}{4}\left(\gamma a^\dagger a\right)^2 dt^2 \approx 1 \text{ to order } dt. \tag{19.1.2}$$

Thus the choices (19.1.1) are correctly normalized in the continuous limit, in which $dt \to 0$.

19.1.1 Master Equation for Continuous Measurement

The density operator after the measurement, assuming the observer does not check whether a photon was detected or not, is the average over both possibilities, that is

$$\rho(t+dt) = E^\dagger(1)\rho(t)E(1) + E^\dagger(2)\rho(t)E(2). \tag{19.1.3}$$

Inserting the explicit forms (19.1.1) and taking the limit $dt \to 0$, this implies that $\rho(t)$ obeys the master equation

$$\frac{\partial \rho(t)}{\partial t} = \tfrac{1}{2}\gamma\left(2a\rho(t)a^\dagger - \rho(t)a^\dagger a - a^\dagger a\rho(t)\right). \tag{19.1.4}$$

Thus the evolution of the density operator as a result of a continuous measurement process is given by the master equation (19.1.4).

a) Inclusion of a Hamiltonian: So far we have ignored what sort of dynamical processes might be influencing the photon mode as the photon counting process proceeds. However, the necessary requirements for measurement operators are also satisfied by a more general form for $E(2)$, namely

$$E(2) = 1 - \left(\frac{i}{\hbar}H + \tfrac{1}{2}\gamma a^\dagger a\right)dt, \tag{19.1.5}$$

where H is any Hermitian Hamiltonian, since to first order in dt this still gives the correct normalization (19.1.2). This Hamiltonian term provides the way of including dynamical processes other than measurement.

We then find that the density operator obeys the general master equation

$$\frac{\partial \rho(t)}{\partial t} = \frac{i}{\hbar}[H, \rho(t)] + \tfrac{1}{2}\gamma\left(2a\rho(t)a^\dagger - \rho(t)a^\dagger a - a^\dagger a\rho(t)\right). \tag{19.1.6}$$

b) Conclusion: The master equation therefore arises naturally as a result of a process of continuous measurement (in this case detection of photons according to (19.1.1) and (19.1.5)) in which the information about the result of a measurement is discarded.

19.1.2 Measurement as a Physical Process

The emergence of the master equation as a consequence of the process of measurement demonstrates the close connection between the concept of measurement, and the concept of elimination of the unobserved modes of the heat bath, which is closely connected to the fact that the physical measurement is a description in terms of macrostates, which themselves are a way of discarding inessential information.

In the derivation of the master equation in Chap. 13, we choose ρ_B to be a stationary solution of the evolution operator $\dot{\rho} = \mathcal{L}_B \rho$, and in particular, we choose a solution corresponding to a thermal state of temperature T, whose correlation functions then take the thermal form as in Sect. 13.2.1, equations (13.2.35) and (13.2.36). We get a master equation of the same kind as that derived above, (19.1.6), when we choose $T = 0$. At non-zero temperature we would also obtain terms in which $a \leftrightarrow a^\dagger$, which would correspond to a hot detector, emitting photons into the mode. In optical situations this is rare, but with microwaves this can be a significant effect.

There is also the possibility that detections which appear like photon detections occur because of processes internal to the detector, yielding a spurious count rate which would give detection of photons even in their absence—this in fact happens (it is called *dark current*) and may lead to the necessity to cool the detector.

19.2 Wavefunction Interpretation of Continuous Measurement

It is possible to develop a simulation method based on the idea of quantum jumps from the formalism of continuous measurement. In this methodology, we develop equations of motion for the *wavefunction* rather than the density operator, and compute physical results as averages over numbers of different stochastic simulations. The procedure was first introduced by *Zoller, Marte* and *Walls* [62], and was later elaborated into a useful and widely employed algorithm by a number of groups [63–65].

19.2.1 Wavefunction Evolution

The evolution of the wavefunction under the process of continuous measurement is straightforwardly worked out. Let us work out the probability of *not* detecting a photon in a finite time interval $(0, t)$, in the case that $E(1)$ is given by (19.1.1), and there is an evolution Hamiltonian, so that $E(2)$ is given by (19.1.5).

If no photon has been detected up to time t, then we can divide the interval $(0, t)$ into infinitesimal subintervals $dt = t/N$, within each of which no photon has been detected. Therefore, the wavefunction at the end of the interval is given by applying the operator

$$E(2) = 1 - \left(\frac{iH}{\hbar} + \tfrac{1}{2}\gamma a^\dagger a\right) dt, \tag{19.2.1}$$

to $|\psi(0)\rangle$ and normalizing the wavefunction at each stage. The end result of this process is easily found;

$$|\psi, t\rangle = \lim_{N\to\infty}\left(\frac{\left(1-\left(\frac{\mathrm{i}H}{\hbar}+\frac{1}{2}\gamma a^\dagger a\right)\frac{t}{N}\right)^N|\psi(0)\rangle}{\left\|\left(1-\left(\frac{\mathrm{i}H}{\hbar}+\frac{1}{2}\gamma a^\dagger a\right)\frac{t}{N}\right)^N|\psi(0)\rangle\right\|}\right), \tag{19.2.2}$$

$$= \frac{\exp\left[-\left(\frac{\mathrm{i}H}{\hbar}+\frac{1}{2}\gamma a^\dagger a\right)t\right]|\psi(0)\rangle}{\left\|\exp(-\left(\frac{\mathrm{i}H}{\hbar}+\frac{1}{2}\gamma a^\dagger a\right)t)|\psi(0)\rangle\right\|}. \tag{19.2.3}$$

19.2.2 The Stochastic Schrödinger Equation

The algorithm for computation of the wavefunction amounts to a description of continuous, but non-unitary evolution of the wavefunction between quantum jumps from one wavefunction to another.

The equation of motion which generates this evolution is called the *stochastic Schrödinger equation.* In *Quantum Noise* this is formulated as a kind of stochastic differential equation, but here we will simply describe the algorithm for generating the evolution. In detail, it takes the form:

i) Compute the wavefunction between photon counts by solving the Schrödinger equation with the *non-Hermitian* Hamiltonian

$$H_{\text{eff}} \equiv \left(H - \tfrac{1}{2}\mathrm{i}\hbar\gamma a^\dagger a\right), \tag{19.2.4}$$

that is, solve

$$\mathrm{i}\hbar\frac{\partial|\varphi, t\rangle}{\partial t} = H_{\text{eff}}|\varphi, t\rangle = \left(H - \tfrac{1}{2}\mathrm{i}\hbar\gamma a^\dagger a\right)|\varphi, t\rangle. \tag{19.2.5}$$

ii) The probability of there being no photon detected up to time t is $\||\varphi, t\rangle\|^2$, and at that time the wavefunction is

$$|\psi, t\rangle_1 = \frac{|\varphi, t\rangle}{\||\varphi, t\rangle\|}. \tag{19.2.6}$$

iii) A simulation procedure using random numbers, similar to that which we gave in Sect. 6.1.4, can be used to determine a random time T at which the first photon is detected.

iv) After the photon has been detected at time T the wavefunction immediately after detection is

$$|\psi, T\rangle_2 = \frac{a|\varphi, T\rangle}{\|a|\varphi, T\rangle\|}. \tag{19.2.7}$$

v) The process continues iteratively—the wavefunction $|\psi, T\rangle_2$ becomes the starting point for a time evolution according to the non-Hermitian Schrödinger equation (19.2.5), which is used to compute the probability distribution of the next photon.

19.2.3 Multiple Detection

To take account of the possibility of multiple kinds of detection happening simultaneously, we introduce instead of (19.1.1)

$$\left.\begin{aligned} E_1(i) &= A_i\sqrt{\gamma_i\,dt}\,, \\ E_2 &= 1-\left(\frac{\mathrm{i}H}{\hbar}+\tfrac{1}{2}\sum_i \gamma_i A_i^\dagger A_i\right)dt\,. \end{aligned}\right\} \tag{19.2.8}$$

This allows the possibility of a number of detections characterized by the operators A_i, whose nature is essentially arbitrary. The kind of detection formulated in Sect. 19.1 envisages only absorption of single photons from a single mode, but this generalization permits us to consider detections from multiple modes, and in ways other than by simple absorption.

a) The Detection Algorithm: This is as formulated in Sect. 19.2.2, but with the modifications:

i) The effective Hamiltonian is

$$H_{\text{eff}} \equiv H - \tfrac{1}{2}\hbar\sum_i \gamma_i A_i^\dagger A_i. \tag{19.2.9}$$

ii) The equation of motion for $|\varphi, t\rangle$ now uses this effective Hamiltonian, and the time T of the next detection is determine from this as above.

iii) The particular kind of detection which takes place is determined with relative probabilities

$$P_i(T) \equiv \gamma_i \langle\varphi, T|A_i^\dagger A_i|\varphi, T\rangle\,. \tag{19.2.10}$$

iv) The particular detection, I, is determined according to these relative probabilities using random numbers, and the new quantum state at time T is

$$|\psi, T\rangle \equiv \frac{A_I|\varphi, T\rangle}{\|A_I|\varphi, T\rangle\|}\,. \tag{19.2.11}$$

v) The process then continues iteratively as above, generating a sequence of detection times and detections of a particular kind.

b) Equivalent Master Equation: The master equation obeyed by the corresponding density operator, analogous to (19.1.6) takes the form

$$\frac{\partial\rho(t)}{\partial t} = \frac{\mathrm{i}}{\hbar}[H,\rho(t)] + \tfrac{1}{2}\sum_i \gamma_i\left(2A_i\rho(t)A_i^\dagger - \rho(t)A_i^\dagger A_i - A_i^\dagger A_i\rho(t)\right). \tag{19.2.12}$$

This density operator can be interpreted as describing the system while it is being measured, and averaging over the results of the measurements. The kinds of operators are arbitrary, and we can make the interpretation that any evolution according to such a master equation can be viewed as the result of measurements being made on the system by an observer who does not communicate to us the results of the measurements.

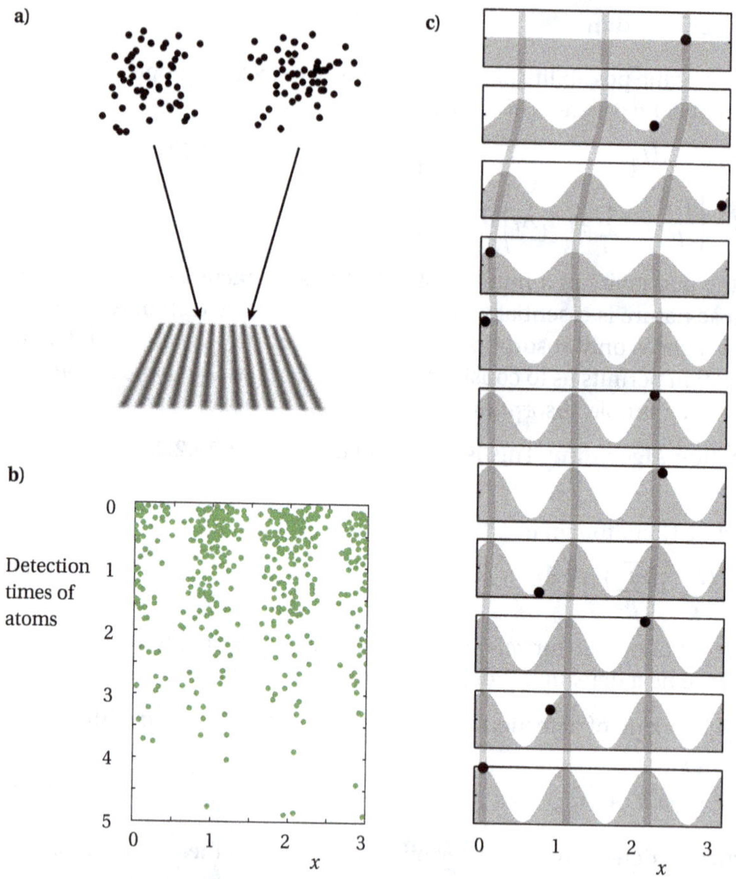

Fig. 19.1. Interference of two Bose–Einstein condensates.

a) Schematic diagram of the experimental setup.

b) Growth of the interference pattern on the detection surface. The rate at which atoms are detected is proportional to the number of undetected atoms, and thus drops off exponentially.

c) Evolution of the probability distribution for the next atom to be detected as the first eleven atoms are detected. The grey lines through the eleven plots show how the position of the first count (marked by a black dot) corresponds precisely to the peak in detection probability for the second count. However, the influence of the detection position of the second count on the probability of the third count is less, although the peak moves in the appropriate direction. As further atoms are detected, the peak position becomes progressively less influenced, and the alignment of the peaks in the probability distribution stabilizes.

19.3 Application to Matter Wave Interference

In Sect. 11.2.2 we considered how quantum fields produce interference patterns in a rather formal manner, in which the observed interference patterns were related to correlation functions of the quantum fields. A more explicit picture of interference can be given by considering the actual process of detection of the field quanta one after the other. We will formulate the description in terms of matter waves, but this is not an essential aspect of the analysis.

This also provides a very nice example of the use of the multiple detection model, where in this case the different detection possibilities correspond to the different possible places where a particle might be detected.

19.3.1 Interference of Independent Bose–Einstein Condensates

Let us now consider dropping two Bose–Einstein condensates in such a way that they expand and eventually overlap, thus producing an interference pattern, as in Fig. 19.1. The particular situation we want to consider is that of *independent condensates*. This is in contrast to textbook descriptions of interference of light beams, where the light which illuminates (for example) a double slit is derived from the same source, so that there is a well-defined phase relation between the light emerging from the two slits.

We will consider ideal condensates in which each consists of the same number, N, of atoms. Since there is no phase associated with a number state, there is no predetermined phase relationship between the condensates. Nevertheless, we will show that as the atoms drop and the clouds overlap, a visible set of interference fringes will emerge, but the location of the fringe pattern is quite random.

a) A Two-Mode Model: We describe the two condensates by destruction operators a and b, and (simplifying to one-dimensional description) write the field operator at the detection surface as

$$\Psi(x) = a\mathrm{e}^{\mathrm{i}k_a x} + b\mathrm{e}^{\mathrm{i}k_b x}. \tag{19.3.1}$$

The model therefore assumes the two condensates are initially well localized and separated, and propagate to the detection surface, giving rise to a superposition with the appropriate phases as written.

b) Detection Model: We will divide the detection surface into intervals of length Δx at positions x_i, and suppose that the length of the surface is L. The operator to detect one particle on the detection surface in the time interval dt, and at position x_i in the space interval is modelled as

$$E_1(i) = \Psi(x_i)\sqrt{\lambda\, dt\, \Delta x/L}. \tag{19.3.2}$$

We can then evaluate

$$\sum_i E_1^\dagger(i)E_1(i) = \lambda\, dt \sum_i \frac{\Delta x}{L}\Big(a^\dagger \mathrm{e}^{-\mathrm{i}k_a x_i} + b^\dagger \mathrm{e}^{-\mathrm{i}k_b x_i}\Big)\Big(a\mathrm{e}^{\mathrm{i}k_a x_i} + b\mathrm{e}^{\mathrm{i}k_b x_i}\Big), \tag{19.3.3}$$

$$\to \lambda\, dt\left(a^\dagger a + b^\dagger b\right), \tag{19.3.4}$$

so that the effective Hamiltonian is

$$H_{\text{eff}} = \left(\hbar\omega - \tfrac{1}{2}\mathrm{i}\hbar\lambda\right)\left(a^\dagger a + b^\dagger b\right). \tag{19.3.5}$$

c) Initial State: The initial state is the two-mode number state $|N,N\rangle$, in which we put, for simplicity, exactly the same number of atoms in each cloud. This does not alter the major features of the modelling.

d) Evolution between Detections: Starting with the initial state $|N,N\rangle$, each detection will reduce the total number of particles by one, so that the system is at all stages in an eigenstate of the total number operator $a^\dagger a + b^\dagger b$, and after K detections, this eigenvalue will be $2N-K$. Hence, the state after K detections, occurring at times T_K, is

$$|\varphi, t\rangle_K = \mathrm{e}^{-(\mathrm{i}\omega+\frac{1}{2}\lambda)(M-K)(t-T_K)}|\varphi, T_K\rangle_K\,. \tag{19.3.6}$$

Since the time development amounts simply to a multiplicative factor, the normalized state does not change as a result of the non-unitary part of the evolution, hence

$$|\psi, t\rangle_K = \mathrm{e}^{-\mathrm{i}\omega(M-K)(t-T_K)}|\psi, T_K\rangle_K\,. \tag{19.3.7}$$

As in Sect. 6.1.4, the times T_K are determined using random numbers Z_K uniformly distributed on $[0,1]$, and the following algorithm:

$$\text{Set} \qquad \tau_K = -\frac{\log Z_K}{\lambda(M-K+1)}, \tag{19.3.8}$$

$$\text{then} \qquad T_K = T_{K-1} + \tau_K. \tag{19.3.9}$$

e) The Sequence of Detections:

i) *First Detection*: For the first detection of an atom in the time interval dt_1, the atom will be detected in the interval (x_1, x_1+dx_1) with probability

$$P_1(x_1) = \langle N,N|\left(a^\dagger \mathrm{e}^{-\mathrm{i}k_a x_1} + b^\dagger \mathrm{e}^{-\mathrm{i}k_b x_1}\right)\left(a\mathrm{e}^{\mathrm{i}k_a x_1} + b\mathrm{e}^{\mathrm{i}k_b x_1}\right)|N,N\rangle \times\lambda\, dx_1\, dt_1, \tag{19.3.10}$$

$$= 2N\lambda\, dx_1\, dt_1\,. \tag{19.3.11}$$

Thus, the position of detection of the first atom is uniformly distributed over space, as is shown in the first plot of Fig. 19.1c.

Let us suppose the atom is actually detected at position x_1, then we will use the notation $|1,x_1\rangle$ for the normalized wavefunction after this detection, and it is given by

$$|1,x_1\rangle \equiv \frac{1}{\sqrt{2N}}\left(a\mathrm{e}^{\mathrm{i}k_a x_1} + b\mathrm{e}^{\mathrm{i}k_b x_1}\right)|N,N\rangle, \tag{19.3.12}$$

$$= \frac{1}{\sqrt{2N}}\left(\mathrm{e}^{\mathrm{i}k_a x_1}\sqrt{N}|N-1,N\rangle + \mathrm{e}^{\mathrm{i}k_b x_1}\sqrt{N}|N,N-1\rangle\right). \tag{19.3.13}$$

This is now an entangled state—the action of detecting a single atom in such a way that it could have come from either condensate has produced this entanglement. After the first detection, the two condensates are no longer independent.

ii) *Second Detection*: We suppose that the second atom is detected in the interval $(x_2, x_2 + dx_2)$; the probability for such a detection is

$$P_2(x_2, x_1) = \langle 1, x_1 | \left(a^\dagger e^{-ik_a x_2} + b^\dagger e^{-ik_b x_2} \right) \left(a e^{ik_a x_2} + b e^{ik_b x_2} \right) | 1, x_1 \rangle \times \lambda\, dx_2\, dt_2 , \tag{19.3.14}$$

$$= 2N\left(2N - 1 + N\cos(k_a - k_b)(x_1 - x_2)\right)\lambda\, dx_2\, dt_2 . \tag{19.3.15}$$

This already displays an interference pattern, with maxima approximately three times as high as the minima, as is shown in the second plot in Fig. 19.1c.

iii) *Subsequent Detections*: It is clear that the detection of the next atom, at a position governed by this probability density, will further modify the probability density governing the detection of the third atom, and so on, until all of the atoms have been detected. The process is straightforward to simulate, with results as shown in Fig. 19.1c. One sees that the sequence of probability densities of detecting atoms rapidly converges to a stable limit with maximum visibility, so that the observed distribution of detected atoms looks like a familiar interference pattern. However, the analytic determination of this interference pattern is not straightforward, and for details the reader should consult the original literature [66].

f) Experimental Realization of the Detection Process: The kind of detection envisaged in this model must be sufficiently fast to enable detection of one atom at a time. This can be done using metastable helium atoms, from which a Bose–Einstein condensate was produced in 2001 [67]. The metastable atoms are in a highly excited state, whose excitation energy is discharged on contact with a multichannel plate detector. This yields a time resolution of the order of nanoseconds.

19.4 Damping of Quantum Coherence

Let us consider a quantum superposition of two coherent states

$$|\psi_{\rm sup}\rangle = |\alpha_0\rangle + |\beta_0\rangle . \tag{19.4.1}$$

If we create such a state, how long can we expect it to survive in the real world, where the oscillator interacts with the rest of the world? An example is a mode of an optical cavity, which cannot avoid interacting with the electromagnetic field on the outside of the cavity. The amplitude of a coherent state will decay at a rate corresponding to the cavity lifetime, and we might hope that the superposition state could be made to live for a similar time.

19.4.1 Stochastic Schrödinger Equation Treatment

Unfortunately, the superposition state is much more fragile than that, and we can use the stochastic Schrödinger equation method to show why. The model of detection given in Sect. 19.2.2 demonstrates the basic principle. Up to the time of the first detection, the coherent state evolution is given

$$|\alpha_0\rangle \to |\alpha_0\rangle_t \equiv \exp\left(-\frac{\mathrm{i}H_{\mathrm{eff}}t}{\hbar}\right)|\alpha_0\rangle, \tag{19.4.2}$$

$$= \exp\left(-(\mathrm{i}\omega + \tfrac{1}{2}\gamma)a^\dagger a\, t\right)|\alpha_0\rangle, \tag{19.4.3}$$

$$= \exp\left(\tfrac{1}{2}|\alpha_0|^2(\mathrm{e}^{-\gamma t}-1)\right)\left|\mathrm{e}^{-(\mathrm{i}\omega+\gamma/2)t}\alpha_0\right\rangle. \tag{19.4.4}$$

To derive this result, we have set the system Hamiltonian $H = \hbar\omega a^\dagger a$, and then used the result of Ex. 10.3 to evaluate the resulting expression (19.4.3). The most important part of this result is that a coherent state remains a coherent state as long as no detection takes place. The amplitude decays and oscillates, and there is a multiplicative factor as well, but this does not affect the conclusion. If the superposition is such that $|\alpha_0| = |\beta_0|$, the multiplicative factor for both components is the same, and this would mean that the superposition of the evolved states would still be a simple sum—thus

$$|\alpha_0\rangle + |\beta_0\rangle \longrightarrow |\alpha_0\rangle_t + |\beta_0\rangle_t. \tag{19.4.5}$$

However when the first detection takes place, at time T, and the state is transformed according to (19.2.7) to

$$\frac{a\big(|\alpha_0\rangle_T + |\beta_0\rangle_T\big)}{\|a\big(|\alpha_0\rangle_T + |\beta_0\rangle_T\big)\|} \propto \alpha_0|\alpha_0\rangle_T + \beta_0|\beta_0\rangle_T. \tag{19.4.6}$$

The two terms of the superposition are now multiplied by different factors, and if α_0 and β_0 are macroscopically different, the superposition has been destroyed.

This happens at the first detection, and the time T at which this happens is only stochastically determined. The typical time of the first detection is given by

$$T \approx \frac{1}{\gamma|\alpha_0|^2}, \tag{19.4.7}$$

since the rate of detection is proportional to γ multiplied by the number of quanta—that is $|\alpha_0|^2$. This means that we can expect a coherent superposition of macroscopically different states to survive only this very small time.

For truly macroscopic states, we would imagine $|\alpha_0|^2$ to possibly of the order of a million, and this would mean that the coherence would be very fragile, being destroyed a million times faster than the decay of the amplitude. For values of the order of magnitude of Avogadro's number, coherence is destroyed essentially instantly.

19.4.2 Interference and the Quantum Characteristic Function

The above argument is only qualitative, and considers only detection of quanta. In this section we will give quantitative results, and consider a system which incorporates gain as well as loss, since amplification is an essential element in quantum measurement theory.

We will use the quantum characteristic function as the basis for the method of calculation, and before doing this, it is worth explicitly writing down how interference is expressed in these terms.

a) Probability Density in Terms of the Quantum Characteristic Function: If we use the relationship (16.3.4) between the destruction operator and the position and momentum operators, then

$$\langle x|\rho|x\rangle = \frac{\kappa}{\pi\sqrt{2\hbar}}\int dv\,\chi_S(\mathrm{i}v,-\mathrm{i}v)\exp\left(-\mathrm{i}\sqrt{\frac{2}{\hbar}}\kappa v x\right). \tag{19.4.8}$$

Exercise 19.1 Probability Density Proof: Prove (19.4.8) by using the definition of the quantum characteristic function (15.1.5) and expressing the trace of any operator R as $\mathrm{Tr}\{R\} = \int dx\,\langle x|R|x\rangle$.

b) Interference Terms: The density operator corresponding to the superposition (19.4.1) is

$$|\alpha_0\rangle\langle\alpha_0| + |\beta_0\rangle\langle\beta_0| + |\beta_0\rangle\langle\alpha_0| + |\alpha_0\rangle\langle\beta_0|. \tag{19.4.9}$$

For simplicity, we will want to consider interference in the case that it is maximal, and this corresponds to the overlapping of two wavepackets in the same position but with opposite momentum. We will also, for simplicity, choose $\kappa = 1$, which means that the coherent state parameters can be written in the form (19.4.1)

$$\alpha_0 = \mathrm{i}\frac{q}{\sqrt{2\hbar}}, \qquad \beta_0 = -\mathrm{i}\frac{q}{\sqrt{2\hbar}}, \tag{19.4.10}$$

which corresponds to a momentum difference $\Delta p = 2q$. In this case, we will compute the quantum characteristic function as the sum of four terms corresponding to those in (19.4.9), and clearly, all we need to do is consider the term $|\alpha_0\rangle\langle\beta_0|$, from which all the other terms can be deduced. For this state, we can use (15.1.1) to show that the normally ordered characteristic function takes the form $\exp(-\lambda^*\alpha_0 + \lambda\beta_0^*)\langle\beta_0|\alpha_0\rangle$, so that the symmetrically ordered characteristic function is

$$\chi_S^{\alpha,\beta}(\lambda,\lambda^*) = \langle\beta_0|\alpha_0\rangle\exp\left(-\lambda^*\alpha_0 + \lambda\beta_0^* - \tfrac{1}{2}|\lambda|^2\right). \tag{19.4.11}$$

The probability distribution can then be evaluated for each of the four terms in (19.4.9) using (19.4.8), and expressed as

$$\langle x|\rho|x\rangle = p^{\alpha,\alpha}(x) + p^{\beta,\beta}(x) + p^{\alpha,\beta}(x) + p^{\beta,\alpha}(x), \tag{19.4.12}$$

$$= \frac{2}{\sqrt{\pi\hbar}}\mathrm{e}^{-x^2/\hbar}\left\{1+\cos\left(\frac{2\mathrm{i}qx}{\hbar}\right)\right\}. \tag{19.4.13}$$

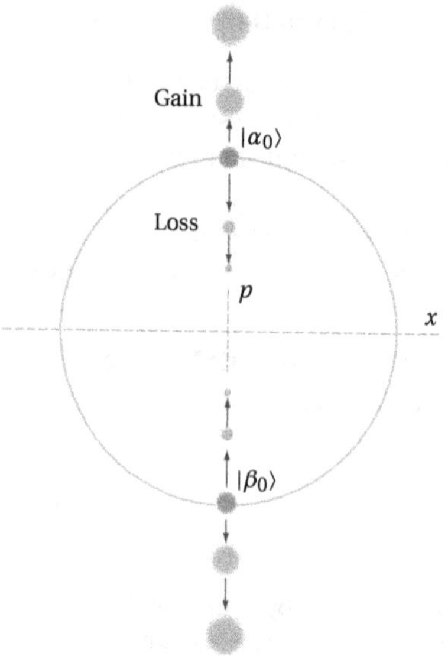

Fig. 19.2. Schematic diagram of loss and gain in a superposition of two coherent states. The harmonic oscillator is viewed in an interaction picture, and the state pictured is one where the wavepackets overlap spatially, but have opposite momenta. The process of *loss* brings the amplitudes $\alpha_0(t)$, $\beta_0(t)$ closer to the origin, and proportionately reduces the spread of the probability distribution. Similarly, the process of *gain* takes the amplitudes $\alpha_0(t)$, $\beta_0(t)$ further from the origin, and proportionately increases the spread of the probability distribution. While the initial distribution corresponds to a quantum superposition of the two coherent states, the distributions at almost all later times are essentially incoherent mixtures. Thus, while interference is initially visible between the two states, this disappears almost immediately.

This shows interference fringes of maximum visibility, arising from the cross terms $p^{\alpha,\beta}(x)$ and $p^{\beta,\alpha}(x)$, with a wavelength corresponding to the momentum difference $\Delta p = 2q$.

19.4.3 The Linear Loss-Gain Model

Let us now consider the time evolution induced by the model of gain or loss given in Sect. 16.2.1. We have in fact already given the solution to this problem in Ex. 16.6, where the time evolution quantum characteristic function was explicitly computed in terms of its initial value; the result is given in (16.2.13), and the behaviour we will discuss is illustrated in Fig. 19.2.

For this situation we choose the initial value to be as given by (19.4.11), so that after a time t, the formula (16.2.13) shows that

$$\chi_S^{\alpha,\beta}(\lambda,\lambda^*,t) = \langle\beta_0|\alpha_0\rangle \exp\left(-\lambda^*\alpha_0(t) + \lambda\beta_0^*(t) - \tfrac{1}{2}|\lambda|^2\right) \times \exp\left(-|\lambda|^2 \frac{\mathcal{N}}{\mathcal{M}-\mathcal{N}}\left(1 - e^{-\gamma(\mathcal{M}-\mathcal{N})t}\right)\right), \tag{19.4.14}$$

$$\alpha_0(t) \equiv \alpha_0 e^{-\gamma(\mathcal{M}-\mathcal{N})t/2}, \qquad \beta_0(t) \equiv \beta_0 e^{-\gamma(\mathcal{M}-\mathcal{N})t/2}. \tag{19.4.15}$$

a) The Case of Pure Attenuation: Here we consider the situation when $\mathcal{N} = 0$, for which the calculation is particularly simple, since the factor in the second line is one. The quantum characteristic function then corresponds to that of

$$\rho^{\alpha,\beta}(t) \equiv g^{\alpha,\beta}(t)|\alpha_0(t)\rangle\langle\beta_0(t)|, \tag{19.4.16}$$

$$\text{where } g^{\alpha,\beta}(t) = \left(\frac{\langle\beta_0|\alpha_0\rangle}{\langle\beta_0(t)|\alpha_0(t)\rangle}\right). \tag{19.4.17}$$

The initial factor in this equation is sufficient to destroy the purity of the quantum state. Notice that this factor evaluates to

$$g^{\alpha,\beta}(t) = \exp\left\{-\left(\tfrac{1}{2}|\alpha_0|^2 + \tfrac{1}{2}|\beta_0|^2 - \beta_0^*\alpha_0\right)\left(1 - e^{-\gamma\mathcal{M}t}\right)\right\}, \tag{19.4.18}$$

and that

$$\left|g^{\alpha,\beta}(t)\right| = \exp\left\{-\tfrac{1}{2}|\alpha_0 - \beta_0|^2\left(1 - e^{-\gamma\mathcal{M}t}\right)\right\}, \tag{19.4.19}$$

$$\longrightarrow \exp\left\{-\tfrac{1}{2}|\alpha_0 - \beta_0|^2\right\} \quad \text{as } t \to \infty. \tag{19.4.20}$$

From this we can see that:

i) If the states $|\alpha_0\rangle$ and $|\beta_0\rangle$ are macroscopically different, the initial superposition is almost totally destroyed when $\gamma\mathcal{M}t \gg 1$, that is

$$\begin{aligned}
&\left(|\alpha_0\rangle + |\beta_0\rangle\right)\left(\langle\alpha_0| + \langle\beta_0|\right) \\
&= |\alpha_0\rangle\langle\alpha_0| + |\beta_0\rangle\langle\beta_0| + |\beta_0\rangle\langle\alpha_0| + |\alpha_0\rangle\langle\beta_0|, && (19.4.21)\\
&\longrightarrow |\alpha_0\rangle\langle\alpha_0| + |\beta_0\rangle\langle\beta_0| + g_\infty^{\alpha,\beta}|\beta_0\rangle\langle\alpha_0| + g_\infty^{\beta,\alpha}|\alpha_0\rangle\langle\beta_0|. && (19.4.22)
\end{aligned}$$

We will consider that the states $|\alpha_0\rangle$ and $|\beta_0\rangle$ are macroscopically different if $|\alpha_0 - \beta_0|^2$ is a number significantly different from 1. So, for example, if $\alpha_0 = \sqrt{10}$ and $\beta_0 = -\sqrt{10}$, then $\left|g_\infty^{\alpha,\beta}\right| = \exp(-20) \approx 2 \times 10^{-9}$.

ii) Alternatively, the typical time τ_c to destroy the superposition is given by that time at which the exponent in (19.4.19) becomes equal to one, namely

$$\tau_c = -\frac{1}{\gamma\mathcal{M}}\log\left(1 - \frac{2}{|\alpha_0 - \beta_0|^2}\right), \tag{19.4.23}$$

$$\approx \frac{1}{\gamma\mathcal{M}}\frac{2}{|\alpha_0 - \beta_0|^2}, \qquad \text{if } |\alpha_0 - \beta_0|^2 \gg 1. \tag{19.4.24}$$

When the states $|\alpha_0\rangle$ and $|\beta_0\rangle$ are macroscopically different, this is considerably shorter than $2/\gamma\mathcal{M}$, the time constant for the decay of the amplitudes $\alpha_0(t)$ and $\beta(t)$. This is a rigorous and more accurate formulation of the result we derived more qualitatively in Sect. 19.4.1.

b) The General Case: When $\mathcal{N} \neq 0$ the analysis is less simple, but we will show that very similar conclusions arise. We will directly compute the interference terms in the same way as we did for the initial state, but using the time-dependent characteristic function (19.4.14). Thus we use the values (19.4.8) for α_0 and β_0, and evaluate the probability density from (19.4.8) as the sum of four terms corresponding to those in (19.4.9).

We find that we can write, for the term arising from $|\alpha_0\rangle\langle\beta_0|$,

$$\langle x|\rho^{\alpha,\beta}(t)|x\rangle = \frac{1}{\sqrt{\pi\hbar\left(r(t)+1\right)}}\exp\left(-\frac{\left(x - u(t) - iv(t)\right)^2}{\hbar\left(r(t)+1\right)}\right)\langle\beta_0|\alpha_0\rangle, \tag{19.4.25}$$

in which we have defined

$$u(t)+\mathrm{i}v(t) \equiv \sqrt{\tfrac{1}{2}\hbar}\,(\alpha_0(t)+\beta_0(t)^*), \tag{19.4.26}$$

$$r(t) \equiv \frac{2\mathcal{N}}{\mathcal{M}-\mathcal{N}}\left(1-\mathrm{e}^{-\gamma(\mathcal{M}-\mathcal{N})t}\right). \tag{19.4.27}$$

Let us now consider the various terms which arise from this expression.

i) *Direct Term*: The explicit values for α_0 and β_0^* in this case are given by

$$\beta_0^* \to \alpha_0^* = -\frac{\mathrm{i}q}{\sqrt{2\hbar}}, \tag{19.4.28}$$

and we find that

$$u(t) \longrightarrow 0, \qquad v(t) \longrightarrow 0, \tag{19.4.29}$$

and the component of the probability distribution is

$$\langle x|\rho^{\alpha,\alpha}(t)|x\rangle = \frac{1}{\sqrt{\pi\hbar\big(r(t)+1\big)}}\exp\left(-\frac{x^2}{\hbar\big(r(t)+1\big)}\right). \tag{19.4.30}$$

ii) *Interference Term*: The explicit values for α_0 and β_0^* in this case are given by

$$\beta_0^* \to -\alpha_0^* = \frac{\mathrm{i}q}{\sqrt{2\hbar}}, \tag{19.4.31}$$

and we find that

$$\langle\beta_0|\alpha_0\rangle \longrightarrow \mathrm{e}^{-q^2/\hbar}, \tag{19.4.32}$$

$$u(t) \longrightarrow 0, \qquad v(t) \longrightarrow \sqrt{2\hbar}\,q\mathrm{e}^{-\gamma(\mathcal{M}-\mathcal{N})t/2}, \tag{19.4.33}$$

and the component of the probability distribution is

$$\langle x|\rho^{\alpha,\beta}(t)|x\rangle = \frac{1}{\sqrt{\pi\hbar\big(r(t)+1\big)}}\exp\left(-\frac{q^2}{\hbar}\right) \times \exp\left(-\frac{x^2-2\mathrm{i}xq\mathrm{e}^{-\gamma(\mathcal{M}-\mathcal{N})t/2}-q^2\mathrm{e}^{-\gamma(\mathcal{M}-\mathcal{N})t}}{\hbar\big(r(t)+1\big)}\right). \tag{19.4.34}$$

This can now be simplified by using the explicit form (19.4.27) for $r(t)$, to yield

$$\langle x|\rho^{\alpha,\beta}(t)|x\rangle = \frac{1}{\sqrt{\pi\hbar\big(r(t)+1\big)}}\exp\left(-\frac{x^2-2\mathrm{i}xq\mathrm{e}^{-\gamma(\mathcal{M}-\mathcal{N})t/2}}{\hbar\big(r(t)+1\big)}\right)\mathcal{F}_{\mathcal{M},\mathcal{N}}(q,t), \tag{19.4.35}$$

where we have defined

$$\mathcal{F}_{\mathcal{M},\mathcal{N}}(q,t) \equiv \exp\left(-\frac{q^2}{\hbar}\frac{(\mathcal{M}+\mathcal{N})\left(1-\mathrm{e}^{-\gamma(\mathcal{M}-\mathcal{N})t}\right)}{\mathcal{M}+\mathcal{N}-2\mathcal{N}\mathrm{e}^{-\gamma(\mathcal{M}-\mathcal{N})t}}\right). \tag{19.4.36}$$

iii) *Probability Distribution*: Adding the four terms gives the final result

$$\langle x|\rho(t)|x\rangle = \frac{2}{\sqrt{\pi\hbar\big(r(t)+1\big)}}\exp\left(-\frac{x^2}{\hbar\big(r(t)+1\big)}\right)$$
$$\times\left\{1+\mathcal{F}_{\mathcal{M},\mathcal{N}}(q,t)\cos\left(\frac{2iqx}{\hbar\big(r(t)+1\big)}\right)\right\}. \tag{19.4.37}$$

c) Behaviour of the Function $\mathcal{F}_{\mathcal{M},\mathcal{N}}(q,t)$: Although the quantitative behaviour of this term depends on the relative values of $\mathcal{M}$ and $\mathcal{N}$, the same qualitative behaviour happens in all cases.

i) *Long Time Behaviour*: This depends on the sign of $\mathcal{M}-\mathcal{N}$:

$$\text{Net Loss}\qquad \mathcal{M}\geq\mathcal{N}:\qquad \mathcal{F}_{\mathcal{M},\mathcal{N}}(q,t)\longrightarrow\exp\left(-\frac{q^2}{\hbar}\right), \tag{19.4.38}$$

$$\text{Net Gain}\qquad \mathcal{M}<\mathcal{N}:\qquad \mathcal{F}_{\mathcal{M},\mathcal{N}}(q,t)\longrightarrow\exp\left(-\frac{q^2}{\hbar}\frac{\mathcal{M}+\mathcal{N}}{2\mathcal{N}}\right). \tag{19.4.39}$$

In all cases the coefficient of the interference term is reduced by a factor similar to that found in the previous subsection for the case of pure attenuation, since $|\alpha_0-\beta_0|^2 = q^2/\hbar$.

ii) *Short Time Behaviour*: When $q^2/\hbar \gg 1$, that is in the case that the two components of the initial superposition are macroscopically different, there is a very rapid initial decay of the form

$$\mathcal{F}_{\mathcal{M},\mathcal{N}}(q,t)\approx\exp\left(-\frac{q^2}{\hbar}(\mathcal{M}+\mathcal{N})t\right),\quad\text{provided }\gamma(\mathcal{M}-\mathcal{N})t\ll 1. \tag{19.4.40}$$

19.4.4 The Stability and Robustness of Coherent States

Individual coherent states display properties of robustness which are in strong contrast to the fragility of a macroscopic superposition of coherent states. We have seen in Sect. 16.2.1 that under both amplification and attenuation, coherent states, or almost-coherent states like Gaussian quantum states, survive amplification and attenuation with quite modest effects on their coherence. Indeed, in Ex. 16.7 we have also shown that amplification and attenuation can be used to make a Gaussian state become more coherent.

On the other hand, the results we have found in the previous subsection show that this robustness is definitely present only for individual coherent states, that a quantum superposition of coherent states is rapidly turned into a mixed state, with essentially no remaining mutual coherence between the two components. The fragility of a superposition is in very strong contrast to the robustness of the individual states.

This tells us that there is something special about coherent states. Our results all depend on the model of loss and gain for the relevant oscillator described by the operators $a^\dagger$ and a, but in fact this is really quite generic—it is hard to imagine any real oscillator which will not experience loss or gain of quanta from the environment, and in fact linear loss and gain are almost universally observed, albeit often in conjunction with other kinds of interaction with the environment, for example phase damping, such as was presented in Sect. 16.2.2. What we see here is the emergence of the macroscopic world, where we are accustomed to see well-defined objects without any effect of quantum superposition. But the most important result is the identification of a particular set of quantum states—coherent states—which can be seen macroscopically, but for which quantum superpositions are essentially forbidden.

19.5 The Emergence of the Macroscopic World

Quantum mechanics itself, as defined by the fundamental postulates such as we have set down in Sect. 18.1.1, does not tell us anything about the structure of the world. To do that requires more than these postulates, it requires a definite specification of the kinds of objects which do exist, and the kinds of interactions between them. The fundamental postulates of quantum mechanics are best seen as defining the way information is structured in the world, but do not specify the world itself. They assert the existence of a Hilbert space and a Hamiltonian, but do not specify the structure of the Hilbert space, or the form of the Hamiltonian.

The emergence of the macroscopic world is not a consequence of the fundamental postulates themselves, but also of the model of the world to which we apply the postulates. In the previous section, we have been able to demonstrate the emergence of a macroscopic world of well-defined coherent states, with no superpositions allowable. The picture that arises is the consequence of the three components present in the model we have chosen, namely:

i) The harmonic oscillator, described by the operators $a^\dagger$ and a.

ii) A generic heat bath, which generates transfer of quanta into and out of the bath.

iii) The kinds of time scales and interaction strengths between the bath and the oscillator, which lead to the master equation describing attenuation and gain.

19.5.1 Modelling Quantum Measurement

The process of quantum measurement is equivalent to the transfer of information about microscopic systems to information about macroscopic systems. As an example, we can consider a model consisting of two harmonic oscillators, one microscopic and one which can become macroscopically occupied.

The model is a kind of quantum mechanical amplifier, described by the Hamiltonian

$$H = -i\hbar a^\dagger a\left(\epsilon^* b - \epsilon b^\dagger\right). \tag{19.5.1}$$

This model has two parts, consisting of two independent sets of harmonic oscillator creation and destruction operators:

i) The *system*, described by the operators a, $a^\dagger$, whose state we wish to measure.

ii) The *meter*, described by the operators b, $b^\dagger$, whose state will become macroscopically occupied, in a way which is correlated with the initial state of the system.

a) Transfer of Information: We take an initial number state $|n\rangle$ of the system, and an initial vacuum state $|0\rangle$ of the meter, and let the joint system evolve according to the Hamiltonian (19.5.1) for a time T_{meas}, so that joint quantum state becomes

$$|\Psi, T_{\text{meas}}\rangle = \exp\left(a^\dagger a(\epsilon^* b - \epsilon b^\dagger) T_{\text{meas}}\right)|n\rangle \otimes |0\rangle, \tag{19.5.2}$$

$$= |n\rangle \otimes |\alpha_n(T_{\text{meas}})\rangle, \tag{19.5.3}$$

$$\text{in which } \alpha_n(T_{\text{meas}}) \equiv n\epsilon T_{\text{meas}}. \tag{19.5.4}$$

Here, $|\alpha_n(T_{\text{meas}})\rangle$ is a coherent state, and we have used the defining formula (10.5.4) for a coherent state to arrive at the result. Thus, the evolution does not change the number state $|n\rangle$ being measured, but produces a coherent state of the meter with amplitude proportional to n.

We could start with a more general state of the system, say $\sum a_n|n\rangle$, or even more generally, an initial density operator

$$\rho(0) \equiv \sum p_{mn}|m\rangle\langle n| \otimes |0\rangle\langle 0|. \tag{19.5.5}$$

Under time evolution, this would produce

$$\rho(T_{\text{meas}}) = \sum_{m,n} p_{m,n}|m\rangle\langle n| \otimes \big|\alpha_m(T_{\text{meas}})\big\rangle\big\langle\alpha_n(T_{\text{meas}})\big|. \tag{19.5.6}$$

b) The Influence of Decoherence: In practice the meter will be influenced by environment induced decoherence, either as a result of amplification or attenuation. If the values of $\alpha_n(T_{\text{meas}})$ are macroscopic, then this will, as we have seen in Sect. 19.4 destroy all coherences. Thus, the final measured density operator will be transformed to

$$\rho(T_{\text{meas}}) \longrightarrow \rho_{\text{meas}} \equiv \sum_n p_{n,n}|n\rangle\langle n| \otimes \big|\alpha_n(T_{\text{meas}})\big\rangle\big\langle\alpha_n(T_{\text{meas}})\big|. \tag{19.5.7}$$

This represents the usual conception of a quantum measurement, where the possible values of the macroscopic variable $\alpha_n(T_{\text{meas}})$ are directly correlated with the microscopic values n.

c) The Measurement Procedure: This can be done in various ways, but one would expect that it would take place in three stages:

i) The system states $|n\rangle$ to be measured would be prepared first, by some method which does not involve the measurement process.

ii) The interaction between the meter and the system would be turned on for a time $T_{\rm meas}$ which is quite short, since the state to be measured may not live very long. The time would only need to be long enough to produce values of $\alpha_n(T_{\rm meas})$ corresponding to a modest number of quanta. Environmental interactions could cause some decoherence during this process.

iii) Next, a strong amplification process would be applied, producing truly macroscopic and fully decoherent states $|\alpha_n(T_{\rm meas})\rangle\rangle$, which are macroscopically measurable.

19.5.2 Zurek's Formulation of Quantum Measurement Theory

In 1981 *Zurek* [68] presented a seminal paper in which he showed, by different means from those we have used, that the essence of quantum measurement lay in the existence of environmentally robust states, such as the coherent states we have been dealing with. He called the variables corresponding to these states, such as $\alpha_n(T_{\rm meas})$, *pointer variables.* Thus he gave a very elegant formalization of the essentials of quantum measurement theory, in a form which can be naturally related to the decoherence which occurs during a measurement.

19.5.3 Relationship to van Kampen's Formulation of Measurement Theory

Van Kampen's down-to-earth formulation of quantum measurement theory that we outlined in Sect. 18.2.4b does not give any well-defined prescription for choosing what kind of physical model can in practice produce a macroscopic result which can then be said to constitute a measurement. In the case of the alpha particle in a cloud chamber, the macroscopic quantities observed are the droplets of water. A droplet of water condensed at a given position, specifiable to some macroscopic accuracy, indicates that an excited atom was produced there. The specification in phase space that this happened is, of course, very imprecise.

However, it is clear that van Kampen's idea of a measured result being a region in phase space, macroscopically tiny, but microscopically enormous, is directly relate to the idea of a macroscopic pointer variable, whose robustness is assured by its interaction with the enormous number of variables in a heat bath. The pointer states are by their very nature microscopically imprecise. Similarly, the idea of metastability is really a requirement that there be a gain medium, to generate the macroscopic variable from the microscopic variable being measured. The two conceptions of measurement theory are therefore really much the same.

20. The Quantum Zeno Effect

Quantum measurement is a physical process, which modifies the state of the system under observation, and a surprising consequence of this is the effect known as the *quantum Zeno effect*, in which continual observation of a physical quantity can prevent the evolution of that quantity with time. That the measurement process can affect the measured value of a physical quantity is not surprising, but the physical consequences can look very counterintuitive. For example, *Misra* and *Sudarshan*, the creators of the terminology "quantum Zeno effect", showed [69] that continual measurement of the state of an unstable nucleus could prevent it ever decaying. The essence of what happens is not mysterious; if the measurement is made very frequently, this amounts to a very strong interaction with the system to be measured, an interaction so strong that it overwhelms other processes, even the ability of a nucleus to decay. The reality of the quantum Zeno effect was demonstrated using ion traps [70] by *Itano, Heinzen, Bollinger* and *Wineland*, and more recently [71] in a microwave context using Rydberg atoms by *Bernu, Deléglise, Sayrin, Kuhr, Dotsenko, Brune, Raimond* and *Haroche*.

20.1 Theoretical Basis for the Quantum Zeno Effect

The effect is quite simple to explain in terms of basic quantum mechanics, and formal measurement theory. Suppose the variable we wish to measure has the operator X, with a complete set of discrete eigenstates $|x_i\rangle$. To measure X we use the complete set of measurement operators

$$\Phi(i) = |x_i\rangle\langle x_i|. \tag{20.1.1}$$

These correspond to measurements of the first kind, and clearly are complete in the sense (18.3.6).

Now suppose that the system Hamiltonian is H, and that we perform repeated measurements of X, in fact we do this N times in a time interval T. Let us suppose also that the system is initially in the eigenstate $|x_0\rangle$ of X. For sufficiently large N, the state immediately before the first measurement is, to second order in $\tau \equiv T/N$,

$$|\psi,1\rangle \equiv \left(1 - \frac{\mathrm{i}H\tau}{\hbar} - \frac{H^2\tau^2}{\hbar^2}\right)|x_0\rangle. \tag{20.1.2}$$

After the measurement, the probability that the system is in the state $|x_0\rangle$ is

$$p_1 = \left|\langle x_0|\psi,1\rangle\right|^2 = 1 - \left(\frac{\tau}{\hbar}\right)^2 \mathrm{var}\,[H]_0, \quad \text{to second order in } \tau. \tag{20.1.3}$$

Here we have used the notation

$$\mathrm{var}\,[H]_0 \equiv \langle x_0|H^2|x_0\rangle - |\langle x_0|H|x_0\rangle|^2. \tag{20.1.4}$$

Now repeat the process N times in the time interval T; that is, let the system evolve unitarily according to the Hamiltonian H between each of N measurements at the ends of the intervals. The probability that the system is in state $|x_0\rangle$ at the end of this process is

$$p_N = \left(1 - \left(\frac{\tau}{\hbar}\right)^2 \mathrm{var}\,[H]_0\right)^N, \tag{20.1.5}$$

and as the number N of measurements becomes larger (with a fixed time interval T) the limiting probability becomes

$$\lim_{N\to\infty} p_N = \lim_{N\to\infty}\left(1 - \frac{T^2 \mathrm{var}\,[H]_0}{N^2\hbar^2}\right)^N = 1. \tag{20.1.6}$$

Mathematically there is no mystery here, the probability to stay in the state depends quadratically on $1/N$, while the number of measurements depends linearly on N. If the system was not measured during the interval, the change in *amplitude* under Hamiltonian evolutions would depend linearly on N^{-1}, and after N time intervals, this would give rise to a finite change in the amplitude to remain in the state $|x_0\rangle$. Thus, the effect of repeated measurements is to suppress the underlying Hamiltonian time evolution.

The naming of this effect is related to the classical paradox of *Zeno*, who proposed that, because time is made up of instants, and an arrow cannot move any distance in an instant, therefore an arrow can never move at all. Zeno's paradox was solved with the invention of calculus and measure theory, and is seen nowadays as a mathematical issue. However, to the extent that mathematics is an abstraction of the real world, it is nevertheless a physical issue also.

The quantum Zeno effect is really of a different nature, being a result of the fact that measurement is a physical process and has physical effects on the system being measured. If the measurement postulates of quantum mechanics, and in particular the collapse of the wavefunction, are seen as being truly additional postulates, rather than consequences of the physics of quantum mechanical time evolution, then the quantum Zeno effect can be seen as a paradox. However, the truth is much simpler—the imposition of a series of rapid instantaneous measurements amounts to a significant change in the dynamics of the system being observed. The original dynamics becomes overwhelmed by the measurement interaction, and ceases to have any effect.

20.1.1 Connection with Continuous Measurement Theory

Our formulation of continuous measurement theory in Chap. 19 is similar to the concept of repeated measurements, but differs in that the measurement operators depend on the time interval dt.

Let us look at a continuous measurement situation analogous to the repeated measurement problem above. In any time interval dt we would have the measurement operators

$$\Phi(1,i,dt) = \sqrt{\gamma_i\,dt}\,|x_i\rangle\langle x_i|, \tag{20.1.7}$$

$$\Phi(2,dt) = 1-\tfrac{1}{2}\sum_i \gamma_i |x_i\rangle\langle x_i|\,dt. \tag{20.1.8}$$

The operators in the first set correspond to succeeding in measuring a result x_i, and in the time interval dt this happens with a probability proportional to dt. The second operator represents the case where no result is found. If we use a notation

$$\Gamma_i = \sqrt{\gamma_i}\,|x_i\rangle\langle x_i|, \tag{20.1.9}$$

then, as in Sect. 19.1.1, the density operator obeys the master equation

$$\frac{\partial\rho(t)}{\partial t} = \frac{\mathrm{i}}{\hbar}[H,\rho(t)] - \tfrac{1}{2}\sum_i \left[\Gamma_i,[\Gamma_i,\rho(t)]\right]. \tag{20.1.10}$$

a) Suppression of Off-Diagonal Elements of the Density Operator: Using this master equation, the equations of motion for the matrix elements $\rho_{lm} \equiv \langle x_l|\rho|x_m\rangle$ of the density operator are

$$\frac{\partial\rho_{lm}}{\partial t} = -\frac{\mathrm{i}}{\hbar}\langle x_l|[H,\rho]|x_m\rangle - \tfrac{1}{2}(\gamma_l+\gamma_m)\rho_{lm}, \qquad \text{for } l\neq m, \tag{20.1.11}$$

$$\frac{\partial\rho_{ll}}{\partial t} = -\frac{\mathrm{i}}{\hbar}\langle x_l|[H,\rho]|x_l\rangle. \tag{20.1.12}$$

If all of the γ_i are sufficiently large then the first equation shows that the off-diagonal terms are rapidly damped to zero, leaving a diagonal density operator. Then, if there are no off-diagonal elements, the right-hand side of the second equation also vanishes, so the result of the measurement process is to set

$$\rho_{lm} \to 0, \qquad l\neq m, \tag{20.1.13}$$

$$\rho_{ll} \to \rho_{ll}(0). \tag{20.1.14}$$

b) Equivalence of a Strong Continuous Measurement to a Standard Measurement: This is exactly what happens with the measurement operators (20.1.1) we used in discussing the quantum Zeno effect, where the density operator after measurement is

$$\rho_{\text{meas}} = \sum \Phi^\dagger(i)\rho\Phi^\dagger(i) = \sum_i \rho_{ii}|x_i\rangle\langle x_i|. \tag{20.1.15}$$

We can therefore conclude that a strong continuous measurement gives the same result as the corresponding standard measurement given by (20.1.1), but acts on a time scale of the order of $1/\gamma_i$. In practice most measurements involving atoms and light are the result of strong continuous measurements which can be switched on and off by some means, such as by applying a strong laser pulse for a short time.

20.2 A Quantum Model of Trapped Atoms

In Sect. 9.2.1 we considered a model of trapped atoms, in which atoms were transferred probabilistically from one state to the other, that is, a classical stochastic model. Here we want to consider the analogous quantum problem.

The model has the same interpretation however. The state $|n\rangle$, where $n = 0, 1$ or 2 is a state with n trapped atoms. The idea is that when there are two atoms in the trap, the strong interactions between the atoms cause the loss of both atoms by means of some irreversible process. This could, for example, be a spin flip to an untrapped state, which leads to the atom moving rapidly away from the trapped position.

The quantum model of trapped atoms differs from the stochastic model in the nature of the processes $0 \leftrightarrow 1$ and $1 \leftrightarrow 2$, which in our quantum mechanical model are *coherent* processes—this could occur in a situation in which there is an optical lattice composed of adjacent traps, and the coherent processes are generated by barrier penetration between adjacent wells. On the other hand, the process $2 \to 0$ is regarded as purely dissipative.

This problem can be treated by almost the same formalism as the classical situation, involving projectors and elimination of the fast variable. However, the description is now in terms of a density operator ρ, and the operators

$$a \equiv |0\rangle\langle 1|, \qquad b \equiv |1\rangle\langle 2|, \qquad d \equiv |0\rangle\langle 2|, \tag{20.2.1}$$

and their Hermitian conjugates, which correspond to the transitions shown in Fig. 20.1. The equation of motion takes the form

$$\dot{\rho} = (L_a + L_b + L_d)\,\rho, \tag{20.2.2}$$

in which

$$L_a\rho = \mathrm{i}g\left[a + a^\dagger, \rho\right], \tag{20.2.3}$$

$$L_b\rho = \mathrm{i}f\left[b + b^\dagger, \rho\right], \tag{20.2.4}$$

$$L_d\rho = w\left(2d\rho d^\dagger - \rho d^\dagger d - d^\dagger d\rho\right). \tag{20.2.5}$$

Thus:

i) The Liouvillians L_a and L_b correspond to coherent transport of atoms from elsewhere, where for notational convenience, the process which changes the atom number from $0 \leftrightarrow 1$ is called L_a, and the process which changes the atom number from $1 \leftrightarrow 2$ is called L_b.

ii) The Liouvillian L_d is a purely dissipative term which describes some incoherent process which results from the interaction of two atoms. This could be caused physically by the formation of a molecule which is itself unstable towards the formation of some other state which is not trapped.

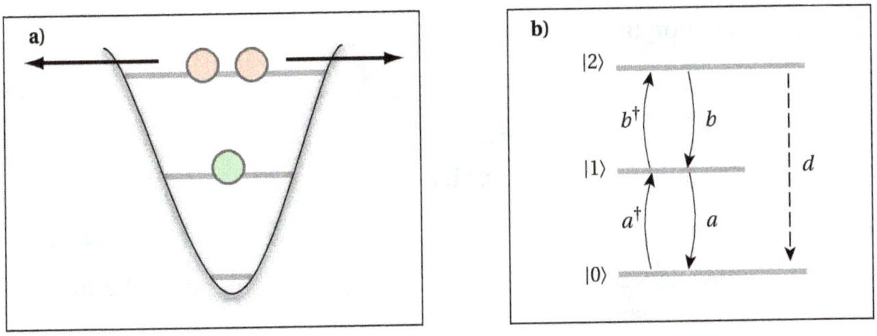

Fig. 20.1. Quantum processes for atoms in a trap. **a)** Repulsive forces between the atoms raise the energy level sufficiently to enable both of them to escape from the trap; **b)** The processes represented by a, $a^\dagger$, b and $b^\dagger$ are the coherent quantum mechanical processes (20.2.3), (20.2.4), while that represented by d is the incoherent process (20.2.5).

20.2.1 Stationary State and Projectors

As in the stochastic model of trapped atoms introduced in Sect. 9.2.1, we want to consider the limit in which the term L_d becomes very strong, and we will solve in terms of the stationary state of L_d, and $\mathcal{P}$, the projector onto this stationary state. As in that section, we introduce the compact notation

$$L_u \equiv L_a + L_b, \tag{20.2.6}$$

and the equation of motion for $v(t) \equiv \mathcal{P}\rho(t)$ is

$$\frac{\partial v(t)}{\partial t} = (\mathcal{P}L_u\mathcal{P})v(t) + (\mathcal{P}L_u\mathcal{Q})\big(s - \mathcal{Q}L_u - wL_1\big)^{-1}(\mathcal{Q}L_u\mathcal{P})v(t), \tag{20.2.7}$$

and in the limit of large w, in which L_d becomes dominant, this becomes

$$\boxed{\frac{\partial v(t)}{\partial t} = (\mathcal{P}L_u\mathcal{P})v(t) - \frac{1}{w}(\mathcal{P}L_u\mathcal{Q})L_1^{-1}(\mathcal{Q}L_u\mathcal{P})v(t).} \tag{20.2.8}$$

We now need to calculate the stationary state, and construct the projector $\mathcal{P}$.

a) Stationary State of L_d: The action of L_d on ρ can be explicitly calculated as

$$L_d\rho\big|_{22} = -2w\rho_{22}, \tag{20.2.9}$$
$$L_d\rho\big|_{21} = -w\rho_{21}, \tag{20.2.10}$$
$$L_d\rho\big|_{12} = -w\rho_{12}, \tag{20.2.11}$$
$$L_d\rho\big|_{20} = -w\rho_{20}, \tag{20.2.12}$$
$$L_d\rho\big|_{02} = -w\rho_{02}, \tag{20.2.13}$$
$$L_d\rho\big|_{00} = 2w\rho_{22}, \tag{20.2.14}$$
$$L_d\rho\big|_{11} = L_d\rho\big|_{10} = L_d\rho\big|_{01} = 0. \tag{20.2.15}$$

The stationary solution then requires

$$\rho_{s,22} = \rho_{s,21} = \rho_{s,12} = \rho_{s,20} = \rho_{s,02} = 0, \tag{20.2.16}$$

$$\rho_{s,11}, \quad \rho_{s,01}, \quad \rho_{s,10}, \quad \text{with arbitrary values.} \tag{20.2.17}$$

To conserve the trace of $\rho_s = 1$, we must have

$$\rho_{s,00} = 1 - \rho_{s,11}. \tag{20.2.18}$$

b) Projector onto the Stationary State: The projector is found by the formula introduced in (8.1.36), namely

$$\mathcal{P} = \lim_{t\to\infty} \exp(L_d t). \tag{20.2.19}$$

To implement this we have to solve the equation $\dot{\rho} = L_d \rho$, and find the stationary state as the limit $t \to \infty$ for an arbitrary initial condition. The result is

$$\mathcal{P}\rho \to \mathcal{P}_{0,1}\rho\mathcal{P}_{0,1} + |0\rangle\langle 2|\rho|2\rangle\langle 0|, \tag{20.2.20}$$

in which

$$\mathcal{P}_{0,1} \equiv |1\rangle\langle 1| + |0\rangle\langle 0|. \tag{20.2.21}$$

Physically what happens is much the same as the classical example; the evolution under L_2 gives a transient behaviour during which the population in level 2 is transferred to level 0.

Exercise 20.1 Time-Dependent Solutions: Solve the equation $\dot{\rho} = L_d \rho$, and find the stationary state as the limit $t \to \infty$. Explicitly show that the projector is given by (20.2.20).

20.2.2 The Master Equation in the Strong Dissipation Limit

a) Evaluation of $\mathcal{P}L_a\mathcal{P}\rho$: The evolution operator L_a involves only the operators a and $a^\dagger$, which operate only within the $[0,1]$ subspace. The projected operator $\mathcal{P}\rho$ has non-zero elements only within this subspace, so it is clear that

$$\mathcal{P}L_a\mathcal{P}\rho = L_a\mathcal{P}\rho, \tag{20.2.22}$$

$$= ig\left[a + a^\dagger, \mathcal{P}\rho\right]. \tag{20.2.23}$$

This then gives the evaluation of the first term in (20.2.8).

b) Other Terms Involving L_a: There are other terms involving L_a in the second (dissipative) part of (20.2.8). From (20.2.22) it follows that $\mathcal{Q}L_a\mathcal{P} = 0$. This means that no term involving L_a will arise from the term $\mathcal{Q}L_u\mathcal{P}$ in (20.2.8). The part involving $\mathcal{P}L_a\mathcal{Q}$ does not *automatically* vanish, but when the explicit forms of the operators a, b, d are inserted, it is in fact zero.

This means that we can now evaluate the dissipative term by setting $L_u \to L_b$.

c) Evaluation of $\mathcal{P}L_b\mathcal{P}\rho$: We first calculate

$$L_b\mathcal{P}\rho = \mathrm{i}f\left[b+b^\dagger,\mathcal{P}\rho\right], \tag{20.2.24}$$

$$= \mathrm{i}f\left(|2\rangle\langle 1|\rho\mathcal{P}_{0,1} - \mathcal{P}_{0,1}\rho|1\rangle\langle 2|\right). \tag{20.2.25}$$

By definition $b|0\rangle = b^\dagger|0\rangle = 0$, so the second term in (20.2.20) has no effect in the evaluation of this term. Since the only non-vanishing parts of this expression are the off-diagonal elements of the rows or columns labelled by 2, applying the projector gives $\mathcal{P}L_b\mathcal{P}\rho = 0$, and hence

$$\mathcal{Q}L_b\mathcal{P}\rho = L_b\mathcal{P}\rho, \tag{20.2.26}$$

$$\mathcal{P}L_b\mathcal{Q}\rho = \mathcal{P}L_b\rho. \tag{20.2.27}$$

d) Action of L_d^{-1}: The matrix represented by

$$L_b\mathcal{P}\rho \propto |2\rangle\langle 1|\mathcal{P}\rho - \mathcal{P}\rho|1\rangle\langle 2| \tag{20.2.28}$$

has non-zero elements only in the positions $(2,1)$, $(2,0)$, $(1,2)$ and $(0,2)$. Using (20.2.27), and the explicit form of the action of L_d given in (20.2.10–20.2.13), we can see that the action of L_d on $L_b\mathcal{P}\rho$ is simply to multiply by $-w$; hence in (20.2.8) we can simply make the replacement $L_d^{-1} \to -1/w$.

e) Evaluation of $\mathcal{P}L_bL_d{}^{-1}L_b\mathcal{P}\rho$: It is now unnecessary to write $\mathcal{Q}$ in the dissipative part. For simplicity in this part let us use the notation

$$\bar\rho = \mathcal{P}\rho. \tag{20.2.29}$$

We now need to evaluate

$$L_bL_b\bar\rho = -f^2\left[b+b^\dagger,\left[b+b^\dagger,\bar\rho\right]\right]. \tag{20.2.30}$$

Since $\bar\rho$ has no elements in either the column or the row labelled 2, we know that $b\bar\rho = \bar\rho b^\dagger = 0$, so that

$$\mathcal{P}\{L_bL_b\bar\rho\} = -f^2\mathcal{P}\left\{2b^\dagger\bar\rho b - bb^\dagger\bar\rho - \bar\rho bb^\dagger\right\}. \tag{20.2.31}$$

Notice now that $bb^\dagger = |1\rangle\langle 1| = a^\dagger a$, and from the form (20.2.20),

$$\mathcal{P}\left\{b^\dagger\bar\rho b\right\} = a\bar\rho a^\dagger. \tag{20.2.32}$$

Thus we conclude in the end that

$$\mathcal{P}\{L_bL_d^{-1}L_b\bar\rho\} = -\frac{f^2}{w}\left\{2a\bar\rho a^\dagger - a^\dagger a\bar\rho - \bar\rho a^\dagger a\right\}. \tag{20.2.33}$$

f) Interpretation as the Quantum Zeno Effect: The resulting equation of motion is then

$$\frac{\partial\bar\rho}{\partial t} = \mathrm{i}g\left[a+a^\dagger,\bar\rho\right] + \frac{f^2}{w}\left\{2a\bar\rho a^\dagger - a^\dagger a\bar\rho - \bar\rho a^\dagger a\right\}. \tag{20.2.34}$$

In the limit that the loss rate w is very large, the second term is negligible, and we see that the system is described by a simple Hamiltonian within the space

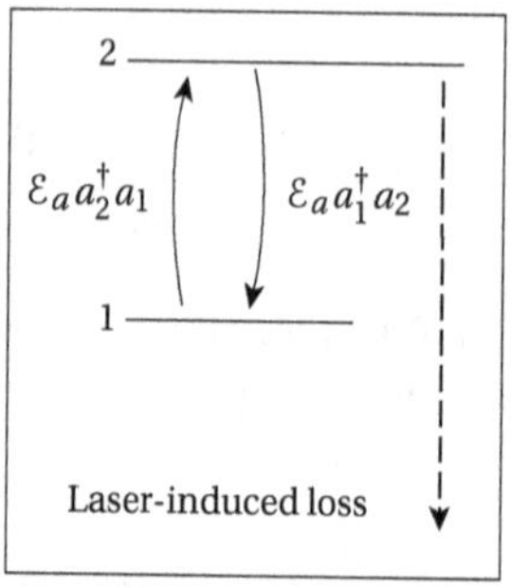

Fig. 20.2. Schematic of the implementation of coherent transfer of atoms within a Bose–Einstein condensate between levels, with dissipative loss from the second level.

spanned by the states $|0\rangle$ and $|1\rangle$—the strong loss process stabilizes this subsystem.

One can view the loss process as a destructive measurement of whether there are two atoms present or not. The constant measurement prevents there ever being more than a single atom or no atoms present, and all time development takes place within the $(0,1)$ subspace.

20.3 Quantum Zeno Effect for a Bose–Einstein Condensate

The Quantum Zeno effect has been investigated experimentally using the hyperfine states of a Bose–Einstein condensate of ^{87}Rb atoms [72]. We consider a Bose–Einstein condensate of magnetically trapped atoms which can exist in two hyperfine states, represented by the Boson destruction operators a_1 and a_2, as illustrated in Fig. 20.2. A continuous measurement of the number of a_2 atoms can be implemented by applying a laser field which transfers atoms from this state to a third untrapped state, so that these atoms are then lost from the system. The two hyperfine states can be driven by a resonant electromagnetic field, so that the resulting Hamiltonian in the interaction picture is

$$H = \hbar g \mathcal{E}_a (a_2^\dagger a_1 + a_1^\dagger a_2), \tag{20.3.1}$$

and the measurement process is implemented by measurement operators of the form (19.1.1), leading to an equation of motion with a dissipative term of the usual form

$$\dot{\rho} = \mathrm{i} g \mathcal{E}_a \left[a_2^\dagger a_1 + a_1^\dagger a_2, \rho \right] + \tfrac{1}{2}\gamma (2 a_2 \rho a_2^\dagger - a_2^\dagger a_2 \rho - \rho a_2^\dagger a_2). \tag{20.3.2}$$

The process is illustrated in Fig. 20.2.

20.3.1 P-Representation Solution

This equation is best solved using a P-representation, as described in Chap. 17. The equations of motion for the P-representation variables α_1 and α_2 have no noise terms, and are simply

$$\dot{\alpha}_1 = -ig\mathcal{E}_a\alpha_2, \tag{20.3.3}$$

$$\dot{\alpha}_2 = -ig\mathcal{E}_a\alpha_1 - \gamma\alpha_2. \tag{20.3.4}$$

a) Solutions of the Equations of Motion: For sufficiently large γ the solutions of these equations can be written in terms of two real eigenvalues ("fast" and "slow")

$$\lambda_f = \frac{-\gamma - \sqrt{\gamma^2 - 4g^2\mathcal{E}_a^2}}{2} \longrightarrow -\gamma + \frac{g^2\mathcal{E}_a^2}{\gamma} \quad \text{as } \gamma \to \infty, \tag{20.3.5}$$

$$\lambda_s = \frac{-\gamma + \sqrt{\gamma^2 - 4g^2\mathcal{E}_a^2}}{2} \longrightarrow -\frac{g^2\mathcal{E}_a^2}{\gamma} \quad \text{as } \gamma \to \infty. \tag{20.3.6}$$

As $\gamma \to \infty$ the "fast" eigenvalue $\lambda_f \to -\infty$, while the "slow" eigenvalue $\lambda_s \to 0$. The corresponding eigenvectors are

$$u_f = \begin{pmatrix} 1 \\ -ig\mathcal{E}_a/\gamma \end{pmatrix}, \qquad u_s = \begin{pmatrix} -ig\mathcal{E}_a/\gamma \\ 1 \end{pmatrix}. \tag{20.3.7}$$

b) Experimental Procedure: The Bose–Einstein condensate is initially in a state in which all atoms are in the state 1. The driving field is first turned on for a time T, but not the measurement field, so that $\mathcal{E}_a \neq 0$ and $\gamma = 0$. In that case there is a Rabi oscillation of the amplitudes at a frequency $g\mathcal{E}_a$, and the state develops non-zero values for both fields. Hence the state can be expressed in terms of the eigenvectors u_s, u_f so that we can write

$$\begin{pmatrix} \alpha_1(T) \\ \alpha_2(T) \end{pmatrix} \equiv Au_s + Bu_f. \tag{20.3.8}$$

c) Measurement Process: At this stage the measurement field can be turned on so that the solution is of the form

$$\begin{pmatrix} \alpha_1(t) \\ \alpha_2(t) \end{pmatrix} \approx Au_s e^{-\lambda_s(t-T)} + Bu_f e^{-\lambda_f(t-T)}. \tag{20.3.9}$$

After a time of the order of $1/\lambda_f$ the component proportional to u_2 is essentially zero, and since $\gamma \gg g\mathcal{E}_a$, this means that the measurement process has detected the presence of some atoms in state $|2\rangle$, and removed almost all of them.

d) Interpretation as the Quantum Zeno Effect: The population in state $|1\rangle$ is almost untouched, and decays only on the slow time scale as long as the measurement field is applied. In the limit $\gamma \to \infty$, this slow time scale becomes infinite, so that the continued measurement to the $|2\rangle$ component makes $\alpha_2 \to 0$, while preserving the value of α_1. This can be seen as the quantum Zeno effect, where strong measurement of α_2 prevents further evolution of α_1 according to the Hamiltonian part of the evolution.

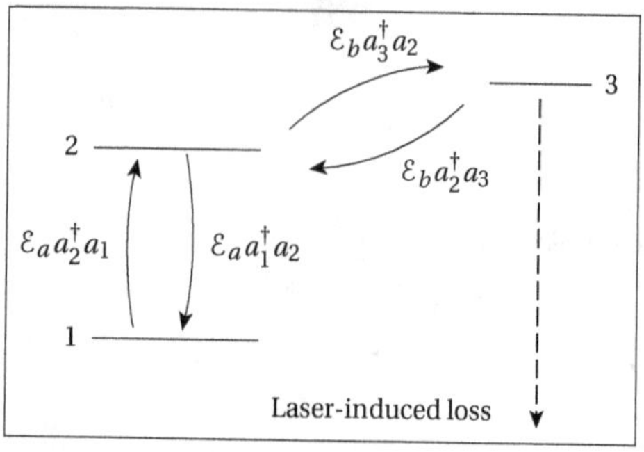

Fig. 20.3. Schematic of the implementation of coherent transfer of atoms between levels 1 and 2. Dissipative loss from level 2 is induced by coherent transfer to level 3, which is antitrapped, and undergoes dissipative loss.

20.3.2 Implementation of Fast Loss Mechanism

The physical way in which the loss is induced is by means of a laser field which makes a coherent transfer $2 \leftrightarrow 3$, where $|3\rangle$ is an antitrapped state, as illustrated in Fig. 20.3. This leads to the atom being ejected from the trap at a rate characteristic of the trap strength.

This can be modelled by the Hamiltonian

$$H = \hbar g \mathcal{E}_a (a_2^\dagger a_1 + a_1^\dagger a_2) + \hbar g \mathcal{E}_b (a_3^\dagger a_2 + a_2^\dagger a_3), \tag{20.3.10}$$

with the loss term similar to that in (20.3.2), but involving the untrapped state $|3\rangle$,

$$\dot{\rho} = \mathrm{i} g \mathcal{E}_a \left[a_2^\dagger a_1 + a_1^\dagger a_2, \rho \right] + \mathrm{i} g \mathcal{E}_b \left[a_3^\dagger a_2 + a_2^\dagger a_3, \rho \right] + \tfrac{1}{2}\gamma (2 a_3 \rho a_3^\dagger - a_3^\dagger a_3 \rho - \rho a_3^\dagger a_3). \tag{20.3.11}$$

In practice the rate of loss will be varied by changing the strength of the driving field $\mathcal{E}_b$, while γ is determined by the nature of the trap and cannot be easily varied.

The P-function equations of motion become

$$\dot{\alpha}_1 = -\mathrm{i} g \mathcal{E}_a \alpha_2, \tag{20.3.12}$$

$$\dot{\alpha}_2 = -\mathrm{i} g \mathcal{E}_a \alpha_1 - \mathrm{i} g \mathcal{E}_b \alpha_3, \tag{20.3.13}$$

$$\dot{\alpha}_3 = -\mathrm{i} g \mathcal{E}_b \alpha_2 - \gamma \alpha_3. \tag{20.3.14}$$

a) Solutions of the Equations for Large γ: If γ becomes very large, in a time scale of order of magnitude $1/\gamma$ we can write

$$\alpha_3 \approx -\frac{\mathrm{i} g \mathcal{E}_b \alpha_2}{\gamma}, \tag{20.3.15}$$

and the equations of motion reduce to

$$\dot{\alpha}_1 = -\mathrm{i} g \mathcal{E}_a \alpha_2, \tag{20.3.16}$$

$$\dot{\alpha}_2 = -\mathrm{i} g \mathcal{E}_a \alpha_1 - \frac{g^2 \mathcal{E}_b^2}{\gamma} \alpha_2. \tag{20.3.17}$$

The effect is to confine the dynamics to the $(1,2)$ subspace, giving Rabi oscillations with a small added damping term, which vanishes in the limit of large γ. This is not the process of interest.

b) Solutions of the Equations for Large $\mathcal{E}_b$: Suppose now that we let $\mathcal{E}_b$ become very large, so that we can neglect the first term in (20.3.13), the equation of motion for α_2. The resulting equations of motion in the (2,3) subspace are of the same form as (20.3.3, 20.3.4), and the eigenvalues can be written for large $\mathcal{E}_b$ as

$$\lambda_\pm = -\tfrac{1}{2}\gamma \pm \mathrm{i}\Omega_b, \tag{20.3.18}$$

$$\Omega_b = \sqrt{g^2\mathcal{E}_b^2 - \gamma^2/4}. \tag{20.3.19}$$

We can therefore write the solution for $\alpha_2(t)$ in the form

$$\alpha_2(t) \approx a\mathrm{e}^{(-\gamma/2+\mathrm{i}\Omega_b)t} + b\mathrm{e}^{(-\gamma/2-\mathrm{i}\Omega_b)t}, \tag{20.3.20}$$

for some constants a, b, and hence

$$\alpha_1(t) = \alpha_1(0) + \mathrm{i}g\mathcal{E}_a\left(a\frac{1-\mathrm{e}^{(-\gamma/2+\mathrm{i}\Omega_b)t}}{\gamma/2-\mathrm{i}\Omega_b} + b\frac{1-\mathrm{e}^{(-\gamma/2-\mathrm{i}\Omega_b)t}}{\gamma/2+\mathrm{i}\Omega_b}\right). \tag{20.3.21}$$

Since we assume that $\mathcal{E}_b \gg \mathcal{E}_a$, the second part of the right-hand side is very small, so that essentially the value of $\alpha_1(t)$ becomes clamped to its initial value, even if the loss term γ is quite finite. However, the time taken to achieve this state is of order $1/\gamma$, so for this clamping to the initial value to be of the quantum Zeno nature, we must have $\gamma \gg g\mathcal{E}_1$.

References

[1] J. Bardeen, *Semiconductor Research Leading to the Point Contact Transistor*, `nobelprize.org` (1956). 3

[2] C. Townes, *Production of Coherent Radiation by Atoms and Molecules*, `nobelprize.org` (1964). 3

[3] H. Haken and H. Sauermann, *Nonlinear Interaction of Laser Modes*, Zeits. für Phys. **173**, 261 (1963). 5

[4] H. Haken and H. Sauermann, *Frequency Shifts of Laser Modes in Solid State and Gaseous Systems*, Zeits. für Phys. **176**, 47 (1963). 5

[5] H. Haken, *Theory of Coherence of Laser Light*, Phys. Rev. Lett. **13**, 329 (1964). 5

[6] M. Lax, *Formal Theory of Quantum Fluctuations from a Driven State*, Phys. Rev. **129**, 2342 (1963). 5

[7] W. E. Lamb, *Theory of an Optical Maser*, Phys. Rev. **134**, A1429 (1964). 5

[8] J. Bardeen, L. N. Cooper, and J. R. Schrieffer, *Theory of Superconductivity*, Phys. Rev. **108**, 1175 (1957). 5

[9] M. H. Anderson, J. R. Ensher, M. R. Matthews, C. E. Wieman, and E. A. Cornell, *Observation of Bose-Einstein Condensation in a Dilute Atomic Vapor*, Science **269**, 198 (1995). 7

[10] D. Jaksch, C. Bruder, J. I. Cirac, C. W. Gardiner, and P. Zoller, *Cold Bosonic Atoms in Optical Lattices*, Phys. Rev. Lett. **81**, 3108 (1998). 7

[11] M. Greiner, O. Mandel, T. Esslinger, T. W. Hansch, and I. Bloch, *Quantum Phase Transition from a Superfluid to a Mott Insulator in a Gas of Ultracold Atoms*, Nature **415**, 39 (2002). 7

[12] D. F. Walls, *Evidence for the Quantum Nature of Light*, Nature **280**, 451 (1979). 7

[13] D. F. Walls, *Squeezed States of Light*, Nature **306**, 141 (1983). 7

[14] A. Einstein, B. Podolsky, and N. Rosen, *Can Quantum-Mechanical Description of Physical Reality Be Considered Complete?*, Phys. Rev. **47**, 777 (1935). 9

[15] E Schrödinger, *Die gegenwärtige Situation in der Quantenmechanik*, Naturwissenschaften **23**, 807 (1935). 9

[16] J. S. Bell, *On the Einstein–Podolsky–Rosen Paradox*, Physics **1**, 195 (1964). 9

[17] A. Aspect, P. Grangier, and G. Roger, *Experimental Tests of Realistic Local Theories via Bell's Theorem*, Phys. Rev. Lett. **47**, 460 (1981). 10

[18] R. P. Feynman, *Simulating Physics with Computers*, International Journal of Theoretical Physics **21**, 467 (1982). 10

[19] R. P. Feynman, *Quantum Mechanical Computers,* Foundations of Physics **16**, 507 (1986). 11

[20] D. Deutsch, *Quantum Theory, the Church–Türing Principle and the Universal Computer,* Proc. Roy. Soc. London Ser. A **400**, 96 (1985). 11

[21] D. Deutsch, *Quantum Computational Networks,* Proc. Roy. Soc. London Ser. A **425**, 73 (1989). 11

[22] P. W. Shor, in *Proceedings of the 35th Annual Symposium on the Foundations of Computer Science* (IEEE Computer Society Press, New York, 1994), p. 124. 11

[23] J. I. Cirac and P. Zoller, *Quantum Computations with Cold Trapped Ions,* Phys. Rev. Lett. **74**, 4091 (1995). 11

[24] F. Schmidt-Kaler, H. Haffner, M. Riebe, S. Gulde, G. P. T. Lancaster, T. Deuschle, C. Becher, C. F. Roos, J. Eschner, and R. Blatt, *Realization of the Cirac–Zoller Controlled-NOT Quantum Gate,* Nature **422**, 408 (2003). 11

[25] D. Leibfried, B. DeMarco, V. Meyer, D. Lucas, M. Barrett, J. Britton, W. M. Itano, B. Jelenkovic, C. Langer, T. Rosenband, and D. J. Wineland, *Experimental Demonstration of a Robust, High-Fidelity Geometric Two Ion-Qubit Phase Gate,* Nature **422**, 412 (2003). 11

[26] J. Benhelm, G. Kirchmair, C. F. Roos, and R. Blatt, *Towards Fault-Tolerant Quantum Computing with Trapped Ions,* Nat. Phys. **4**, 463 (2008). 11

[27] M. D. Barrett, J. Chiaverini, T. Schaetz, J. Britton, W. M. Itano, J. D. Jost, E. Knill, C. Langer, D. Leibfried, R. Ozeri, and D. J. Wineland, *Deterministic Quantum Teleportation of Atomic Qubits,* Nature **429**, 737 (2004). 11

[28] H. Dehmelt, in *Advances in Laser Spectroscopy,* edited by F. T. Arecchi, F. Strumia, and H. Walther (Plenum, New York, 1983), p. 153. 11

[29] W. Paul, *Electromagnetic Traps for Charged and Neutral Particles,* Rev. Mod. Phys. **62**, 531 (1990). 11, 12

[30] J. T. Barreiro, M. Muller, P. Schindler, D. Nigg, T. Monz, M. Chwalla, M. Hennrich, C. F. Roos, P. Zoller, and R. Blatt, *An open-system quantum simulator with trapped ions,* Nature **470**, 486 (2011). 11

[31] R. Blatt and D. Wineland, *Entangled States of Trapped Atomic Ions,* Nature **453**, 1008 (2008). 12

[32] J. Clarke and F. K. Wilhelm, *Superconducting Quantum Bits,* Nature **453**, 1031 (2008). 13

[33] C. Santori and Y. Yamamoto, *Quantum Dots: Driven to Perfection,* Nat. Phys. **5**, 173 (2009). 13

[34] M. P. Hedges, J. J. Longdell, Y. Li, and M. J. Sellars, *Efficient Quantum Memory for Light,* Nature **465**, 1052 (2010). 13

[35] M. J. Collett and C. W. Gardiner, *Squeezing of Intracavity and Travelling Wave Light Fields Produced in Parametric Amplification,* Phys. Rev. A **30**, 1386 (1984). 19

[36] C. W. Gardiner and M. J. Collett, *Input and Output in Damped Quantum Systems—Quantum Stochastic Differential Equations and the Master Equation,* Phys. Rev. A **31**, 3761 (1985). 19

[37] H. J. Carmichael, *Quantum Trajectory Theory for Cascaded Open Systems*, Phys. Rev. Lett. **70**, 2273 (1993). 19

[38] C. W. Gardiner, *Driving a Quantum System with the Output Field from Another Driven Quantum System*, Phys. Rev. Lett. **70**, 2269 (1993). 19

[39] R. Brown, *A Brief Account of Microscopical Observations … on the Particles Contained in the Pollen of Plants…*, Phil. Mag. **4**, 121 (1828). 27

[40] R. Clausius, *Über die Art der Bewegung, die wir Wärme nennen*, Annalen der Physik **100**, 353 (1857). 27

[41] P. Langevin, *Sur la Théorie du Mouvement Brownien*, Comptes Rendues **146**, 530 (1906). 29

[42] A. Khinchin, *Korrelationstheorie der stationären stochastischen Prozesse*, Math. Annalen **109**, 604 (1934). 37

[43] *Higher Transcendental Functions*, edited by A. Erdélyi (McGraw-Hill, New York, Toronto, London, 1953), Vol. 2. 54

[44] R. P. Feynman and A. R. Hibbs, *Quantum Mechanics and Path Integrals* (McGraw-Hill, New York, 1965). 57, 249

[45] M. Abramowitz and I. A. Stegun, *Handbook of Mathematical Functions* (Dover, New York, 1965). 59

[46] N. G. van Kampen, *Stochastic Processes in Physics and Chemistry*, 1st ed. (North Holland, Amsterdam, New York, Oxford, 1981). 90, 91

[47] G. C. Wick, *The Evaluation of the Collision Matrix*, Phys. Rev. **80**, 268 (1950). 118

[48] R. Hanbury Brown and R. Q. Twiss, *A Test of a New Type of Stellar Interferometer on Sirius*, Nature **178**, 481 (1956). 137

[49] V. Weisskopf and E. Wigner, *Berechnung der natürlichen Linienbreite auf Grund der Diracschen Lichttheorie*, Zeitschrift für Physik **63**, 54 (1930). 148

[50] T. Grunzweig, A. Hilliard, M. McGovern, and M. F. Andersen, *Near-Deterministic Preparation of a Single Atom in an Optical Microtrap*, Nat. Phys. **6**, 951 (2010). 158

[51] E. Jaynes and F. Cummings, *Comparison of Quantum and Semiclassical Radiation Theories with Application to the Beam Maser*, Proc IEEE **51**, 89 (1963). 159

[52] W. J. Munro and C. W. Gardiner, *Non-Rotating-Wave Master Equation*, Phys. Rev. A **53**, 2633 (1996). 167

[53] R. H. Koch, D. J. Van Harlingen, and J. Clarke, *Measurements of Quantum Noise in Resistively Shunted Josephson Junctions*, Phys. Rev. B **26**, 74 (1982). 180

[54] F. Bloch, *Nuclear Induction*, Phys. Rev. **70**, 460 (1946). 186

[55] R. J. Glauber, *The Quantum Theory of Optical Coherence*, Phys. Rev. **130**, 2530 (1963). 207

[56] E. Wigner, *On the Quantum Correction For Thermodynamic Equilibrium*, Phys. Rev. **40**, 749 (1932). 210

[57] P. B. Blakie, A. S. Bradley, M. J. Davis, R. J. Ballagh, and C. W. Gardiner, *Dynamics and Statistical Mechanics of Ultra-Cold Bose Gases Using C-Field Techniques*, Advances in Physics **57**, 363 (2008). 219, 233

[58] N. Lütkenhaus and S. M. Barnett, *Nonclassical Effects in Phase Space*, Phys. Rev. A **51**, 3340 (1995). 231

[59] A. Einstein, *Über einen die Erzeugung und Verwandlung des Lichtes betreffenden heuristischen Gesichtspunkt*, Annalen der Physik **17**, 132 (1905). 250

[60] N. F. Mott, *The Wave Mechanics of α-Ray Tracks*, Proc. Roy. Soc. London. Ser. A **126**, 79 (1929). 251

[61] N. G. van Kampen, *Ten Theorems about Quantum Mechanical Measurements*, Physica A **153**, 97 (1988). 256

[62] P. Zoller, M. Marte, and D. F. Walls, *Quantum Jumps in Atomic Systems*, Phys. Rev. A **35**, 198 (1987). 265

[63] C. W. Gardiner, A. S. Parkins, and P. Zoller, *Wave-Function Quantum Stochastic Differential Equations and Quantum-Jump Simulation Methods*, Phys. Rev. A **46**, 4363 (1992). 265

[64] K. Mølmer, Y. Castin, and J. Dalibard, *Monte Carlo Wave-Function Method in Quantum Optics*, J. Opt. Soc. Am. B **10**, 524 (1993). 265

[65] H. J. Carmichael, *An Open Systems Approach to Quantum Optics* (Springer, Berlin Heidelberg New York, 1993). 265

[66] J. I. Cirac, C. W. Gardiner, M. Naraschewski, and P. Zoller, *Continuous Observation of Interference Fringes from Bose Condensates*, Phys. Rev. A **54**, R3714 (1996). 271

[67] A. Robert, O. Sirjean, A. Browaeys, J. Poupard, S. Nowak, D. Boiron, C. I. Westbrook, and A. Aspect, *A Bose–Einstein Condensate of Metastable Atoms*, Science **292**, 461 (2001). 271

[68] W. H. Zurek, *Pointer Basis of Quantum Apparatus: Into What Mixture Does the Wave Packet Collapse?*, Phys. Rev. D **24**, 1516 (1981). 280

[69] B. Misra and E. C. G. Sudarshan, *The Zeno's Paradox in Quantum Theory*, J. Math. Phys. **18**, 756 (1977). 281

[70] W. M. Itano, D. J. Heinzen, J. J. Bollinger, and D. J. Wineland, *Quantum Zeno Effect*, Phys. Rev. A **41**, 2295 (1990). 281

[71] J. Bernu, S. Deléglise, C. Sayrin, S. Kuhr, I. Dotsenko, M. Brune, J. M. Raimond, and S. Haroche, *Freezing Coherent Field Growth in a Cavity by the Quantum Zeno Effect*, Phys. Rev. Lett. **101**, 180402 (2008). 281

[72] E. W. Streed, J. Mun, M. Boyd, G. K. Campbell, P. Medley, W. Ketterle, and D. E. Pritchard, *Continuous and Pulsed Quantum Zeno Effect*, Phys. Rev. Lett. **97**, 260402 (2006). 288

Author Index

Subject Index

www.ingramcontent.com/pod-product-compliance
Lightning Source LLC
Chambersburg PA
CBHW061624220326
41598CB00026BA/3872
* 9 7 8 1 7 8 3 2 6 4 6 1 2 *